Plant Genes and Genetics

Plant Genes and Genetics

Editor: Evan Adams

Callisto Reference

www.callistoreference.com

Callisto Reference,
118-35 Queens Blvd., Suite 400,
Forest Hills, NY 11375, USA

Visit us on the World Wide Web at:
www.callistoreference.com

ISBN: 978-1-64116-524-2 (Hardback)

Cataloging-in-Publication Data

Plant genes and genetics / edited by Evan Adams.
p. cm.
Includes bibliographical references and index.
ISBN 978-1-64116-524-2
1. Plant gene expression. 2. Plant genetics. 3. Plant genetic engineering. I. Adams, Evan.
QK981 .P53 2022
572.862--dc23

Table of Contents

Preface

Plant genes determine the qualitative or quantitative traits a plant exhibits. Plant genetics is a field of botany which studies plant genes, variation and heredity. The DNA sequence in plants is studied to understand the role of specific plant genes. Desired traits, such as disease and pest resistance, increased yields and improved nutritional value can be produced in plants by altering the genes and loci encoded by the DNA sequence. Advances in biotechnology have led to a greater understanding of plant genes and genetics, and improved plant breeding and hybridization. Modern plant genetics delves into the sequencing of plant genomes to further the techniques of genetic modification. Agrobacterium method and gene gun method are modern ways of modifying plants. This book is compiled in such a manner, that it will provide in-depth knowledge about plant genes and genetics. The topics covered in this extensive book deal with the core aspects of genetic engineering of plants. For someone with an interest and eye for detail, this book covers the most significant topics in plant genetics.

The information shared in this book is based on empirical researches made by veterans in this field of study. The elaborative information provided in this book will help the readers further their scope of knowledge leading to advancements in this field.

Finally, I would like to thank my fellow researchers who gave constructive feedback and my family members who supported me at every step of my research.

Editor

1

Cycle, canopy architecture and yield of common bean genotypes (*Phaseolus vulgaris*) in Santa Catarina State

Marcio Zilio[1], Clovis Arruda Souza[2*], Cileide Maria Medeiros Coelho[2], David José Miquelluti[2] and Adelina Ferreira Michels[1]

*[1]Programa de Pós-graduação em Ciências Agrárias (Produção Vegetal), Centro de Ciências Agroveterinárias, Universidade do Estado de Santa Catarina, Lages, Santa Catarina, Brazil. [2]Departamento de Agronomia, Centro de Ciências Agroveterinárias, Universidade do Estado de Santa Catarina, Av. Luís Camões, 2090, 88520-000, Lages, Santa Catarina, Brazil. *Author for correspondence. E-mail: souza_clovis@cav.udesc.br*

ABSTRACT. The objective of this study was to identify the phenotypic and genotypic correlations between the plant cycle and plant habit and the effect on yield in landrace bean genotypes. The experiment was conducted in Joaçaba and Lages, Santa Catarina State for the 2008/2009 crop using 26 bean genotypes: 22 landrace and 4 commercial genotypes obtained from the UDESC. We evaluated the number of days from emergence to flowering, number of days from flowering to physiological maturity and number of days from emergence to physiological maturity in relation to the genotypes' cycle. The aerial plant architecture characteristics evaluated were the growth habit, plant habit, plant height, stem diameter, height of the first pod insertion and number of nodes on the main stem. Trail analysis showed that there was a positive correlation between the emergence-to-flowering period and yield and that the emergence-physiological maturity; flowering-physiological maturity showed a negative correlation to the yield in both locations. Therefore, short-cycle genotypes, especially those with a reduced post-flowering period, produced increased yields. The aerial plant architecture characteristics showed phenotypic and genotypic positive correlations with the yield in both environments. To increase yield, the reproductive period needs to coincide with periods of the greatest photosynthetically active radiation.

Keywords: trail analysis, genetic variability, short cycle, upright habit.

Introduction

The planting of short-cycle bean genotypes presents various advantages: making better use of the cultivation area; allowing for a more favorable period for sowing and harvesting; reducing climatic stress, disease occurrence and harvest losses; lowering water consumption and the soil 'use' time under intensive and irrigated cultivation; and facilitating crop rotation. Furthermore, due to the early harvest, early-maturing beans can provide a better profit for the farmer (DALLA CORTE et al., 2003).

The number of days from emergence to flowering has been used by researchers to define early-maturing characteristics (RAMALHO et al., 2005; SILVA et al., 2007). The flowering initiation characteristic shows a high heritability and a high and positive correlation with physiological maturity (SILVA et al., 2007). However, genotypes with the shortest emergence-flowering period may not be the earliest to mature, suggesting that it is not always the early-flowering genotypes that will have the shortest cycles (RIBEIRO et al., 2004).

Early maturity can be related to bean yield. It has been reported that a reduction of one day of the cycle caused a 33 kg ha^{-1} increase in yield during the dry season in Minas Gerais State (RAMALHO et al., 1993) and that the yield of black and carioca bean genotypes increase with a reduction in the cycle in the central basin region of Rio Grande do Sul (RIBEIRO et al., 2004). Traka-Mavrona et al. (2002) showed that advanced families had a high degree and stability of early bean cultivars, a high yield, and normal seeds and pods of high quality. Therefore, the use of short-cycle genotypes can lead to increased yields.

The use of more productive cultivars, which have a plant architecture that facilitates cultural operations and mechanical harvest, has increased recently (SILVA et al., 2009): the modern cultivars tend to have a type-II growth habit, grow more upright, resist lodging and mature uniformly. However, Alves et al. (2001) argued that there is a negative correlation between more upright plant crops and yield, which is likely to be a problem for the modern cultivars.

According to Souza et al. (2010a and b), plants with greater hypocotyl and epicotyl diameters and a greater average plant height at harvest have better architecture. However, a negative linear association between the hypocotyl and epicotyl diameters and grain yield has been observed, indicating that plants with a more upright architecture demonstrate lower yields. In addition, it must be considered that there is a high negative genotypic correlation between the epicotyl diameter and the number of pods and branch internodes, indicating that plants with greater epicotyl diameters have fewer branches.

The cycle and aerial plant architecture (growth habit) are highly influenced by the environment, however the choice of a short-cycle genotype with ideal architecture is possible, especially if various genotypes are tested in diverse environments (SILVA et al., 2009; DAWO et al., 2007). Therefore, studies concerned with plant architecture, bean genotype diversity and growth cycle should address the effect of the environment on these characteristics and also concentrate on identifying productive plants with early maturity; a large, upright habit; and a higher point of first pod insertion, traits that are of great importance to meet new market demands.

The objective of this work was to determine the effect of the direct and indirect relationships between the cycle and aerial plant architecture on the bean grain yield through genotypic and phenotypic correlations.

Material and methods

The experiment was conducted under field conditions in Joaçaba, Santa Catarina State (27°10' S, 51°30' W, 522 meters above sea level) and Lages, Santa Catarina State (27°52' S, 50°18' W, 930 meters above sea level) for the 2008/2009 crop using 26 genotypes, including 22 landrace genotypes and 4 commercial genotypes, obtained from the common bean germplasm bank, named 'Banco Ativo de Feijão' (BAF) of the 'State University Santa Catarina' (UDESC). The following BAF genotypes used were used: 3; 4; 7; 13; 23; 36; 42; 44; 46; 47; 50; 55; 57; 60; 68; 75; 81; 84; 97; 102; 108; 112; 115; 120; 121 and 192. The 4 commercial cultivars were BAF 112 (IPR88-Uirapurú), BAF 115 (BRS-Valente), BAF 121 (Iapar-81) and BAF 192 (BRS-Radiante).

A random experimental block layout was used, comprising three repetitions in plots with four 3 meter-long lines, with 0.5 m between the lines. Fifteen seeds were sown per meter, with the two external lines being considered as border strips, thus making the two internal lines the useful area but excluding 0.5 m from the extremities of each.

Base fertilizer was applied next to the lines before sowing and in accordance with a soil analysis and the recommendations of the 'Comissão de Química e Fertilidade do Solo' - CQFS-RS/SC (2004). Top dressing was applied twice, at the three-lobed-leaf stage and at the start of flowering (V4 and R5) using 30 kg ha^{-1} of N for each application. Weed, disease and pest control was performed according to necessity using chemical products recommended for this culture.

The grain yield was estimated in kg ha^{-1}, corrected to a 13% moisture content (fresh weight), based on the grain yield obtained from the useful area of each plot. As for the cycle, the genotypes were evaluated during the emergence-flowering (number of days from emergence to flowering), flowering-physiological maturity (number of days from flowering to physiological maturity), and emergence-physiological maturity (number of days from emergence to physiological maturity) periods. In each plot, the phenological stage was considered as follows: emergence - 50% plus one seedling emergence; flowering - 50% plus one plant with an open flower; physiological maturity - 50% plus one plant with yellow pods (CIAT, 1983).

The following aerial plant architecture characteristics were evaluated: growth habit (type I, type II, type III and type IV); plant habit (upright, semi-upright, and prostrate); plant height (cm); stem diameter (cm); height of the first pod insertion (cm); and number of nodes on the main stem. The growth habit, plant habit, plant height and stem diameter were determined in R6, and the height of the first pod insertion and number of nodes were determined in R8. The determination of these characteristics was made for 10 plants, randomly selected from the useful area in each plot (IPGRI, 2001). The field genotype germination percentage was also estimated by counting the number of emerged plants in the useful area of each plot.

All of the variables studied were subjected to univariate variance analysis according to the fixed-effects linear model (LITTEL et al., 2006; ZILIO et al., 2011) for the components, location, genotype, and genotype response to the cultivation environment. The comparison between the average values of each variable analyzed for the different genotypes in each location was performed using the Scott-Knott test and for each genotype between locations using the Tukey test. The association between the characteristics was estimated through phenotypic and genotypic trail analysis. Trail analysis consists of a study of the direct and indirect effects of independent explanatory characteristics on a principal, basic dependent variable for which the estimates are obtained using regression equations using previously standardized variables (CABRAL et al., 2011). The minimum level of significance for all of the tests was 5%.

Results and discussion

The variance analysis indicated a genotype x environment interaction for the plant height characteristic only. For the other characteristics, the effect of genotype was only observed between the locations, suggesting that the genotypes do not respond differently to environmental influences in terms of the stem diameter, height of the first pod insertion and number of nodes. However, a significant difference was observed between the genotypes for all of the characteristics evaluated within the two locations, except for the height of the first pod insertion in the Lages environment. These results indicate the presence of variability for the majority of the characteristics in the two locations, thus facilitating the identification of the genotypes with ideal architecture.

Different genotype responses to cultivation environments were observed in terms of the cycle and yield, proving the existence of genetic diversity over the duration of the cycle and the effect of the location on each genotype. For this reason, the data obtained for the different genotypes are presented by the cultivation location.

After the phenotypic trail analysis (Table 1), it was revealed that the emergence-flowering period showed a positive correlation with the yield for both locations, i.e., the greater the number of days between emergence and flowering, the greater was the yield. According to Fageria et al. (2006), the pre-flowering period is necessary for the adequate development of the leaves to supply photosynthate to the grains during the grain-filling stage.

However, the emergence-physiological maturity and flowering-physiological maturity periods showed negative correlations in both of the environments, indicating that grain yield increases were obtained when the number of days between emergence and physiological maturity was reduced. Therefore, short-cycle genotypes, especially those with a reduced cycle post-flowering, give greater yields.

Similar results were found by Ramalho et al. (1993) in Minas Gerais State and by Ribeiro et al. (2004) in Rio Grande do Sul State in that reductions in the cycle led to yield increases. In a study conducted by Ribeiro et al. (2001), carioca bean genotypes showed a negative genetic correlation between the cycle and grain yield, weight of 100 grains, number of pods and grains per plant, and number of grains per legume. Cabral et al. (2011) also observed a negative correlation between the yield and emergence-flowering and emergence-harvest periods.

Table 1. Estimate of the phenotypic and genotypic correlation coefficients and their direct and indirect effects on the yield characteristics and cycle in common beans.

Characteristics	Estimate of correlation			
	Phenotypic		Genotypic	
Emergence – Flowering	Joaçaba	Lages	Joaçaba	Lages
Direct effect on yield	-2.79	8.88	-2.32	-26.1
Indirect effect via emergence-physiological maturity	2.57	-4.01	2.51	11.25
Indirect effect via flowering-physiological maturity	-0.24	-5.16	-0.24	14.02
Indirect effect via plant height	0.26	-0.05	0.41	-0.37
Indirect effect via stem diameter	0.17	0.05	0.28	-0.01
Indirect effect via first pod height	0.05	0.30	-0.05	0.82
Indirect effect via internodes number	0.17	0.24	-0.31	0.66
Indirect effect via emergence	-0.04	0.01	-0.12	0.01
Total	0.14	0.25	0.16	0.29
Emergence - Physiological Maturity				
Direct effect on yield	3.10	-15.28	3.02	36.44
Indirect effect via emergence- physiological maturity	-2.31	2.33	-1.93	-8.06
Indirect effect via flowering-physiological maturity	-1.42	12.76	-1.40	-28.56
Indirect effect via plant height	0.20	-0.01	0.32	-0.01
Indirect effect via stem diameter	0.13	-0.04	0.21	0.02
Indirect effect via first pod height	0.05	0.07	-0.05	0.31
Indirect effect via internodes number	0.11	-0.11	-0.21	-0.41
Indirect effect via emergence	-0.07	0.02	-0.19	0.03
Total	-0.20	-0.24	-0.24	-0.24
Flowering - Physiologic Maturity				
Direct effect on yield	-2.19	15.52	-2.18	-37.31
Indirect effect via emergence-physiological maturity	-0.30	-2.95	-0.26	9.81
Indirect effect via flowering-physiological maturity	2.00	-12.56	1.94	27.89
Indirect effect via plant height	0.01	0.03	0.01	0.25
Indirect effect via stem diameter	-0.01	-0.07	0.01	0.03
Indirect effect via first pod height	0.03	-0.11	-0.02	-0.27
Indirect effect via internodes number	-0.03	-0.25	0.05	-0.84
Indirect effect via emergence	-0.07	0.02	-0.18	0.02
Total	-0.55	-0.38	-0.64	-0.42
Plant Height				
Direct effect on yield	0.37	-0.14	0.57	-0.93
Indirect effect via emergence-physiological maturity	-1.98	3.53	-1.69	-10.42
Indirect effect via flowering-physiological maturity	1.69	-0.02	1.71	0.44
Indirect effect via plant height	-0.02	-3.66	-0.03	9.89
Indirect effect via stem diameter	0.12	0.07	0.20	-0.02
Indirect effect via first pod height	0.03	0.37	-0.03	0.91
Indirect effect via internodes number	0.13	0.22	-0.25	0.65
Indirect effect via emergence	-0.02	-0.01	-0.06	0.01
Total	0.34	0.36	0.41	0.51
Stem Diameter				
Direct effect on yield	0.33	0.14	0.50	-0.04
Indirect effect via emergence-physiological maturity	-1.45	2.81	-1.30	-10.30
Indirect effect via flowering-physiological maturity	1.22	4.69	1.29	-13.94
Indirect effect via plant height	0.01	-7.56	-0.01	23.75
Indirect effect via stem diameter	0.14	-0.06	0.23	-0.60
Indirect effect via first pod height	0.04	0.20	-0.05	0.81
Indirect effect via internodes number	0.07	0.27	-0.14	1.03
Indirect effect via emergence	-0.02	-0.01	-0.08	-0.03
Total	0.34	0.48	0.45	0.68
First Pod Height				
Direct effect on yield	0.13	0.50	-0.11	1.02
Indirect effect via emergence- physiological maturity	-1.02	5.43	-1.00	-21.21
Indirect effect via flowering- physiological maturity	1.25	-2.27	1.44	11.29
Indirect effect via plant height	-0.49	-3.41	-0.57	9.81
Indirect effect via stem diameter	0.10	-0.10	0.14	-0.84
Indirect effect via first pod height	0.11	0.06	0.22	-0.03
Indirect effect via internodes number	-0.01	0.12	0.01	0.35
Indirect effect via emergence	0.01	0.01	-0.02	0.01
Total	0.09	0.33	0.10	0.40
Internodes Number				
Direct effect on yield	0.21	0.59	-0.32	1.49
Indirect effect via emergence-physiologic maturity	-2.26	3.64	-2.21	-11.58
Indirect effect via flowering-physiologic maturity	1.69	2.87	1.96	-9.93
Indirect effect via plant height	0.30	-6.56	0.35	21.06
Indirect effect via stem diameter	0.24	-0.05	0.45	-0.40
Indirect effect via first pod height	0.12	0.07	0.22	-0.03
Indirect effect via internodes number	-0.01	0.10	0.01	0.24
Indirect effect via emergence	-0.02	-0.01	-0.10	-0.01
Total	0.27	0.64	0.34	0.84

Continue...

...continuation

Characteristics	Estimate of correlation			
	Phenotypic		Genotypic	
Emergence	Joaçaba	Lages	Joaçaba	Lages
Direct effect on yield	0.16	-0.03	0.38	-0.02
Indirect effect via emergence-physiologic maturity	0.75	-1.76	0.74	12.87
Indirect effect via flowering-physiologic maturity	-1.35	8.84	-1.53	-53.11
Indirect effect via plant height	0.91	-6.90	1.02	40.6
Indirect effect via stem diameter	-0.04	-0.01	-0.09	0.01
Indirect effect via first pod height	-0.05	0.07	-0.10	-0.06
Indirect effect via internodes number	0.01	-0.02	0.01	-0.35
Indirect effect via emergence	-0.03	0.21	0.09	1.00
Total	0.35	0.39	0.52	0.94

According to Rocha and Vello (1999), soybean grain yield varied with maturity cycle and cultivation location: short-cycle, semi-short-cycle and medium-cycle genotypes demonstrated greater grain yields than late-cycle varieties. These authors also reported that late-cycle genotypes were more susceptible to fungal diseases due to their longer exposure to field conditions than short-cycle genotypes; the latter were exposed to field conditions for less time and were, thus, less affected or not affected at all by phytopathogenic fungi. The authors further indicated that medium-cycle genotypes demonstrated a greater stability to environmental factors compared to late-maturing genotypes due to their greater chance of not being adversely affected by fungi.

Other authors have found a relationship between the cycle and bean yield components. According to Cabral et al. (2011), plants with a lower number of days between emergence and flowering and between flowering and physiological maturity produce a greater number of pods per plant, seeds per pod and weight of 100 seeds.

In the present study, the aerial plant architecture characteristics showed a positive phenotypic correlation with the yield for both of the locations. The characteristic that contributed most to the increased grain yield in beans was the plant height (+0.34) and stem diameter (+0.34) in Joaçaba and the number of nodes (+0.64) and stem diameter (+0.48) in Lages. Similar results were found by Vale et al. (2009) who also found a positive correlation between the yield, plant height and height of the first pod insertion. In contrast, Cabral et al. (2011) found that the plant height and height of the first pod insertion did not present a correlation with the yield.

According to this, environmental conditions drastically affect the degree of association between cycle and yield. In general, the yield, rate of maturity, greater plant height, stem diameter and number of nodes were efficient for the selection of high-yielding genotypes by the cultivation location.

Our genotypic trail analysis results showed that the aerial plant architecture characteristics presented positive correlation coefficients with the yield for both of the locations (Table 1). Most notable were the stem diameter (+0.45) and plant height (+0.41) in Joaçaba and the number of nodes (+0.84) and stem diameter (+0.68) in Lages. These results indicate that elevated yields can be achieved when genotypes with greater aerial structure, number of nodes, height of first pod insertion and, especially, stem diameter are used.

The genotype correlation coefficient estimates show a positive association between the emergence-flowering period and yield for both of the locations, whereas the emergence-physiological maturity and flowering-physiological maturity periods show negative correlation coefficients for the environments (Table 1).

When the phenotypic and genotypic correlation coefficient values are compared, in general, the genotypic correlations are superior to the corresponding phenotypic values; when the correlation magnitudes are not similar, the genotypic correlations are more useful than the phenotypic correlations for selection strategies. However, genotypically correlated characteristics that are not phenotypically correlated may not be of practical use in selection, as selection is generally based on phenotype (CARVALHO et al., 2002). Thus, both types of correlation were considered in this study to facilitate the decision about the efficiency of these indirect selection criteria.

For characteristics related to the aerial plant architecture, the genotypes studied showed different behavior between the two locations in terms of the plant height only. However, there was variability within each location for the characteristics, with the exception of the height of the first pod insertion in Lages, Santa Catarina State (Table 2).

The height of the genotypes in Joaçaba varied between 100.5 cm (BAF 108) and 45.8 cm (BAF 4), with an average of 70.6 cm. Despite BAF 108 presenting an elevated height, this genotype exhibited a prostrate undetermined climber growth habit, which does not permit mechanical harvest. BAF 4 presented an upright determined growth habit. In Lages, the tallest genotype was BAF 7 (90.1 cm), which displayed a prostrate undetermined growth habit and was semi-upright, and the shortest was BAF 120 (35.5 cm), with a determined, upright growth habit. The genotypes with determined growth habits presented shorter plants in Joaçaba (60.0 cm) and Lages (42.9 cm), compared to the undetermined-habit genotypes (77.8 cm to 64.8 cm in Joaçaba and Lages, respectively) (Table 2).

The stem diameter of the genotypes cultivated in Joaçaba varied from 0.8 cm (BAF 7) to 0.6 cm (BAF 47), with an average of 0.7 cm. BAF 7 presented a type III growth habit, and BAF 47 showed a type II habit and an upright growth habit. In Lages, the genotypes varied from 0.7 cm (BAF 121) to 0.5 cm (BAF 47), with an average of 0.6 cm. BAF 121 (Iapar-81) was upright with a type II growth habit (Table 2).

The height of the first pod insertion for the Joaçaba genotypes was from 35.5 cm (BAF 13) to 14.0 cm (BAF 47), with an average of 20.7 cm. In Lages, the genotype with the greatest height of first pod insertion was BAF 7 (32.7 cm) and the lowest was BAF 192 (11.5 cm), with an average of 20.1 cm. The number of nodes on the genotypes in Joaçaba varied between 18.1 (BAF 50) and 5.6 (BAF 97), with an average of 9.3 nodes per plant. In Lages, the genotypes showed a variation of between 13.3 (BAF 121) and 6.2 (BAF 4), with an average of 9.6 nodes (Table 2).

Among the landrace genotypes, BAF 7 was notable, presenting elevated values for all of the architecture characteristics evaluated at both locations, whereas BAF 47 presented low characteristic values at both locations. Among the commercial genotypes, BAF 121 was prominent, presenting elevated plant height, height of the first pod insertion, stem diameter, and number of nodes values in Lages and increased stem diameter and number of nodes in Joaçaba (Table 2).

For characteristics related to the culture cycle and yield, the genotypes studied presented different behavior within and between each location (Table 3). Various studies have reported a genotype x environment interaction and genetic diversity in the cycle (BARELLI et al., 1999; BURATTO et al., 2007; DALLA CORTE et al., 2002; RIBEIRO et al., 2004).

Table 2. Growth habit, plant height, stem diameter, height of the first pod insertion and number of internodes of common bean genotypes cultivated in Joaçaba and Lages, Santa Catarina State, during the 2008/2009 crop season.

BAF	(1)Growth habit		(2)Habit		(3)/Plant height (cm)		Stem diameter (cm)		First pod height (cm)		Internodes number		Emergence (%)	
	Joaçaba	Lages	Joaçaba	Lages	Joaçaba	Lages	Joaçaba	Lages	Joaçaba	Lages	Joaçaba	Lages	Joaçaba	Lages
3	I	I	2	1	A 55.4 e	A 43.8 b	0.8a	0.6a	19.3b	21.9a	7.2a	7.1b	81.1b	69.5a
4	I	I	1	1	A 45.9 f	A 39.0 b	0.6b	0.5b	17.0b	17.4a	7.2a	6.2b	78.9b	54.6a
7	III	III	2	2	A 80.2 b	A 90.1 a	0.8a	0.6a	33.4a	32.7a	9.8a	12.4	75ba	87.9a
13	II	II	1	1	A 83.6 b	A 65.6 a	0.7a	0.6a	35.5a	26.2a	7.3a	9.8a	88.9a	83.9a
23	I	I	1	1	A 57.0 e	A 42.7 b	0.6b	0.6b	19.8b	18.0a	8.2a	6.4b	74.2b	57.7a
36	II	II	1	1	A 70.4 c	A 73.8 a	0.7b	0.6a	14.8b	25.4a	8.4a	8.9a	80b	86.3a
42	I	I	1	1	A 75.9 c	A 45.7 b	0.8a	0.6a	15.5b	16.0a	10.2a	9.9a	76.7b	67.4a
44	III	II	2	1	A 71.6 c	A 69.9 a	0.6b	0.5b	15.8b	20.3a	9.3a	10.9	85a	80.9a
46	II	III	1	2	A 74.0 c	A 77.4 a	0.6b	0.6a	20.8b	23.1a	9.0a	9.1a	93.9a	84.1a
47	II	II	1	1	A 68.7 d	A 49.7 b	0.6b	0.5b	14.0b	16.7a	7.9a	8.1b	88.6a	63.6a
50	II	II	2	1	A 83.0 b	A 51.2 b	0.7b	0.5b	21.2b	18.7a	18.1a	12.7a	88.3a	62.2a
55	II	II	1	1	A 66.9 d	A 47.1 b	0.8a	0.6a	20.1b	18.9a	10.6a	9.4a	90.6a	69.1a
57	II	II	1	1	A 66.4 d	A 54.7 b	0.6b	0.5b	23.1b	16.1a	7.2a	8.1b	91.1a	69.8a
60	I	I	1	1	A 66.3 d	A 49.1 b	0.7b	0.6a	22.7b	17.3a	10.8a	9.8a	84.2a	66.5a
68	I	I	1	1	A 50.3 f	A 47.3 b	0.6b	0.5b	16.8b	24.1a	6.9a	7.7b	84.7a	59.6a
75	II	II	1	1	A 80.6 b	A 85.7 a	0.7b	0.7a	24.6b	23.8a	9.7a	9.1a	85.8a	88.6a
81	II	II	1	1	A 87.3 b	B 38.4 b	0.8a	0.6a	20.7b	17.6a	10.4a	11.5a	82.8a	57.9a
84	I	I	1	1	A 72.5 c	A 40.5 b	0.7b	0.5b	18.3b	22.2a	10.9a	10.8a	84.4a	67.9a
97	I	I	1	1	A 62.3 d	A 47.1 b	0.6b	0.5b	23.3b	18.1a	5.6a	6.5b	89.7a	63.3a
102	II	III	1	2	A 72.8 c	A 71.0 a	0.7a	0.6a	16.0b	22.5a	9.8a	10.4a	84.2a	81.7a
108	IV	IV	3	3	A 100.5 a	A 85.8 a	0.6b	0.6b	20.7b	21.2a	12.7a	10.2a	70.8b	76.8a
112	II	II	1	1	A 71.8 c	A 43.6 b	0.8a	0.6a	23.6b	16.8a	9.5a	12.1a	90.3a	66.6a
115	II	II	1	1	A 83.3 b	B 44.9 b	0.8a	0.6a	18.5b	14.1a	10.0a	11.0a	86.4a	66.1a
120	I	I	1	1	A 57.4 e	A 35.6 b	0.8a	0.5b	23.7b	16.2a	7.3a	6.9b	85a	66.9a
121	II	II	1	1	A 74.6 c	A 88.1 a	0.8a	0.7a	19.2b	26.1a	11.2a	13.3a	78.3b	97.7a
192	I	I	1	1	A 57.2 e	A 38.8 b	0.7b	0.5b	20.6b	11.6a	6.5a	11.1a	86.1a	63.4a
Average					70.6	56.4	0.7	0.6	20.7	20.1	9.3	9.6	84.0	81.3
C.V.					6.1	17.7	6.1	8.0	18.4	23.5	9.3	6.6	5.2	8.0

(1) I (determined); II (undetermined shrubby); III (undetermined prostrate); IV (undetermined climber). (2) 1 (upright); 2 (semi–upright); 3 (prostrate). (3)/Averages followed by the same capital letter do not significantly differ between the cultivation locations (rows) using the Tukey test at 5% probability. Averages followed by the same lowercase letter do not significantly differ in each cultivation location (columns) using the Scott-Knott test at 5% probability.

Table 3. Duration in the number of days of the emergence-physiological maturity, emergence-flowering and flowering-physiological maturity periods of bean genotypes cultivated in Joaçaba and Lages, Santa Catarina State, during the 2008/2009 crop season.

BAF	V1–R9 (days)		V1–R6 (days)		R6–R9 (days)		Yield (kg ha^{-1})	
	Joaçaba	Lages	Joaçaba	Lages	Joaçaba	Lages	Joaçaba	Lages
3	[1/]A 68 c	A 79 b	B 31 g	A 36 b	A 37 a	A 43 b	A 2617 b	A 1220 a
4	A 68 c	A 81 b	B 30 h	A 34 c	A 38 a	A 47 b	A 2011 b	A 1656 a
7	A 84 a	A 90 a	B 42 a	A 48 a	A 42 a	A 43 b	A 2776 b	AB 1938 a
13	A 68 c	A 79 b	A 34 d	A 34 c	A 34 b	A 45 b	A 3768 a	A 3065 a
23	B 68 c	A 85 a	B 30 h	A 33 c	B 38 a	A 52 a	A 2788 b	B 571 a
36	B 67 d	A 86 a	B 30 h	A 34 c	A 37 a	A 53 a	A 3327 b	A 1605 a
42	A 67 c	A 79 b	A 34 e	A 34 c	A 33 b	A 45 b	A 4452 a	AB 2470 a
44	B 68 c	A 90 a	B 34 e	A 37 b	B 34 b	A 53 a	A 2282 b	A 1957 a
46	B 66 e	A 81 b	A 30 h	A 31 d	A 36 b	A 50 a	A 3724 a	B 1541 a
47	B 66 e	A 81 b	B 30 h	A 33 c	A 36 b	A 48 a	A 2661 b	A 1107 a
50	A 77 b	A 86 a	A 42 a	B 36 b	B 35 b	A 51 a	A 3665 a	A 2646 a
55	A 68 c	A 82 b	A 34 e	A 36 b	A 34 b	A 46 b	A 3766 a	A 1806 a
57	A 65 e	A 78 b	B 30 h	A 33 c	A 35 b	A 45 b	A 5096 a	B 492 a
60	A 67 c	A 82 b	A 34 e	A 36 b	A 33 b	A 46 b	A 3941 a	B 1714 a
68	B 68 c	A 90 a	B 30 h	A 37 b	B 38 a	A 54 a	A 2761 b	A 750 a
75	A 68 c	A 73 b	A 34 f	A 36 b	A 34 b	A 37 c	A 3956 a	B 1519 a
81	A 77 b	A 73 b	A 35 c	A 34 c	A 42 a	A 39 c	A 3693 a	A 2330 a
84	A 68 c	A 73 b	B 34 f	A 37 b	A 34 b	A 36 c	A 3949 a	B 1430 a
97	B 66 e	A 85 a	A 29 i	A 31 d	B 37 a	A 54 a	A 3104 b	A 1058 a
102	A 67 d	A 73 b	A 34 d	A 34 c	A 33 b	A 39 c	A 5018 a	B 2673 a
108	A 77 b	A 78 b	A 38 b	A 36 b	A 39 a	A 42 b	A 2750 b	A 1888 a
112	A 68 c	A 78 b	A 34 f	A 34 c	A 34 b	A 44 b	A 4386 a	B 1931 a
115	A 68 c	A 81 b	A 34 f	A 36 b	A 34 b	A 45 b	A 3856 a	A 1871 a
120	B 68 c	A 86 a	B 30 h	A 34 c	A 38 a	A 53 a	A 2822 b	A 982 a
121	A 68 c	A 78 b	A 34 f	A 36 b	A 34 b	A 42 b	A 4019 a	A 2524 a
192	B 65 e	A 81 b	A 30 h	A 31 d	A 35 b	A 50 a	A 3247 b	B 706 a
Average	68.8	80.8	33.1	34.9	35.7	45.9	3478.3	1671.1
C.V.	0.4	2.3	0.1	1.2	0.8	4.2	15.0	33.5

[1/]Averages followed by the same capital letter do not significantly differ between the cultivation locations (rows) using the Tukey test at 5% probability. Averages followed by the same lowercase letter do not significantly differ in each cultivation location (columns) using the Scott-Knott test at 5% probability. V1-R9=emergence-physiological maturity; V1-R6=emergence-flowering; R6-R9=flowering-physiological maturity.

BAFs 23, 36, 44, 46, 47, 68, 97, 120 and 192 presented a shorter cycle period between emergence and physiological maturity in Joaçaba in comparison to Lages. The reduction in the number of days varied from 15 days (BAF 46 and BAF 47) to 22 days (BAF 44 and BAF 68). For the period between emergence and flowering, BAFs 3, 4, 7, 23, 36, 44, 47, 57, 68, 84 and 120 presented reduced cycles in Joaçaba compared to Lages, with the observed variation being from 3 days (BAFs 23, 44, 47 and 57) to 6 days (BAF 7). The exception was for BAF 50, which presented a longer cycle (6 days) in Joaçaba than in Lages. In the flowering-physiological maturity period, BAFs 36, 44, 50, 68 and 97 had shorter cycles in Joaçaba, with variation amplitude of 16 days (BAFs 36, 50 and 68) to 19 days (BAF 44) (Figure 1a and b).

The duration of the emergence-flowering period presented a distinct variation for the genotypes at both locations. In Joaçaba, the genotypes' cycle duration was from 84 to 65 days, with BAFs 46, 47, 57, 97 and 192 being the shortest. In Lages, the variation in the total cycle was 90 to 73 days, with BAFs 75, 81, 84 and 102 all having a cycle of 73 days. BAF 7 presented the longest cycle in both locations.

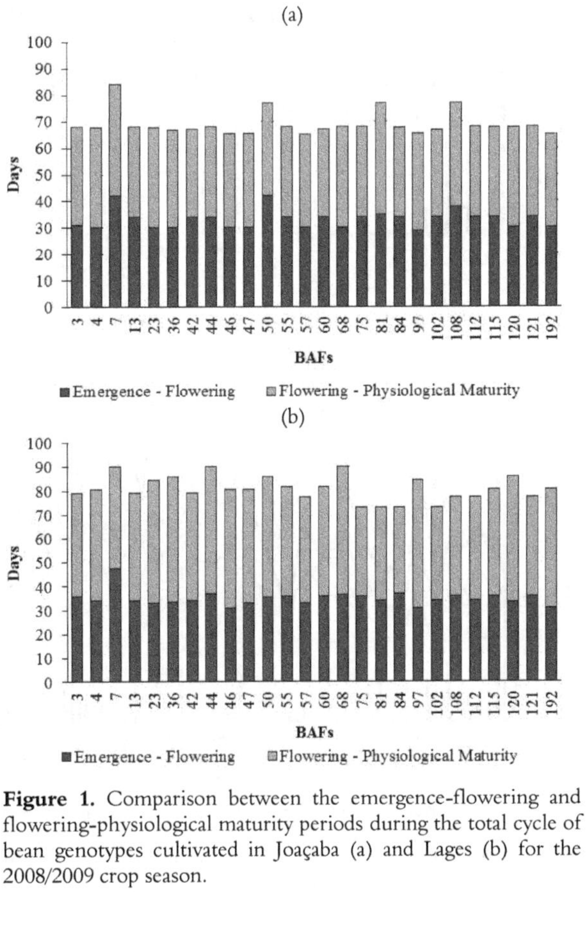

Figure 1. Comparison between the emergence-flowering and flowering-physiological maturity periods during the total cycle of bean genotypes cultivated in Joaçaba (a) and Lages (b) for the 2008/2009 crop season.

The number of days to flowering in Joaçaba varied from 42 (BAF 7) to 29 days (BAF 97) and varied from 48 (BAF 7) to 31 days (BAF 192) in Lages, with BAFs 46, 97 and 192 presenting the earliest flowering at both locations. BAF 7 also presented the longest number of days to flowering. The flowering-physiological maturity period in Joaçaba varied from 42 (BAF 7) to 33 days (BAF 102), with BAFs 42, 60 and 102 having 33 days; in Lages, the variation was from 54 (BAF 68) to 36 days (BAF 84).

The shortest cycle genotypes have been defined as those with the lowest average number of days necessary to initiate flowering (RAMALHO et al., 1993; SILVA et al., 2007). However, in the current study, the genotypes with the shortest emergence-flowering period duration are different from those labeled as short-cycle varieties (RIBEIRO et al., 2004).

The results show that the greatest differences between the genotypes' total cycle and location, principally in the flowering-physiological maturity period, must be due to pluviometric precipitation during this period, which was greater in Lages than in Joaçaba (Figure 2a and b).

Regions suitable for common bean cropping must present temperatures between 15 and 29.5°C and an average rainfall of 300 mm (FANCELLI; DOURADO NETO, 2007). During the emergence-physiological maturity period, the average temperature in Joaçaba was 20.3°C, whereas the average temperatures registered were 19.5 and 21.2°C during the emergence-flowering and flowering-physiological maturity periods, respectively. In Lages, the temperatures were similar to those registered in Joaçaba, with the emergence-physiological maturity period average temperature being 20.2°C and 19.3 and 20.1°C during the emergence-flowering and flowering-physiological maturity periods, respectively. Therefore, the average temperatures that occurred during the 2008/2009 crop period in Joaçaba and Lages were ideal for the cultivation of beans. These thermal conditions indicate that the lower yields observed in Lages were not due to flower and pod abortion caused by elevated temperatures during the flowering-physiological maturity period.

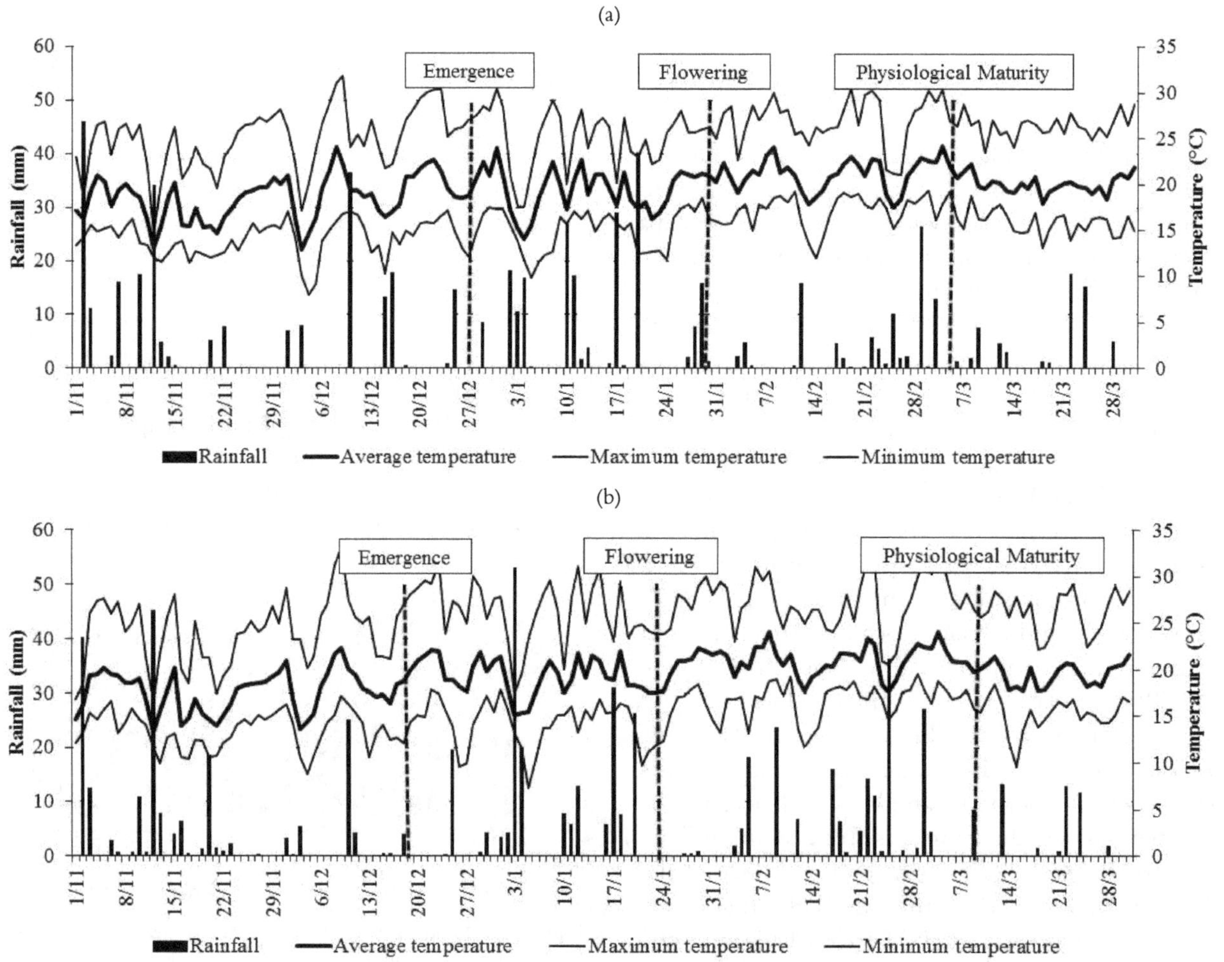

Figure 2. Maximum, average and minimum temperatures and pluvial precipitation for the months of November 2008 through March 2009 in Joaçaba (a) and Lages (b).

In Joaçaba, the total pluvial precipitation during the bean cycle was 296 mm, whereas it was 387 mm in Lages; thus, there was no hydric deficiency for the crops in either location. The pluvial precipitation during the period between emergence and flowering was similar in both locations (206 mm in Joaçaba and 201 mm in Lages). However, during the period between flowering and physiological maturity, an elevated pluvial precipitation was recorded in Lages (181 mm) compared to Joaçaba (95 mm). These facts explain the observation that, on average, the genotypes in Lages presented a greater number of days between flowering and physiological maturity than in Joaçaba.

The lower yield of the Lages genotypes could be due to the lower luminosity associated with an excess of rainfall during the period between flowering and physiological maturity. While studying the problem of low luminosity, Liu et al. (2010) and Liu et al. (2006) concluded that the occurrence of long periods of greater cloud cover can cause photosynthetic deficiency and, as a consequence, a reduction in yield. According to Santos et al. (2003), in common beans, soybeans and weed plants, the yield is generally limited by photosynthetic capacity from the start of pod formation and as a result, a reduction in the rate of photosynthesis causes variations in the grain yield or biomass production.

Comparing the year with the least precipitation to the year with the highest amount of precipitation, while evaluating bean genotypes over three years of cultivation in terms of the cycle duration and yield, Ribeiro et al. (2004) observed that there was an increase in the length of the cycle during the flowering-physiological maturity period. In the same study, it was observed that the reduction in the cycle length was associated with greater yield. According to Albayrak and Töngel (2006), vetch genotypes grown under different amounts of precipitation presented a negative correlation between the number of days to maturity and yield, with the harvest index and biological yield being the principal characteristics that determine the correlation value. Türk et al. (2008) observed a negative correlation between the yield and the number of days to flowering in *Vicia narbonensis* L. and described the harvest index and biological yield as the most important characteristics in determining the magnitude and direction of the correlation.

Conclusion

The common bean canopy characteristics were not affected by environmental influences, except plant height. The genotypes showed diversity of the cycle, and there was a cultivation location effect on each genotype, particularly, the differences on reproductive phase. The genotypes with the longest cycles presented the lowest yields. The phase from flowering to physiological maturity must coincide with a period of regular rainfall associated with a minimum of cloudy days leading to a decrease in the cycle length. Considering the 26 genotypes studied, Joaçaba/SC presents better climatic conditions and, thus, a higher potential yield for beans in comparison to Lages, Santa Catarina State.

References

ALBAYRAK, S.; TÖNGEL, Ö. Path analyses of yield and yield-related traits of common vetch (*Vicia sativa* L.) under different rainfall conditions. **Journal of Faculty of Agriculture of Ondokuz Mayis University**, v. 21, n. 1, p. 27-32, 2006.

ALVES, G. F.; RAMALHO, M. A. P.; ABREU, Â. F. B. Desempenho de cultivares antigas e modernas de feijão avaliadas em diferentes condições ambientais. **Ciência e Agrotecnologia**, v. 25, n. 4, p. 863-870, 2001.

BARELLI, M. A. A.; GONÇALVES-VIDIGAL, M. C.; AMARAL JÚNIOR, A. T.; VIDIGAL FILHO, P. S.; SILVÉRIO, L. Genetic control on number of days to flowering and yield components in common bean (*Phaseolus vulgaris* L.). **Acta Scientiarum. Agronomy**, v. 21, n. 3, p. 423-427, 1999.

BURATTO, J. S.; MODA-CIRINO, V.; FONSECA JÚNIOR, N. S.; PRETE, C. E. C.; FARIA, R. T. Adaptabilidade e estabilidade produtiva em genótipos precoces de feijão no estado do Paraná. **Semina: Ciências Agrárias**, v. 28, n. 3, p. 373-380, 2007.

CABRAL, P. D. S.; SOARES, T. C. B.; LIMA, A. B. P.; SOARES, Y. J. B.; SILVA, J. A. Análise de trilha do rendimento de grãos de feijoeiro (*Phaseolus vulgaris* L.) e seus componentes. **Revista Ciência Agronômica**, v. 42, n. 1, p. 132-138, 2011.

CARVALHO, C. G. P.; ARIAS, C. A. A.; TOLEDO, J. F. F.; OLIVEIRA, M. F.; VELLO, N. A. Correlações e análise de trilha em linhagens de soja semeadas em diferentes épocas. **Pesquisa Agropecuária Brasileira**, v. 37, n. 3, p. 311-320, 2002.

CIAT-Centro Internacional de Agricultura Troppical. FERNÀNDEZ, F.; GEPTS, P.; LÓPEZ, M. (Ed.). **Etapas de desarollo de la planta de frijol común**. Cali: CIAT, 1983.

CQFS-Comissão de Química e Fertilidade do Solo - RS/SC. **Manual de adubação e de calagem para os estados do Rio Grande do Sul e Santa Catarina**. 10. ed. Porto Alegre: SBCS - Núcleo Regional Sul/UFRGS, 2004.

DALLA CORTE, A.; MODA-CIRINO, V.; DESTRO, D. Adaptability and phenotypic stability in early common bean cultivars and lines. **Crop Breeding and Applied Biotechnology**, v. 2, n. 4, p. 525-534, 2002.

DALLA CORTE, A.; MODA-CIRINO, V.; SCHOLZ, M. B. S. Enviroment effect on grain quality in early common bean cultivars and lines. **Crop Breeding and Applied Biotechnology**, v. 3, n. 3, p. 193-202, 2003.

DAWO, M. I.; SANDERS, F. E.; PILBEAM, D. J. Yield, yield components and plant architecture in the F3 generation of common bean (*Phaseolus vulgaris* L.) derived from a cross between the determinate cultivar 'Prelude' and an indeterminate landrace. **Euphytica**, v. 156, n. 1, p. 77-87, 2007.

FAGERIA, N. K.; BALIGAR, V. C.; CLARK, R. **Physiology of crop production**. New York: Food Products Press, 2006.

FANCELLI, A. L.; DOURADO NETO, D. **Produção de feijão**. 2. ed. Piracicaba: Livroceres, 2007.

IPGRI. **Descritores para *Phaseolus vulgaris***. Rome: International Plant Genetic Resources Institute, 2001.

LITTEL, R. C.; MILLIKEN, G. A.; STROUP, W. W.; WOLFINGER, R. D.; SCHABENBERGER, O. **SAS® for mixed models**. 2nd ed. Cary: SAS Institute Inc., 2006.

LIU, X. B.; HERBERT, S. J.; BAATH, K.; HASHEMI, A. M. Soybean (*Glycine max*) seed growth characteristics in response to light enrichment and shading. **Plant, Soil and Environment**, v. 52, n. 4, p. 178-185, 2006.

LIU, B.; LIU, X. B.; WANG, C.; LI, Y. S.; JIN, J.; HEBERT, S. J. Soybean yield and yield component distribution across the main axis in response to light enrichment and shading under different densities. **Plant, Soil and Environment**, v. 56, n. 8, p. 384-392, 2010.

RAMALHO, M. A. P.; ABREU, A. F. B.; SANTOS, J. B. Genetics progress after four cycles of recurrent selection for yield and grain traits in common bean. **Euphytica**, v. 144, n. 1, p. 23-29, 2005.

RAMALHO, M. A. P.; ABREU, A. F. B.; SANTOS, J. B. Desempenho de progênies precoces de feijoeiro (*Phaseolus vulgaris* L.) em diferentes locais e épocas de plantio. **Revista Ceres**, v. 40, n. 229, p. 272-280, 1993.

RIBEIRO, N. D.; HOFFMANN JUNIOR, L.; POSSEBON, S. B. Variabilidade genética para ciclo em feijão dos grupos preto e carioca. **Revista Brasileira de Agrociência**, v. 10, n. 1, p. 19-29, 2004.

RIBEIRO, N. D.; MELLO, R. M.; DALLA COSTA, R.; SLUSZZ, T. Correlações genéticas de caracteres agromorfológicos e suas implicações na seleção de genótipos de feijão carioca. **Revista Brasileira de Agrociência**, v. 7, n. 2, p. 93-99, 2001.

ROCHA, M. M.; VELLO, N. A. Interação genótipos e locais para rendimento de grãos de linhagens de soja com diferentes ciclos de maturação. **Bragantia**, v. 58, n. 1, p. 69-81, 1999.

SANTOS, J. B.; PROCÓPIO, S. O.; SILVA, A. A.; COSTA, L. C. Captura e aproveitamento da radiação solar pelas culturas da soja e do feijão e por plantas daninhas. **Bragantia**, v. 62, n. 1, p. 147-153, 2003.

SILVA, C. A.; ABREU, A. F. B.; RAMALHO M. A. P. Associação entre arquitetura de planta e produtividade de grãos em progênies de feijoeiro de porte ereto e prostrado. **Pesquisa Agropecuária Brasileira**, v. 44, n. 12, p. 1647-1652, 2009.

SILVA, F. B.; RAMALHO, M. A. P.; ABREU, A. F. B. Seleção recorrente fenotípica para florescimento precoce de feijoeiro 'Carioca'. **Pesquisa Agropecuária Brasileira**, v. 42, n. 10, p. 1437-1442, 2007.

SOUZA, C. A.; COELHO, C. M. M.; GUIDOLIN, A. F.; ENGELSING, M. J.; BORDIN, L. C. Influência do ácido giberélico sobre a arquitetura de plantas de feijão no início de desenvolvimento. **Acta Scientiarum. Agronomy**, v. 32, n. 2, p. 325-332, 2010a.

SOUZA, C. A.; COELHO, C. M. M.; STEFEN, D. L. V.; SACHS, C.; FIGUEIREDO, B. P. Atributos morfométricos e componentes da produção do feijoeiro sob efeito de redutores de crescimento. **Científica**, v. 38, n. 1-2, p. 30-38, 2010b.

TRAKA-MAVRONA, E.; GEORGAKIS, D.; KOUTSIKA-SOTIRIOU, M. Phenology and quality studies on a snap bean cultivar deviation. **Journal of New Seeds**, v. 4, n. 4, p. 41-56, 2002.

TÜRK, M.; ÇELÝK, N.; BAYRAM, G.; BUDAKLI, E. Relationships between seed yield and yield components in narbon bean (*Vicia narbonensis* L.) by path analysis. **Bangladesh Journal of Botany**, v. 37, n. 1, p. 27-32, 2008.

VALE, N. M.; BARILI, L. D.; ROCHA, F.; ROZZETTO, D. S.; PEREIRA, T. P.; COIMBRA, J. L. M.; GUIDOLIN, A. F.; BERTOLDO, J. G. Métodos de semeadura empregados como critério de avaliação fenotípica em melhoramento de feijão. **Biotemas**, v. 22, n. 4, p. 73-81, 2009.

ZILIO, M.; COELHO, C. M. M.; SOUZA, C. A.; SANTOS, J. C. P.; MIQUELLUTI, D. J. Contribuição dos componentes de rendimento na produtividade de genótipos crioulos de feijão (*Phaseolus vulgaris* L.). **Revista Ciência Agronômica**, v. 42, n. 2, p. 429-438, 2011.

2

History of northern corn leaf blight disease in the seventh cycle of recurrent selection of an UENF-14 popcorn population

Rodrigo Moreira Ribeiro[1*], Antônio Teixeira do Amaral Júnior[1], Guilherme Ferreira Pena[1], Marcelo Vivas[1], Railan Nascimento Kurosawa[1] and Leandro Simões Azeredo Gonçalves[2]

*[1]Centro de Ciências e Tecnologia Agropecuária, Universidade Estadual do Norte Fluminense Darcy Ribeiro, Av. Alberto Lamego, 2000, 28013-602, Parque Califórnia, Campos dos Goytacazes, Rio de Janeiro, Brazil. [2]Departamento de agronomia, Universidade Estadual de Londrina, Londrina, Paraná, Brazil. *Author for correspondence. E-mail: rodrigo.moreira85@yahoo.com.br*

ABSTRACT. To investigate the genetic resistance to northern corn leaf blight (*Exserohilum turcicum*) disease in the UENF-14 popcorn population and the inclusion of this trait in the analysis, two hundred ten popcorn half-sib families were evaluated using an incomplete block design with treatments arranged in replicates "Sets". Regarding the Families within "Set" (F/S) source of variation, enough genetic variability was found for northern corn leaf blight disease resistance to be explored in the next cycles of the UENF popcorn breeding program. The open-pollinated UENF-14 variety has a source of resistance to this disease in its genetic basis, given that it originated from crossing with an American cultivar with yellow grains that is resistant to disease. These results strengthen the possibility of selecting resistant families across the cycles, so it is concluded that the selection was effective at maintaining the balance in the occurrence of the disease in the study population and that the source of resistance is not lost with advancing cycles. The addition of the foliar disease variable in the analysis is of the utmost importance for the improvement of popcorn as it makes it possible to aggregate genes for resistance to this disease along with agronomic traits of interest.

Keywords: *Exserohilum turcicum*, half-sib families, *Zea mays*.

Introduction

Popcorn (*Zea mays everta* Sturt.) is a special type of maize and a popular snack food in Brazil, where its consumption has increased over the years, so it has become an economically attractive crop for farmers across the country (Mendes de Paula et al., 2010; Silva, Amaral Júnior, Gonçalves, Freitas Júnior, & Ribeiro, 2011; Moterle et al., 2012; Gonçalves et al., 2014). However, official data from Brazilian government institutions have revealed that popcorn production remains limited relative to the potential market for the crop, and the primary limiting constraint is obtaining cultivars with multiple favorable agronomic traits (Arnhold, Mora, Silva, Good-God, & Silva, 2009; Moterle et al., 2012; Ribeiro et al., 2012; Silva et al., 2013). There are 48 registered cultivars in Brazil (Ministério da Agricultura Pecuária e Abastecimento [MAPA], 2013), and most of them belong to the popcorn industry, which is represented by packing companies that make the seed stock available to only a few partner producers.

In contrast to common corn, popcorn plants generally have smaller, thinner and weaker stems; are very precocious in the maturation and drying of grains; produce tillers more often; have higher susceptibility to diseases and pests; are more prolific, i.e., have a greater number of ears per plant; and have fewer and narrower leaves and a small grain size (Li et al., 2008; Viegas Neto et al., 2012). Popcorn plants also have a less developed root system and suffer greater damage by attack from curcubit beetle worms (*D. speciosa*) and nematodes, which makes the plant more susceptible to lodging and desiccation (Li et al., 2008). This maize crop is further affected by several leaf diseases, which can cause significant damage to yield and grain quality. In particular, popcorn is susceptible to the northern corn leaf blight, which is caused by *Exserohilum turcicum* (Pass.) Leonard & Suggs (sin. *Helminthosporium turcicum* Pass.) and is characterized as one of the main foliar diseases (Carson, 2006; Harlapur et al., 2008; Scapin, Carnelossi, Vieira, Schwan-Estrada, & Cruz, 2010; Ishfaq et al., 2014).

According to Ishfaq et al. (2014), northern corn leaf blight disease is characterized by long elliptical, greyish green or tan leaf lesions that first appear on the lower leaves and increase in size and number until very little living tissue remains, and yield is reduced due to a lack of carbohydrates for grain filling. Northern corn leaf blight is a disease that occurs widely in all regions where susceptible corn, sweetcorn and popcorn are grown, and it has great potential to cause damage and has been studied for years (Rossi & Reis, 2014). Damage to the green leaf area during grain filling can cause up to a 40% reduction in grain yield in susceptible hybrids (Ferguson & Carson, 2007; Wang et al., 2010; 2012). The use of resistant cultivars has been the primary control measure of corn leaf diseases (Casela, Ferreira, & Pinto, 2006; Ferguson & Carson, 2007; Vieira et al., 2009; Ishfaq et al., 2014; Ayiga-Aluba, Edema, Tusiime, Asea, & Gibson, 2015), since it reduces production costs and minimizes management activities and environmental risks.

In this context, recurrent selection is an excellent breeding strategy for developing disease resistance since the goal of this method is to gradually increase the frequency of favorable alleles in a population with no loss in genetic variability. Recurrent selection consists of three steps: the development of progeny, their evaluation, and their recombination. These steps are carried out cyclically until the frequency of favorable alleles reaches satisfactory levels in the population (Hallauer, Carena, & Miranda Filho, 2010). Jenkins, Robert, and Findley Junior (1954) tested the efficiency of recurrent selection to concentrate polygenic resistance genes into nine groups of progenies. The results revealed that, in most groups, two cycles of recurrent selection were efficient for concentrating resistance genes that promoted good control of northern corn leaf blight disease, but in some cases, there was a reduction in the level of resistance.

As most of the gene action is additive and the level of resistance is related to the number of lesions, Hooker (1973) believes that a simple selection procedure can be effective in isolating maize lines with polygenic resistance to *E. turcicum*. According to the author, recurrent selection has been shown to be an effective means to concentrate resistance genes, and Miles, Dudley, Dudley, and Lambert (1981) found that it is possible to increase resistance to *E. turcicum* by any of several recurrent selection methods. Quantitative resistance to disease has been rapidly obtained through recurrent selection, and resistance to *E. turcicum* can be achieved with only two or three cycles of selection.

Thus, the Darcy Ribeiro North Fluminense State University (UENF) has developed a popcorn breeding program that employs a recurrent selection strategy, and it is currently in the seventh cycle of selection of an open-pollinated UENF-14 population. In its genetic base, the UENF-14 popcorn population has been crossed with the American popcorn variety, which is resistant to northern corn leaf blight (Resh et al., 2015), so the main purpose of this study was to evaluate the severity of the disease over seven recurrent selection cycles (C_0 to C_6) to verify that the alleles that confer resistance to *E. turcicum* have been maintained over the cycles.

Material and methods

Population origin

The UNB-2U open-pollinated population, currently named UENF-14, originated from the UNB-2 variety after two cycles of mass selection at Campos dos Goytacazes, Rio de Janeiro State, Brazil. The UNB-2 was originated from the UNB-1 variety, which came from a 'Composto Indígena' (Indian compound) selection donated to UNB (University of Brasília, Brazil) by ESALQ/USP, Piracicaba, São Paulo State, Brazil. The UNB-1 was crossed with an American popcorn variety, and selected plants from that cross were crossed with yellow grain popcorn, with a high-yield and *Exserohilum turcicum*-resistant genotype. After the second crossing, mass selection was applied to form a population of resistant, high yielding plants with yellow grains. This population was then backcrossed three times with the American popcorn variety, eventually originating the UNB-2 open-pollinated variety (Pereira & Amaral Júnior, 2001; Daros, Amaral Júnior, & Pereira, 2002).

The summary of the population trajectory is shown in Table 1, which presents the strategies used during the seven cycles of recurrent selection in the UENF-14 popcorn population. The selection indices and the characteristics used for selection are also found in Table 1.

Plant material

To obtain the half-sib families, ten lines of each cycle (C_0 to C_6) interspersed with 10 lines of the tester cycle (C_0) of the UENF-14 population were used so that the half-sib families were obtained from crossings of the recurrent selection cycles (C_0, C_1, C_2, C_3, C_4, C_5 and C_6) with the tester cycle (C_0). As they arose, the tassels were eliminated from the line used as a female parent (C_0, C_1, C_2, C_3, C_4, C_5 and C_6) so that only the lines containing the tester (C_0) were able to produce pollen and pollinate the experimental field without contamination. Plants were spaced 0.20 m; there were 0.90 m between rows, and the lines were 5.0 m long. Three seeds were planted per hole at a depth of 0.05 m, and at 21 days after emergence, thinning was carried out, leaving one plant per hole.

Thus, this procedure produced 210 half-sib families, i.e., 30 half-sib families for each cycle. In other words, 30 families were obtained from the crossings between C_0 x C_0, 30 additional families from the crossings between C_0 x C_1 and so on until the 30 families from the crossings between the C_0 x C_6 intersection. These 210 half-sib families were evaluated in trials in Campos (Latitude: 21 44 '47' 'S, Longitude: 41 18' 24 " W and Altitude: 11 m), Rio de Janeiro, Brazil. The off-season planting was performed on April 18, 2013 (Environment 1) and repeated for the second crop, for which planting occurred on September 4, 2013 (Environment 2).

Experimental design

The experimental design was a randomized block that was repeated in "Sets". Three "Sets" with three replications were used, and each "Set" contained 73 treatments, i.e., 70 families of half-sib and three controls (Beija-Flor, RS-20 and Barão Viçosa). Each "Set" was composed of families from each cycle that were numbered from 1 to 30. Therefore, in "Set" 1, 1 to 10 randomly selected families were clustered with more controls; families 11 to 20 were grouped in the "Set" 2 with more controls; and families from 21 to 30 were placed in "Set" 3 along with the controls.

The lines were cultivated in rows that were 2.40 m long and spaced 0.90 m from each other with 13 plants spaced 0.20 m apart in each row. Three seeds were sown in each hole at a depth of 0.05 m, and at 21 days after emergence, thinning was carried out leaving one plant per hole for a combined population of 60,185 plants per hectare. Fertilization planting was carried out according to the soil analysis, and the topdressing was held for approximately 30 days after planting. The cultural treatments were performed according to the needs of the culture.

Table 1. Strategies, selection indexes and evaluated characteristics in the seven cycles of recurrent selection of the UENF-14 popcorn population at UENF.

Cycle	Strategy Selection	Selection index	Evaluated features	Reference
C_0	Mass selection	-	SILK, PH, GY, W100, SV, PE	Pereira and Amaral Júnior, (2001)
C_1	Full-sib	-	SILK, PH, GY, PE	Daros et al. (2002)
C_2	S_1	Smith (1936) and Hazel (1943)	NSE, PE, NP, BP, PHE, NE, GY	Daros et al. (2004)
C_3	Half-sib	Mulamba and Mock (1978)	PE, GY, W100, NSE, EP, PH, HE, NP, NBP, LP, PHE, SILK	Santos, Amaral Júnior, Freitas Júnior, Rangel, and Pereira (2007)
C_4	Full-sib	Mulamba and Mock (1978)	NE, NSE, EP, WP, GY, PE, W100, SILK, PH, HE, NP, NBP, LP, PHE	Freitas Júnior, Amaral Júnior, Rangel, and Viana (2009)
C_5	Full-sib	Mulamba and Mock (1978)	PH, HE, NSE, EP, GY, PE	Rangel, Amaral Júnior, Gonçalves, Freitas Júnior, and Candido (2011)
C_6	Full-sib	Mulamba and Mock (1978)	NP, NBP, PH, HE, NE, NSE, WE, GY, W100, PE	Ribeiro et al. (2012)

NP = number of plants per plot; SV = volume of one hundred seeds; NBP = average number of broken plants; PH = average plant height; HE = average height of the insertion of the first ear; NE = average number of ears; NSE = average number of sick ears; WE = mean weight of ears; GY = grain yield; W100 = mean weight of 100 grains; PE = popping expansion; SILK = silk emergence; PHE = poorly hulled ears; EP = average number of ears attacked by pests; LP = average number of lodged plants; and WP = average weight of ears with pests.

Evaluated characteristics

Given that the experimental area had been successively planted with corn for many years, there was no pathogen inoculation, so the disease occurred spontaneously in the course of crop development.

The reaction of the genotypes to the foliar disease was monitored by estimating the severity of symptoms. Two estimation modes were adopted: measuring the percentage of the entire plant showing symptoms and measuring the percentage of the leaf immediately below the first ear showing symptoms, which were termed the severity in the plant and the severity in the leaf, respectively.

The severity of the disease based on the plant was estimated by modifying the scale notes adopted by Agroceres (1996), where the scale ranged from 1 to 9, and note 1: 0.5% severity; note 2: 1% severity; note 3: 10% severity; note 4: 30% severity; note 5: 50% severity; note 6: 70% severity; note 7: 80% severity; note 8: 90% severity; and note 9: 100% severity. The evaluation was performed by an appraiser once each season using six competitive plants per plot after the plants flowered.

The assessment of northern corn leaf blight severity on the leaf was performed once in both the 1st and 2nd seasons using six competitive plants per plot after flowering. The diagrammatic scale proposed by Lazaroto, Santos, Konflanz, Malagi, and Camochena (2012) was used, which included severity ranges according to the following estimated percentages of the leaf area affected by the disease: 0.5, 1.0, 2.5, 6.5, 15.5, 30.0, and 54.0%. These percentages were attributed to notes from 1 to 7, according to the observed severity.

Genetic x Statistical analysis

The variance analysis of the data was performed according to the statistical model $Y_{ijkl} = \mu + A_i + S_j + AS_{ij} + R/AS_{ijk} + F/S_{jl} + AF/S_{ijl} + e_{ijkl}$, where μ is the average; A_i is the fixed effect of the ith environment; S_j is the effect of the jth "Set"; AS_{ij} is the effect of the interaction between the environments and the "Sets"; R/AS_{ijk} is the effect of the kth repetition within the interaction between the ith environment and jth "Set"; F/S_{jl} is the random effect of the ith family within the jth "Set" (NID, 0, $\sigma^2_{F/S}$); AF/S_{ijl} is the effect of the interaction between the environments and families within the jth "Set"; and e_{ijkl} is the experimental error (NID, 0, σ^2). Based on the proposed model, the variance analysis was performed using SAS® (SAS 9.1, SAS Institute, 2002, Cary, NC, USA).

The genetic, phenotypic and environment components were obtained, where

$$\hat{\sigma}_G^2 = \frac{QMF/S - QMR}{ar}$$

is the estimator of the genotypic variance among families;

$$\hat{\sigma}_F^2 = \frac{QMF/S}{ar}$$

is the estimator of the phenotypic variance between families;

$$\hat{\sigma}_R^2 = QMR/ar$$

is the estimator of the residual average variance;

$$\hat{\sigma}_{GA}^2 = \frac{QM(AxF)/S - QMR}{r} \cdot \frac{a-1}{a}$$

is the estimator of the variance in the genotype-versus-environment interaction;

$$\hat{h}_{\bar{x}}^2 = \frac{\hat{\sigma}_G^2}{\hat{\sigma}_F^2} = \frac{QMF/S - QMR}{QMF/S}$$

x 100 is the percentage heritability based on the family averages;

$$\hat{I}_v = \frac{CVg}{CVe}$$

is the estimator of the variation index, where CVg is the genetic variation coefficient, and CVe is the coefficient of experimental variation.

Results and discussion

Through analysis of variance, significant differences were found in the Environment (E) source of variation between the two traits (Table 2), and the significance of this source of variation shows that the environments were distinct enough to promote differences in the evaluated characteristics. This result was expected since the pathogen has good survivability in crop residues, and its dissemination occurs by transport conidia over long distances by wind. Furthermore, moderate to high temperatures and high humidity favor the proliferation of the disease (Palaversic et al., 2012); the infection caused by *Exserohilum turcicum* is favored by mild temperatures ranging from 20 to 25°C and relative humidity above 90%, conditions that are ideal for the development of epidemics (Cota, Silva, & Costa, 2013). From Table 2, it can be seen that the mean observed for the first environment was higher for the disease severities on both the plant and the leaf. This result corroborates the prediction, considering that average temperatures were lower at the time of the experiment. In contrast, during the period corresponding to Environment 2, higher temperatures were prevalent, which are not considered optimal for the development of the disease, so the disease severity was lower.

The Families within "Set" (F/S) source of variation was found to be significant for both of the evaluated characteristics ($p < 0.01$), demonstrating sufficient genetic variability to be explored in the next cycle of the UENF popcorn breeding program in regard to northern corn leaf blight disease. The UENF-14 open-pollinated variety has a source of resistance to this disease in its genetic basis, given that it originated from the crossing of an American cultivar with yellow grains that is resistant to disease (Resh et al., 2015). In this context, the population appears to have been stable for this variable throughout the recurrent selection process, which was confirmed by means of *boxplots* (Figures 1 and 2).

Further, intuitive selection for greater disease resistance was conducted once the highly affected plants were not selected for recombination. Moreover, there is no denying that the families that were greatly affected by the disease had lower productivity, so this variable was selected for indirect selection. Because it is a characteristic of great importance for the improvement of this population, it would be interesting to include productivity in the model/selection index for new cycles to select families that add favorable alleles for this trait. Another possibility is the selection of families that add favorable alleles for grain yield and expansion capacity, which are achieved over the recurrent selection cycles, and exhibit a considerable level of resistance to northern corn leaf blight.

For the Environment versus Families within "Set" (ExF)/S source of variation, there were significant differences in the evaluated characteristic at the 5% level of probability. The significance of the (ExF)/S interaction indicates that the evaluated families behaved distinctly in both of the environments, but the significance is a question of probability. With greater degrees of freedom, there is also greater sensitivity, allowing for the detection of significant differences. As the cycles unfold, the set of treatments in each cycle is much smaller, which reduces the statistical power because higher interaction values are required to determine significance. Thus, greater attention should be paid to the unfolding as it can determine if the significance of the main effects and interactions occurred within each level. Significance was not observed for most characteristics in the unfolding, except for C_4, so it is possible to select for superior families independent of the environment because the interactions over the cycles were not significant.

Table 2. Estimates of the mean squares, averages and percentages of the experimental coefficients of variation from two characteristics evaluated in 210 half-sib families in the UENF-14 popcorn population at Campos dos Goytacazes, Rio de Janeiro State in 2013.

FV	GL	Mean Square	
		SL[/1]	SP[/1]
Environment (E)	3	124.94**	33.11**
Set (S)	2	0.0624[ns]	0.1987[ns]
E x S	6	0.0488[ns]	0.2127[ns]
Replications (R)/E x S	24	0.1290**	0.5030**
Families (F)/ S	207	0.0391**	0.1825**
C_0	27	0.0329[ns]	0.2195**
C_1	27	0.0419**	0.1561[ns]
C_2	27	0.0351[ns]	0.1755[ns]
C_3	27	0.0305[ns]	0.1049[ns]
C_4	27	0.0446**	0.2200*
C_5	27	0.0282[ns]	0.1429[ns]
C_6	27	0.0413**	0.1957*
Contrast	18	0.0688**	0.2764**
(E x F)/ S	621	0.0279*	0.1532*
C_0	81	0.0264[ns]	0.1483[ns]
C_1	81	0.0236[ns]	0.1434[ns]
C_2	81	0.0218[ns]	0.1186[ns]
C_3	81	0.0317[ns]	0.1280[ns]
C_4	81	0.0424**	0.2377*
C_5	81	0.0258[ns]	0.1359[ns]
C_6	81	0.0304[ns]	0.1471[ns]
Contrast	54	0.0057[ns]	0.0576[ns]
Residue	1618	0.0226	0.1211
Original average (Oa)		2.58	0.53
Oa (Env 1)		3.65	1.21
Oa (Env 2)		1.50	0.12
CV (%)		8.74	37.50

[/1] SL = disease severity on the leaf; SP = disease severity on the plant; ** = Significant at the 1% level of probability by F test; * = Significant at 5% the level of probability by F test; [ns] = Not significant

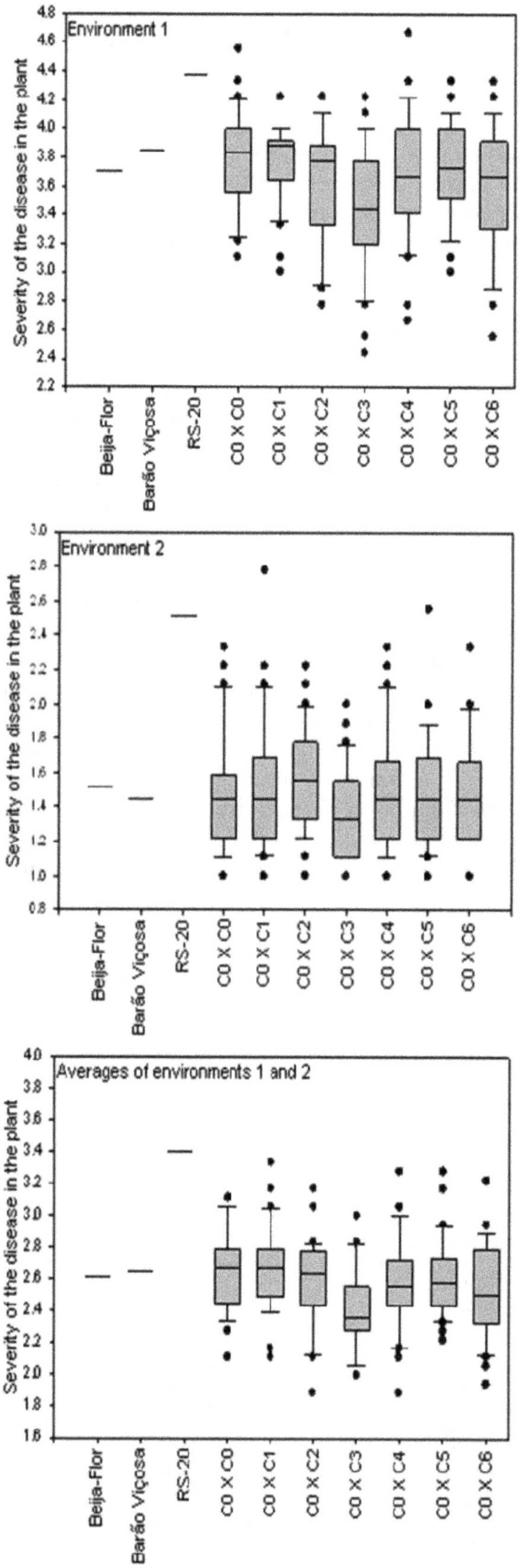

Figure 1. Boxplot – Severity of the disease on the plant in environments 1and 2 and the averages of environments 1 and 2.

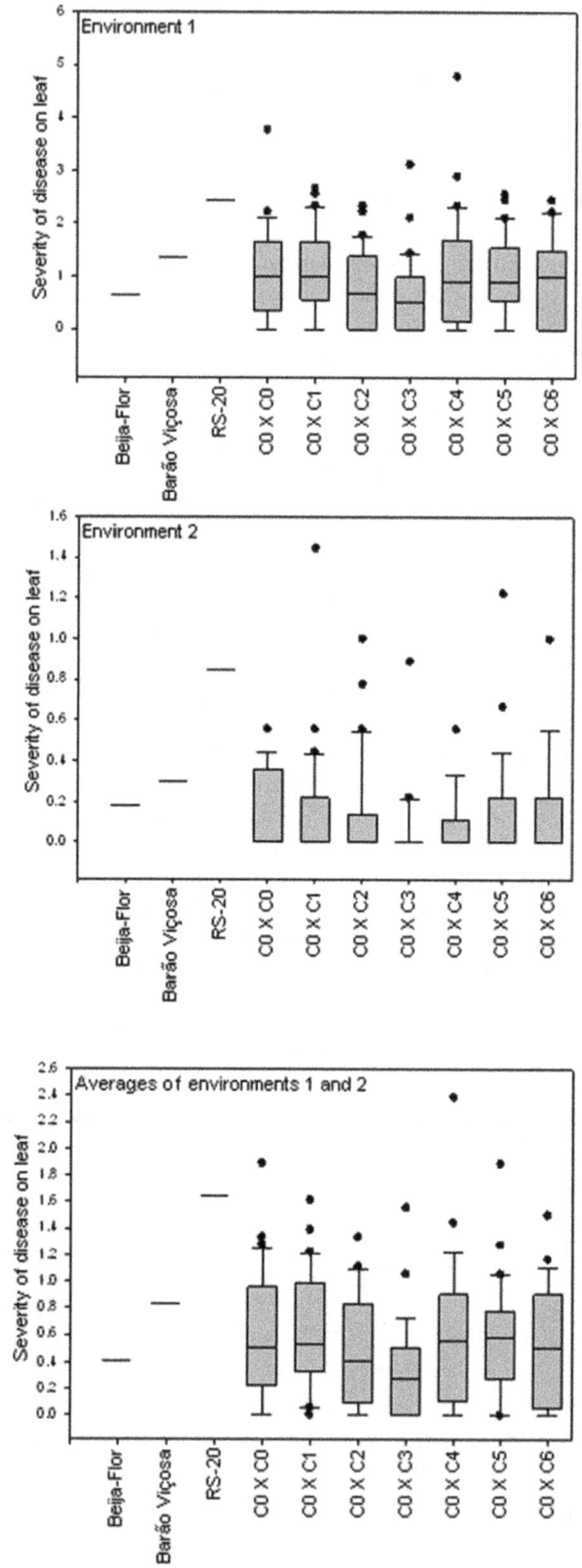

Figure 2. Boxplot – Severity of disease on the leaf in environments 1 and 2 and the averages of environments 1 and 2.

Table 3 shows that there is heritability for disease severity on the leaf ranging from 19.86 to 49.33%, and the severity of the disease on the plant ranges from 0 to 44.95%. The zero heritability value obtained in this experiment was due to the lack of genotypic variance between individuals in the C_3 cycle. It is highly probable that this event is due to the use of inbred S_1 families in the previous cycle (C_2), which may have led to the narrowing of the genetic basis of the selected individuals. In the same table, it can be seen that the values of the variation index are greater than one unit. These results strengthen the possibility of selecting resistant families across the cycles so that the inclusion of this variable in the analysis would add additional information and result in the release of the variety with the highest level of resistance to northern corn leaf blight.

Ayiga-Aluba et al. (2015) assessed the response to two cycles of S_1 recurrent selection for northern corn leaf blight in a population of an open-pollinated maize variety. They found that moderate heritabilities, desirable selection differentials and significant improvements in northern corn leaf blight disease resistance indicate that recurrent S_1 selection was effective in improving that study population. This result confirms the importance of assessing the severity of this disease in breeding programs through recurrent selection.

Furthermore, in Table 3, we see that the heritability values fluctuate greatly, which reveals that the target characteristic suffers from significant environmental effects, so more robust methods to improve selection of superior families are necessary. According to the *boxplot* graphics, the same consideration should be made as noted above; there is variation due to disparate individuals (outliers) with both higher and lower levels of disease.

Therefore, it is concluded that the selection was effective at maintaining the balance in the occurrence of the disease in the study population and that the source of resistance is not lost with advancing cycles. The addition of a foliar disease variable in the analysis is of the utmost importance for the improvement of popcorn as it would be possible to aggregate genes for resistance to this disease along with agronomic traits of interest, allowing the selection of resistant and productive families in advanced cycles. Thus, the development of a new variety of popcorn will be possible so that farmers will be able to find high productivity and resistance to one of the major diseases that affect the culture in the same cultivar.

Conclusion

The results obtained in this study strengthens the possibility of selecting resistant families across the cycles, so it is concluded that the selection was effective at maintaining the balance in the occurrence of the disease in the study population and that the source of resistance is not lost with advancing cycles. Thus, the addition of the foliar disease variable in the analysis is of the utmost importance for the improvement of popcorn as it makes it possible to aggregate genes for resistance to this disease along with agronomic traits of interest.

Table 3. Genetic parameters: estimates of phenotypic variance ($\hat{\sigma}_F^2$), genotypic variance ($\hat{\sigma}_G^2$), residual variance ($\hat{\sigma}_r^2$), variance in the genotype-versus-environment interaction ($\hat{\sigma}_{GA}^2$), heritability based on the family averages ($\hat{h}^2\bar{x}$), genetic variation coefficient ($CV\hat{g}$), variation index ($\hat{I}_v$) and additive variance (σ_a^2) in the severity of the disease on the plant and on the leaf. Evaluated in 210 families in the UENF-14 popcorn population at Campos dos Goytacazes, Rio de Janeiro State 2013/2014.

SL[1/]	$\hat{\sigma}_F^2$	$\hat{\sigma}_G^2$	$\hat{\sigma}_r^2$	$\hat{\sigma}_{GA}^2$	$\hat{h}^2\bar{x}$	$CV\hat{g}$	$\hat{I}_v$	σ_a^2
C_0	0.0055	0.0017	0.0038	0.0006	31.31	21.21	2.43	0.0069
C_1	0.0070	0.0032	0.0038	0.0002	46.06	11.82	1.35	0.0129
C_2	0.0059	0.0021	0.0038	0.0000	35.61	19.74	2.26	0.0083
C_3	0.0051	0.0013	0.0038	0.0015	25.90	0.00	0.00	0.0053
C_4	0.0074	0.0037	0.0038	0.0033	49.33	21.60	2.47	0.0147
C_5	0.0047	0.0009	0.0038	0.0005	19.86	10.64	1.22	0.0037
C_6	0.0069	0.0031	0.0038	0.0013	45.28	21.82	2.50	0.0125
SP[1/]	$\hat{\sigma}_F^2$	$\hat{\sigma}_G^2$	$\hat{\sigma}_r^2$	$\hat{\sigma}_{GA}^2$	$\hat{h}^2\bar{x}$	$CV\hat{g}$	$\hat{I}_v$	σ_a^2
C_0	0.0366	0.0164	0.0202	0.0045	44.83	2.51	0.07	0.0656
C_1	0.0260	0.0058	0.0202	0.0037	22.42	3.41	0.09	0.0233
C_2	0.0293	0.0091	0.0202	0.0000	31.00	2.86	0.08	0.0363
C_3	0.0175	0.0000	0.0202	0.0012	0.00	2.58	0.07	0.0000
C_4	0.0367	0.0165	0.0202	0.0194	44.95	3.83	0.10	0.0659
C_5	0.0238	0.0036	0.0202	0.0025	15.26	1.90	0.05	0.0145
C_6	0.0326	0.0124	0.0202	0.0043	38.12	3.62	0.10	0.0497

[1/] SL = Severity of the disease on the leaf; SP = severity of the disease on the plant.

Acknowledgements

The authors are grateful to the Graduate Program in Genetics and Plant Breeding and the Foundation for Research Support of the State of Rio de Janeiro (FAPERJ) for the scholarship provided to the first author and to the Universidade Estadual do Norte Fluminense Darcy Ribeiro (UENF) for their financial support of this research.

References

Agroceres. (1996). *Guia agroceres de sanidade,* (p. 72). São Paulo, SP: Sementes Agroceres.

Arnhold, E., Mora, F., Silva, R. G., Good-God, P. I. V., & Silva, R. G. (2009). Evaluation of top-cross popcorn hybrids using mixed linear model methodology. *Chilean Journal of Agricultural Research, 69*(1), 46-53.

Ayiga-Aluba, J., Edema, R., Tusiime, G., Asea, G., & Gibson, P. (2015). Response to two cycles of S1 recurrent selection for turcicum leave blight in an open pollinated maize variety population (Longe 5) *Advances in Applied Science Research, 6*(12), 4-12.

Carson, M. L. (2006). Response of a maize synthetic to selection for components of partial resistance to Exserohilum turcicum. *Plant Disease, 90(7*), 910-914.

Casela, C. R., Ferreira, A. S., & Pinto, N. F. J. (2006). *Doenças na cultura do milho* (Circular Técnica 83, p. 14). Sete Lagoas, MG: Embrapa Milho e Sorgo.

Cota, L. V., Silva, D. D., & Costa, R. V. (2013). *Helmintosporiose Causada por Exserohilum turcicum na Cultura do Milho*, (Circular Técnica 195, p. 8,). Sete Lagoas, MG: Embrapa Milho e Sorgo.

Daros, M., Amaral Júnior, A. T., & Pereira, M. G. (2002). Genetic gain for grain yield and popping expansion in full-sib recurrent selection in popcorn. *Crop Breeding and Applied Biotechnology*, *2*(3), 339-344.

Daros, M., Amaral Júnior, A. T., Pereira, M. G., Santos, F. S., Grabiel, A. P. C., Scapim, C. A., Freitas Júnior, S. P., & Silvério, L. (2004). Recurrent selection in inbred popcorn families. *Scientia Agricola*, *61*(6), 609-614.

Ferguson, L. M., & Carson, M. L. (2007). Temporal variation in Setosphaeria turcica between 1974 and 1994 and origin of races 1, 23, and 23N in the United States. *Phytopathology, 97*(11), 1501-1511.

Freitas Júnior, S. P., Amaral Júnior, A. T., Rangel, R. M., & Viana, A. P. (2009). Predição de ganhos genéticos na população de milho pipoca UNB- 2U sob seleção recorrente utilizando-se diferentes índices de seleção Genetic gain prediction on UNB-2U popcorn population under recurrent selection by using different selection indexes. *Semina: Ciências Agrárias*, *30*(4), 803-814.

Gonçalves, L. S. A., Freitas Jr, S. D. P., Teixeira do Amaral-Jr, A., Scapim, C. A., Rodrigues, R., Dornelles Marinho, C., & Stefani Pagliosa, E. (2014). Estimating combining ability in popcorn lines using multivariate analysis. *Chilean Journal of Agricultural Research*, *74*(1), 10-15.

Hallauer, A. R., Carena, M. J., & Miranda Filho, J. B. (2010). *Quantitative genetics in maize breeding*, (3a ed., p. 500). New York, USA: Springer.

Harlapur, S. I., Kulkarni, M. S., Wali, M. C., Srikant, K., Yashoda, H., & Patil, B. C. (2008). Status of turcicum leaf blight of maize in Karnataka. *Journal of Agricultural Science*, *21*(1), p. 55-60.

Hooker, A. L. (1973). Maize. In R. R. Nelson (Ed.), *Breeding plants for disease resistance* (p. 132-154). Pennsylvania, PA: Pennsylvania State University Press, University Park.

Ishfaq, A., Dar, Z. A., Lone, A. A., Ali, G., Gazal, A., Hamid, B., & Mohiddin, F. A. (2014). Disease reaction studies of maize (*Zea mays* L.) against turcicum leaf blight involving indigenously identified cytosterile source. *African Journal of Microbiology Research*, *8*(27), 2592-2597.

Jenkins, M. T., Robert, A. L., & Findley Junior, W. R. (1954). Recurrent selection as a method for concentrating genes for resistance to Helminthosporium turcicum leaf blight in corn. *Agronomy Journal*, *46*(2), 89-94.

Lazaroto, A., Santos, I., Konflanz, V. A., Malagi, G., & Camochena, R. C. (2012). Escala diagramática para avaliação de severidade da helmintosporiose comum em milho. *Ciência Rural, 42*(12) 2131-2137.

Li, Y., Dong, Y., Niu, S., Cui, D., Wang, Y., Liu, Y., Wei, M., & Li, X. (2008). Identification of agronomically favorable quantitative trait loci alleles from a dent corn inbred Dan232 using advanced backcross QTL analysis and comparison with the F 2:3 population in popcorn. *Molecular Breeding*, *21*, 1–14.

Ministério da Agricultura Pecuária e Abastecimento [MAPA]. (2013). Retrieved from www.agricultura.gov.br.

Mendes de Paula, T. O. P., Gonçalves, L. S. A., Amaral Júnior, A. T., Oliveira, E. C., Silva, V. Q. R., & Scapim, C. A. (2010). Magnitude of the genetic base of commercial popcorn and recommendation in Brazil. *Crop Breeding and Applied Biotechnology*, *10*, 289-297.

Miles, J. W., Dudley, J. W., Dudley, D. G., & Lambert, R. J. (1981). Response to selection for resistence to four disease in two corn populations. *Crop Science*, *21*(6), 980-983.

Moterle, L. M., Braccini, A. L., Scapim, C. A., Pinto, R. J. B., Gonçalves, L. S. A., Rodrigues R., & Amaral Júnior, A. T. (2012). Combining ability of popcorn lines for seed quality and agronomic traits. *Euphytica*, *185*(3), 337-347.

Palaversic, B., Jukic, M., Jukic, K., Zivkovic, I., Buhinicek, I., Jozinovic, T., Vragolovic, A., Kozic, Z., & Jemenarstvo, S. (2012). Breeding maize for resistance to Northern leaf blight (*Exserohilumturcicum* Pass.) [Croatian]. *Sjemenarstvo*, *29*(3/4), 111-120.

Pereira, M. G., & Amaral Júnior, A. T. (2001). Estimation of genetic componentes in popcorn based on the nested design. *Crop Breeding and Applied Biotechnology*, *1*(1), 3-10.

Rangel, R. M., Amaral Júnior, A. T., Gonçalves, L. S. A., Freitas Júnior, S. P., & Candido, L. S. (2011). Análise biométrica de ganhos por seleção em população de milho pipoca de quinto ciclo de seleção recorrente. *Revista Ciência Agronômica., 42*(2), 473-481.

Resh, F. S., Scapim, C. A., Machado, M. F. P. S., Mangolin, C. A., Amaral Junior A. T., Ramos, H. C. C., & Vivas, M. (2015). Genetic diversity of popcorn genotypes using molecular analysis. *Genetics and Molecular Research, 14*(3), 9829-9840.

Ribeiro, R. M., Amaral Júnior, A. T., Gonçalves, L. S. A., Candido, L. S., Silva, T. R. C., & Pena, G. F. (2012). Genetic progress in the UNB-2U population of popcorn under recurrent selection in Rio de Janeiro, Brazil. *Genetic Molecular Research, 11*(2), 1417-1423.

Rossi, R. L., & Reis, E. M. (2014). Semi-selective culture medium for Exserohilum turcicum isolation from corn seeds. *Summa Phytopathol, 40*(2), 163-167.

Santos, F. S., Amaral Júnior, A. T., Freitas Júnior, S. P., Rangel, R. M., & Pereira, M. G. (2007). Predição de ganhos genéticos por índices de seleção na população de milho-pipoca UNB-2U sob seleção recorrente. *Bragantia, 66*(3), 389-396.

SAS Institute. (2002). *Getting started with the SAS learning edition* (p. 200). Cary, NC: SAS Institute Inc.

Scapin, C. R., Carnelossi, P. R., Vieira, R. A., Schwan-Estrada, K. R. F., & Cruz, M. E. S. (2010). Fungitoxidade in vitro de extratos vegetais sobre *Exserohilum turcicum* (Pass.) Leonard & Suggs. *Revista Brasileira de Plantas Medicinais, 12*(1), 57-61.

Silva, T. C. R., Amaral Júnior, A. T., Gonçalves, L. S. A., Candido, L. S., Vittorazzi, C., & Scapim, C. A. (2013). Agronomic performance of popcorn genotypes in Northern and Northwestern Rio de Janeiro State. *Acta Scientiarum. Agronony, 35*(1), 57-63.

Silva, V. Q. R., Amaral Júnior, A. T., Gonçalves, L. S. A., Freitas Júnior, S. P., & Ribeiro, R. M. (2011). Heterotic parameterizations of crosses between tropical and temperate lines of popcorn. *Acta Scientiarum. Agronony, 33*(2), 243-249.

Vieira, R. A., Scapim, C. A., Moterle, L. M., Tessmann, D. J., Conrado, T. V., & Amaral Júnior, A. T. (2009). Diallel analysis of leaf disease resistance in inbred Brazilian popcorn cultivars. *Genetics and Molecular Research, 8*(4), 1427-1436.

Viegas Neto, A. L., Heinz, R., Gonçalves, M. C., Correia, A. M. P., Mota, L. H. S., & Araújo, W. D. (2012). Milho pipoca consorciado com feijão em diferentes arranjos de plantas. *Pesquisa Agropecuária Tropical, 42*(1), 28-33.

Wang, P., Souma, K., Kobayashi, Y., Iwabuchi, K., Sato, C., & Masuko, T. (2010). Influences of Northern Leaf Blight on corn silage fermentation quality, nutritive value and feed intake by sheep. *Animal Science Journal, 81*(4), 487-493.

Wang, H., Xiao, Z. X., Wang, F. G., Xiao, Y. N., Zhao, JR R., Zheng, Y. L., Qiu, F. Z. (2012). Mapping of HtNB, a gene conferring nonlesion resistance before heading to Exserohilum turcicum (Pass.), in a maize inbred line derived from the Indonesian variety Bramadi. *Genetics and Molecular Research, 11*(3), 2523-2533.

3

Estimates of genetic parameter in maize commercial variety IPR 114 at Paraná State, Brazil

Sara Regina Silvestrin Rovaris[1*], Pedro Mário de Araújo[2], Deoclécio Domingos Garbuglio[2], Cássio Egídio Cavenaghi Prete[1], Valquíria Sheila Zago[1] and Lucas José Ferreira da Silva[1]

*[1]Departamento de Agronomia, Centro de Ciências Agrárias, Universidade Estadual de Londrina, Rod. Celso Garcia Cid, PR-445, Km 380, 86051-990, Londrina, Paraná, Brazil. [2]Área de Melhoramento Genético Vegetal, Diretoria Técnico Cientifica, Instituto Agronômico do Paraná, Londrina, Paraná, Brazil. *Author for correspondence. E-mail: sara_rsr@yahoo.com.br*

ABSTRACT. The objective of this study was to estimate the genetic parameters of the IPR 114 maize commercial variety for breeding purpose and the efficiency of the method. It was perfomed two cycles of half-sib progenie, between 2004 and 2007. The progenies were evaluated in a randomized complete block desing with two replications per environments and the plots were composed of one 4 m length row spaced on 0.8 m, and five plants per meter. The trials were carried out in Londrina and Ponta Grossa, Paraná State, Brazil. There were evaluated 173 and 154 progenies for selection cycles 1 and 2, respectively, analyzing the following traits: plant height, ear height, days to flower, prolificacy and predict grain yield. In both cycles verified significant genetic variability between progenies. Significant selection gains were obtained for predict grain yield, showing 6.33 and 5.03% for cycles 1 and 2, respectively. The estimative for genetic parameters obtained in this work justify the selection program continuity of the IPR 114 maize population

Keywords: *Zea mays* L., progeny of maize, genetic gain, intrapopulacional selection.

Introduction

Maize (*Zea mays* L.) is cultivated in all States of Brazil. Paraná State produced 15.37 milion ton in 2008, standing out as the largest national producer (CONAB, 2008).

This cereal presents great genetic plasticity and therefore is grown in several different environments and in different production systems. The selection of more productive cultivars with better adaptation to local or regional conditions consists in a simple and essential technology to increase crop yield, especially by not burdening the costs of production.

According to Sawazaki and Paterniani (2004), before the 60's decade, the cultivated maize cultivars, besides being low productive, were excessively high, lodged easily and did not tolerate high planting density. With the development of breeding programs, expressive changes were obtained not only in productivity, but also in reductions of plant height and life cycle, and increase of plant health. This resulted in better adaptation to water stress conditions, lower lodging, higher resistance to pests and diseases and enhancement in nutrient quality of the grain.

The Agronomic Institute of Paraná (Instituto Agronômico do Paraná - IAPAR) has been developing breeding in maize populations using different methods of recurrent selection, with emphasis in the method named selection among and within half-sib progenies with remant seeds. According Paterniani (1967), this method is extensively used due to its facility of execution and efficiency in acquiring genetic gains to populations, presenting as an advantage the possibility of estimating the additive genetic variation in intrapopulational level.

Thus, breeding of maize populations aims the enhancement of frequency of favorable alleles inside the population. According to Vencovsky (1987), the probability of obtaining a superior crop line depends on the frequency of favorable alleles of this population. The frequency of those alleles can be gradually increased through successive selection cycles, which are an efficient way of enhancing varieties aiming both commercial use and line extraction.

Several works using half-sib progenies have been reported in literature. Proving the efficiency of this method, Carvalho et al. (2002) emphasize that, in addition of being practical, it is capable of maintaining sufficient genetic variability to promote progress during the selection cycles.

The objective of this work was to obtain estimatives of genetic parameters of IPR 114 maize variety and to evaluate the population potential for breeding purposes.

Material and methods

The IPR 114 commercial maize variety was developed at IAPAR through diallelic cross between a group of commercial hybrids resulting in a synthetic variety (its genetic constitution involves less than 16 inbreeds), being subsequently submitted to two recombination cycles. It is a population of open pollination, presenting the following characteristics: early cycle, low plant height and ear insertion, good tolerance to lodging and stalking, good ear coverage, wide adaptation, good productive potential and semi-flint yellow-orange colored grains.

This population was subsequently submitted to two selection cycles among and within half-sib progenies with remant seeds. The following traits were evaluated: plant and ear height (PH and EH, respectively) in centimeters, female flowering (FF) in days, prolificacy (PRO) in units and grain yield (GY) in kg ha^{-1}, correcting the grain humidity to 14,5%. In order to obtain the genetic parameter estimative, GY data were converted to g $plant^{-1}$. This conversion was necessary due to the fact that most data presented in literature use that unit, which allows comparison with results obtained by other authors.

The experiment was conducted at IAPAR, at the main Londrina Station and at the Experimental Station of Ponta Grossa, Paraná State, Brazil.

Two selection cycles among and within half-sib progenies with remant seeds were performed using a selection intensity of about 20%. In both cycles, the recombination of selected progenies was done in isolated field. The method consists in bulked of a sample of the selected progenies seeds to generate the male lines. This lines were planted in two periods with 10 days of difference, to effectively guarantee the recombination among all selected progenies. At the flowering time, individualized progenies were detasselled, constituting in this way the female lines. At the end of the cycle, the best plants within the lines were selected for further agronomic evaluations.

The evaluation of selection cycle 1 of this population started at 2004/2005 season, using as checks the following cultivars: IPR 115 Simple Hybrid synthesized at IAPAR; BRS 1010 Simple Hybrid of Embrapa Maize and Sorghum Research Center; and P30F33 Simple Hybrid by Pioneer, this completing 182 entries.

Based on this evaluation, it was performed the selection among progenies. Were selected 40 progenies with the best yield means, which were recombined at 2005/2006 season in isolated field. Simultaneously, the selection within progenies was performed, thus completing one selection cycle. At this stage, 154 ears were selected to constitute the progenies for new evaluations in field trial, starting this way the selection cycle 2. Along with this evaluation, the genetic parameters and the selection gain estimative were also obtained.

The evaluation of second cycle 2 occurred at 2006/2007 season, in which the treatments were constituted by 154 selected progenies using as checks: IPR 115, IPR 114-SB – Basic Seed; IPR114-SG – Genetic Seed; DKB 390 – Simple Hybrid by Monsanto; DOW 2B 710 – Simple Hybrid by Dow Agroscience; and P30F53 – Simple Hybrid by Pioneer, totalizing 160 treatments. At 2007/2008 season the selected progenies were recombined, finishing the second selection cycle.

In both selection cycles 1 and 2, progenies were evaluated in completely randomized complete block design with two replication at each locations (Ponta Grossa e Londrina, Paraná State). The experimental plots were representedby single rows 4 m long spaced 0.80 m apart with 25 plants per plot after thinning.

After evaluation assays between progenies of cycles 1 and 2, single and joint variance analyses were performed, considering that the effects of genotypes and environments were randomized. Analyses were carried out using the Genes v. 2007.0.0 software (CRUZ, 2007).

The variance components were estimated with the expected means square of variance analysis (VENCOVSKY, 1987). It was estimated the heritability coefficient at average level (h^2_m), the additive genetic variance for half-sib progenies, which corresponds with $4\sigma^2_p$, and variation index b that is estimated by the relation between the genetic variation coefficient and environment variation coefficient (CRUZ et al., 2003).

The selection gains were estimated in g plant^{-1} and in percentage, according to the following expressions, respectively:

$$(\overline{X}_1 - \overline{X}_0)h^2_m \text{ and } \frac{SG}{\overline{X}_0}.100$$

where:

$\overline{X}_1$ = means of selected progenies;
$\overline{X}_0$ = means of all tested progenies;
σ^2_p = variance among progenies;
σ^2_F = phenotypic variation;
GY = predict selection gain.

Results and discussion

The joint variance analyses of progenies for both selection cycles revealed significant differences for: FF, PH, EH and GY, showing persistence of genetic variation at the IPR 114 population (Table 1).

It was also observed that mean square of locations for FF presented highly significant differences at 1% probability level in both selection cycles. This result was observed due to the great difference grouch degree day acumulation between the locations interfering in the progenies cycle. Ponta Grossa is located at a mild climate environment, where the accumulation of degree days happens in a slower manner, resulting in longer vegetative cycle.

The mean square of P x L interaction for GY was not significant in both selection cycles, evidencing that besides location presented edafoclimatic differences, progenies obtained similar performances. The populations will have the best progenys recombined, searching a population with major adeptness for all regions of the Paraná State. The experimental variation coefficients were considered low for FF and PRO and intermediate for the other traits, demonstrating a good performace of the trial (PIMENTEL-GOMES, 1976; SCAPIM et al., 1995).

Average GY for progenies were of 7,648 kg ha^{-1} on cycle 1 and 8,500 kg ha^{-1} on cycle 2, equivalent to 78.13% of the IPR 115 control productivity used on both cycles.

The genetic parameters estimatives obtained on both selection cycles of IPR 114 population allow evidencing a decrease in genetic variability from cycle 1 to cycle 2 for all five traits analyzed (Table 2). Comparable data have been observed in similar breeding works that utilize this kind of progenies (CARVALHO et al., 2000; CARVALHO; SOUZA, 2007; PATERNIANI, 1967; SANTOS; NASPOLINI FILHO, 1986).

As emphasized by Ramalho (1977 apud CARVALHO; SOUZA, 2007), this happens due to the maximum utilization of free variability existing among polygenic blocks. From cycle 2 on, it is used the latent genetic variability present within polygenic blocks and which is gradually dispensed, throughout genetic crossing-over.

Table 1. Mean squares of joint variance analyses and variation coefficients on two selection cycles of progenies of maize IPR 114 population, in Londrina and Ponta Grossa, 2004 to 2007.

			1st Cycle			
			Mean Square			
Variation Source	F	FF	PH	EH	PRO	GY[1]
Progenies (P)	172	16**	285.3**	194.6**	0.1781**	566.95**
Locations (L)	1	2177.4**	27754*	9501.2**	0.452	3189.90
P x L	172	44.9**	122.8	109.6*	0.0089	276.31
Mean error	362	3.2	102.7	85.2	0.0081	322.39
Means	-	74.3	211.2	104.7	1.08	122.37
C.V.(%)	-	2.4	4.8	8.8	8.52	14.82
			2nd Cycle			
			Mean Square			
Variation Source	F	FF	PH	EH	PRO	GY[1]
Progenies (P)	153	4.9**	206.5**	186.4*	0.0113	635.5**
Locations (L)	1	12843**	214988**	121467**	2.4267*	221445**
P x L	153	2.3	114.5	135.9**	0.0088	363.2
Mean error	318	1.8	94.2	91.5	0.0074	303.4
Means	-	65.8	212.5	113.3	0.9664	136.0
C.V.(%)	-	2.0	4.5	8.4	8.92	12.8

[1]g plant^{-1}; **, *: Significant by F test t 1 at 1 and 5% pobability level.

Table 2. Obtained estimatives referred to additive variance of progenies, variance of progenies x locations interaction, mean heritability coefficient, genetic variation coefficient, and genetic gains within progenies in % and g $plant^{-1}$.

Genetic Parameters	FF Days	PH ------cm------	EH	PRO un	YG $g\ pl^{-1}$
1st Cycle					
$\hat{\sigma}^2_A$	11.600	162.500	85.060	0.0088	290.610
$\hat{\sigma}^2_{pl}$	0.5861	9.26	11.938	0.0002	-15.968
CV_g	2.29	3.02	4.40	4.31	7.20
b	0.93	0.62	0.49	0.50	0.49
h^2%	72.06	56.95	43.70	49.60	51.26
Selection Gains (SG)	0.23	1.70	-0.87	0.02	7.49
Selection Gains % (SG%)	0.31	0.83	-0.83	2.04	6.33
2nd Cycle					
Genetic Parameters	FF Days	PH ------cm------	EH	PRO un	PG[1] $g\ pl^{-1}$
$\hat{\sigma}^2_A$	2.720	92.000	50.542	0.0248	272.380
$\hat{\sigma}^2_{pl}$	0.2050	11.775	21.517	0.0006	29.222
CV_g	1.25	2.25	3.13	2.58	6.12
B	0.60	0.50	0.36	0.28	0.47
h^2%	54.84	44.55	27.10	21.83	42.85
Selection Gains (SG)	0	1.13	0.71	0.008	6.82
Selection Gains % (SG%)	0	0.53	0.61	0.833	5.06

Analyzing Table 2 we can observe that the estimatives for additive genetic variances obtained on cycle 1 for PH and EH traits were of 162.5 and 85.05 (cm $plant^{-1})^2$, respectively, and on cycle 2 were of 91.20 and 50.52 (cm $plant^{-1})^2$, being considered high when compared to values obtained by Silva et al. (2001) in the ESAM-1 maize composite. According to Paterniani (1967), it is of interest that the additive genetic variance is maintained as high as possible to allow the achievement of substantial gains by selection. The same occurs with values obtained for the estimatives for FF and PRO, which were low on both cycles. However, these are expected values, since the IPR 114 variety was synthesized from hybrids that present low height, early cycle and small prolificacy.

In breeding of population of open pollination, the additive genetic variance contribute significantly in answer of selection of population. Many works (CARVALHO et al., 2002; HALLAUER; MIRANDA FILHO, 1988; PATERNIANI, 1967; WEBEL; LONQUIST, 1967) reports, in that populations there are existence of additive genetic variance, that is the principal component of genetic variance of grain yield.

For GY, additive genetic variance was of 290,61 (g $plant^{-1})^2$ and can be considered high when compared to the value obtained by Carvalho and Souza (2007), which was of 147,7 (g $plant^{-1})^2$ from BR 5011 Sertanejo variety.

The negative variance value of the progenies x locations interaction obtained on cycle 1 for GY was due to its reduced mean square value, which was lower than the mean square of the error. This still lead to an increase in progenies variance, considering that the estimative was obtained through the expected mean squares of progenies and errors, causing the enhancement of average heritability coefficient for GY, on selection cycle 1. However, for the other traits the value of $\hat{\sigma}^2_{pl}$ was positive, but it is noticed a decrease from cycle 1 to cycle 2.

Due to the fact that the progenies evaluations were performed in two locations, the magnitudes of the obtained estimatives presented less influenced by the progenies x location interaction. Hallauer and Miranda Filho (1988) mention that the estimatives of genetic and phenotypic parameters obtained in only one location can be overestimated, as being influenced by the progenies x location interaction.

The b index was close to 1.0 only for FF on both selection cycles, presenting values of 0.93 and 0.60 on cycles 1 and 2, respectively. According to Vencovsky and Barriga (1992), when b coefficient reaches a value equal or superior than 1.0 in experiments with half-sib progenies, there is a very favorable situation for selection.

Nevertheless, in this work, the genetic variation coefficients presented for both selection cycles reflect the wide variation existing among half-sib progenies, the same as the values for selection gains indicate the viability for the recurrent selection program with IPR 114 population, especially for yield enhancement.

In general, from cycle 1 to cycle 2 it was observed that heritability coefficient at average level decreased. The same happened for selection gains for GY in percentage and in g $plant^{-1}$. This decrease can be explained by the reduction in genetic variability that occurs through selection from cycle 1 to cycle 2.

The heritability coefficients at average level estimated in IPR 114 population for GY on both selection cycles presented superior values than the ones obtained by Silva et al. (2001), which showed 31.55%, and Cardoso et al. (2003) for the CMS 47 and BR 5039 population, that was of 18.57 and 2.56%, respectively.

The selection gain estimated in percentage to GY was 6.33% in cycle 1 and 5.03% in cycle 2. This estimated gains can be considered intermediate, however, they are superior to the values obtained by Cardoso et al. (2003), that were of 4.88 and 0.78% for CMS 47 and BR 5039, respectively, and inferior to the value observed by Carvalho and Souza (2007), which was of 27% for BR 5011 Sertanejo population. Still, the values here estimated are in agreement with the mean reported by Hallauer and Miranda Filho (1988) in a survey done with several works that utilized this selection method, in which the gain varied from 2 to 13%.

For EH, on cycle 1 it was obtained a negative selection gain, meaning that the height of ear insertion decreased with the selection process, which is a characteristic of great agronomic interest.

Conclusion

The magnitude of genetic parameters estimatives obtained indicates the possibility of continuous gains for grain yield in the IPR 114 population, without greater alterations in the other characteristics, thus justifying the continuity of the selection process.

Acknowledgements

We thank CNPq for this work's financial support.

References

CARDOSO, M. J.; CARVALHO, H. W. L.; LEAL, M. L. S.; SANTOS, M. X. Melhoramento genético das cultivares de milho CMS 47 e BR 5039 (São Vicente) na Região Meio-Norte do Brasil. **Revista Brasileira de Milho e Sorgo**, v. 2, n. 3, p. 88-96, 2003.

CARVALHO, H. W. L.; SOUZA, E. M. Ciclos de seleção de progênies de meios-irmaõs do milho BR 5011 Sertanejo. **Pesquisa Agropecuária Brasileira**, v. 42, n. 6, p. 803-809, 2007.

CARVALHO, H. W. L.; LEAL, M. L. S.; SANTOS, M. X. Estimativas de parâmetros genéticos na população de milho CMS 35 no Estado de Sergipe. **Pesquisa Agropecuária Brasileira**, v. 37, n. 10, p. 1399-1405, 2002.

CARVALHO, H. W. L.; LEAL, M. L. S.; SANTOS, M. X.; CARVALHO, P. C. L. Três ciclos de seleção entre e dentro de progênies de meios-irmãos na população de milho CMS-52. **Pesquisa Agropecuária Brasileira**, v. 35, n. 8, p. 1621-1628, 2000.

CONAB-Companhia Nacional de Abastecimento. **Dados de produção**. Available from: <http://www.conab.gov.br/conabweb/download/safra/3graos_08.09.pdf>. Access on: 20 Dec. 2008.

CRUZ, C. D. **Programa Genes-Versão Windows**: Aplicativo computacional em Genética e Estatística. Versão 2007.0.0. Viçosa: UFV, 2007.

CRUZ, C. D.; REGAZZI, A. J.; CARNEIRO, P. C. S. **Modelos biométricos aplicados ao melhoramento genético**. Viçosa: UFV, 2003. v. 2.

HALLAUER, A. R.; MIRANDA FILHO, J. B. **Quantitative genetics in maize breeding**. 2nd ed. Ames: Iowa State University Press, 1988.

PATERNIANI, E. Selection among and within half-sibs families in a Brazilian population of maize (*Zea mays* L.) **Crop Science**, v. 7, n. 3, p. 212-216, 1967.

PIMENTEL-GOMES, F. P. **Curso de estatística experimental**. São Paulo: Nobel, 1976.

SANTOS, M. X.; NASPOLINI FILHO, W. Estimativas de parâmetros genéticos em três ciclos de seleção entre e dentro de famílias de meios-irmãos no milho (*Zea mays* L.) dentado composto Nordeste. **Revista Brasileira de Genética**, v. 9, n. 2, p. 307-319, 1986.

SAWAZAKI, E.; PATERNIANI, M. E. A. G. Z. Evolução dos cultivares de milho no Brasil. In: GALVÃO, J. C. C.; MIRANDA, G. V. (Ed.). **Tecnologias de produção do milho**. 20. ed. Viçosa: Universidade Federal de Viçosa, 2004. p. 13-53.

SCAPIM, C. A.; CARVALHO, C. G. P.; CRUZ, C. D. Uma proposta de classificação dos coeficientes de variação para a cultura do milho. **Pesquisa Agropecuária Brasileira**, v. 30, n. 5, p. 683-686, 1995.

SILVA, E. S.; SILVA, P. S. L.; NUNES, G. H. S.; SILVA, K. M. B. Estimação dos parâmetros genéticos no composto de milho ESAM-1. **Caatinga**, v. 14, n. 1/2, p. 43-52, 2001.

VENCOVSKY, R. Herança quantitativa. In: PATERNIANI, E.; VIEGAS, G. P. (Ed.). **Melhoramento e produção de milho**. Campinas: Fundação Cargill, 1987. p. 137-214.

VENCOVSKY, R.; BARRIGA, P. **Genética biométrica aplicada ao fitomelhoramento**. Ribeirão Preto: Sociedade Brasileira de Genética, 1992.

WEBEL, O. D.; LONQUIST, J. H. Na evaluation of modifified ear-to-row selection in a population of corn (*Zea mays* L.). **Crop Science**, v. 7, p. 651-655, 1967.

4

Genetic dissimilarity for resistance to *Mononychellus tanajoa* (bondar) (Acari, Tetranychidae) among domesticated and wild *Manihot* species

Verônica de Jesus Boaventura, Rudiney Ringenberg and Carlos Alberto da Silva Ledo*

*Embrapa Mandioca e Fruticultura, Parque Estação Biológica, s/n.º, 70770-901, Brasília, Distrito Federal, Brasil. *Author for correspondence. E-mail: carlos.ledo@embrapa.br*

ABSTRACT. The aim of this study was to evaluate the genetic dissimilarity among wild and domesticated species of *Manihot* for resistance to cassava green mite during the insect life cycle. Nine accessions of wild *Manihot* species, *M. esculenta* ssp. *flabellifolia*, *M. esculenta* ssp. *peruviana*, and *M. carthaginensis* ssp. *glaziovii*, and two clones of *M. esculenta* (Cigana Preta and Sacaí) were evaluated under laboratory conditions at 25 ± 1°C, 70 ± 10% RH, and a 12-h photophase. Daily observations during the mite life cycle stages (larva-adult) were recorded. The data were subjected to an analysis of variance, a Scott-Knott test (5%), and Singh criterion, cluster, and principal component analyses. The larval-adult period ranged from 5.53 to 7.01 days: the longest period was observed on an *M. glaziovii* accession (GLA-19-DF) and the shortest on an *M.* flabellifolia accession (FLA-025V). The UPGMA method allowed the division of the genotypes into six groups, with the greatest distance between the FLA-025V and GLA-19-DF accessions. The first two main components explained 77.50% of the total accumulated variation. The association of the longest cycle duration of M. tanajoa with the lowest larval-adult viability suggests that GLA-19-DF is less favorable to mite development compared to the other accessions. Significant variability among the genotypes was observed.

Keywords: biology, cassava, biotic stress, green mite.

Introduction

In Brazil, cassava is cultivated in all biomes due to the great diversity of varieties adapted to different ecosystems, resulting in a great genetic diversity among species. This diversity is due to natural selection and is favored by cross-pollination, a high degree of heterozygosity, and abrupt fruit dehiscence, leading to a plethora of new genotypes (FUKUDA; SILVA 2002).

Historically, cassava is a culture characterized by low supplies of inputs and agrochemicals with a high tolerance to drought, which can remain in the soil until its consumption, playing an important role in feeding the population (CAMARGO, 2003). However, due to its long growing cycle, the crop is subject to attack by a wide range of feeding arthropods. It is estimated that there are over 200 mite species associated with cassava in the Americas among which the green mite *Mononychellus tanajoa* (Bondar, 1938) (Acari, Tetranychidae) is considered a major pest affecting cassava cultivation in northeastern Brazil, particularly in the semiarid region (FUKUDA, 2006).

Green mite infestations develop largely under conditions of high temperature and low relative humidity, beginning at the top of the plants, attacking the buds and young leaves. Such infestation affects the formation of leaves, which become reduced in severely attacked plants and exhibit deformation, with the shortening of internodes, potentially leading to the death of the branch apices. Losses are estimated to be 10-50%, depending on the intensity of infestation (MORAES; FLECHTMANN, 2008).

The potential use of wild species has recently been reported in studies conducted at the International Center for Tropical Agriculture (CIAT). The results of these experiments include the following: 1) interspecific hybrids of *M. esculenta* ssp. *flabellifolia* show moderate to high levels of resistance to mite, whitefly, and scale insects (BURBANO et al., 2007); 2) an interspecific hybrid between *M. esculenta* and *M. walkerae* Croizat exhibits reduced post-harvest physiological spoilage (CIAT, 2003); 3) *M. glaziovii* Muell, *M. catingae* Ule, and *M. carthaginensis* Mull. Arg. are adapted to semiarid conditions and are potential sources of genes for drought tolerance; 4) the roots of interspecific hybrids of *M. esculenta* ssp. *flabellifolia* contain a high protein content (AKINBO et al., 2012a); 5) *M. flabellifolia* accessions and interspecific crosses of *M. esculenta* ssp. *flabellifolia* exhibit different levels of resistance to the whitefly *Aleurotrachelus socialis* Bondar (CARABALI et al. 2010); and 6) resistance to *A. socialis* can be introgressed in interspecific hybrids of commercial cassava (AKINBO et al., 2012b).

Considering the characteristics of agronomical interest found in wild and domesticated cassava species, it is important to select cassava genotypes that are resistant to green mite for use in integrated mite control programs (ARGOLO et al., 2005).

This study was aimed at assessing the genetic dissimilarity among wild and domesticated *Manihot* species for resistance to *M. tanajoa* during the life cycle of the insect (larva-adult period).

Material and methods

Location of bioassays

The study was performed at the Laboratory of Entomology - Embrapa Cassava and Fruits, Cruz das Almas, Bahia State, Brazil.

Mites and *Manihot* accessions

Cassava green mite specimens (*M. tanajoa*) were obtained from colonies established in the cassava clone "Cigana Preta" maintained in a greenhouse. The wild *Manihot* genotypes are part of the Work Collection of Wild *Manihot* Species - Embrapa Cassava and Fruits, comprising 628 accessions belonging to 28 species, and were installed in the field in 2005. This study evaluated nine accessions belonging to three wild cassava species, *M. esculenta* ssp. *flabellifolia* (FLA-025V, FLA-026V, and FLA-029V), *M. esculenta* ssp. *peruviana* (PER-006V, PER-007V, and PER-010V), and *M. carthaginensis* ssp. *glaziovii* (GLA-03-DF, GLA-10-DF, and GLA-19-DF), and clones of a domesticated species (*M. esculenta*), Cigana Preta (BGM 0116) and Sacaí (BGM 0384), all from the experimental area belonging to Embrapa Cassava and Fruits. These genotypes were selected according to characteristics of interest: *M. esculenta* ssp. *flabellifolia* is the ancestor of *M. esculenta* and provides resistance to such pests as mites; *M. esculenta* ssp. *peruviana* is close to *M. flabellifolia*; *M. carthaginensis* ssp. *glaziovii* presents drought tolerance; and the *M. esculenta* clone Cigana Preta is considered susceptible and Sacaí resistant to *M. tanajoa*.

Mite eggs and bioassays

To obtain eggs, eighty female *M. tanajoa* obtained from maintenance breeding were allowed to lay eggs in the lobes of fully developed young leaves of the tested accessions. The lobes were positioned with the top (adaxial surface) in contact with nylon foam (1 cm thick) that was moistened with distilled water and surrounded with strips of cotton wool inside trays (20 x 15 x 5 cm). The trays were sealed with transparent polyvinylchloride film (PVC), which was perforated to allow aeration. After 24 hours, the females were removed, and the eggs were allowed to develop on the lobes until hatching. After hatching, the larvae were isolated on leaf disks (2.5 cm in diameter) of each accession tested; the disks were deposited on nylon foam (1 cm thick) moistened with distilled water inside Petri dishes (14 cm in diameter x 2 cm deep) and sealed with PVC film perforated to allow aeration, according to the methodology of Noronha et al. (1995). The Petri dishes were maintained in a BOD-type incubator at 25 ± 1°C, a relative humidity of 70 ± 10%, and a photoperiod of 12 hours. Except when they were in quiescent stage (protochrysalis, deutochrysalis, and teleiochrysalis stages), the mites were transferred to new disks every two days with the aid of a stereoscopic microscope and a brush with fine bristles. The development of *M. tanajoa* was followed to adulthood, with daily observations of the larval, protochrysalis, protonymph, deutochrysalis, deutonymph, teleiochrysalis, and adult stages to verify the succession to the next stage, as characterized by the presence of exuviae. A completely randomized design with 30 replicates per genotype was used. The first 30 hatched larvae were captured. Each plot consisted of one mite.

Statistical analysis

The data were subjected to an analysis of variance, and the means of the genotypes were grouped by the Scott-Knott test (SCOTT; KNOTT, 1974) at 5% significance using SISVAR software (FERREIRA, 2011). Cluster and principal components analyses were performed. For the cluster analysis, the mean Euclidean distance was considered as a dissimilarity measure. The hierarchical groupings from the dissimilarity matrix were obtained by the Unweighted Pair Group Method with Arithmetic Mean (Upgma) method (SNEATH; SOKAL, 1973). The Singh (1981) criterion was also used to quantify the relative contribution of the duration of the life cycle stages for the genetic dissimilarity analyses performed using the Genes (CRUZ, 2001) and Statistica 7.1 software (STATSOFT, 2005).

Results and discussion

The biological life cycle (larva-adult period) of *M. tanajoa* was grouped according to the means of the Scott-Knott test at 5% significance (Table 1). The egg stage lasted for five days and did not vary between the accessions. The larvae used in the study were of the same age. A variation from 1.07 to 1.53 days was observed for the larval stage, whereas the protonymph stage ranged from 0.96 to 1.29 days, with the formation of only one group, which did not statistically differ from *Manihot* accessions studied in the two cycle stages. In contrast, two groups were distinguished for the other stages.

For the protochrysalis stage, the variation ranged from 0.98 to 1.62 days, with significant differences between the accessions. The longest mite protochrysalis, protonymph, and larva-adult stage durations were found for the accessions of *M. peruviana* species; the shortest protochrysalis stage was observed on the *M. carthaginensis* ssp. *glaziovii* accessions.

For the deutochrysalis and deutonymph phases, the variation ranged from 0.93 to 1.28 days and 0.90 to 1.27 days, respectively. The *M. peruviana* accessions and *M. esculenta* clones Cigana Preta and Sacaí of were among the group showing the shortest deutochrysalis stage, whereas the *M. esculenta* clones were among the group showing the longest deutonymph stage. The teleiochrysalis stage varied from 0.64 to 1.09 days, with the *M. flabellifolia* accessions and *M. esculenta* clones exhibiting the longest durations and the *M. peruviana* accessions and two *M. glaziovii* accessions exhibiting the shortest duration.

The larva-adult period ranged from 5.53 to 7.01 days. The group that showed the longest larva-adult period was formed by FLA-026V, GLA-03-DF, GLA-019-DF, PER-006V, PER-007V, PER-010V, Cigana Preta, and Sacaí. The accessions that formed the group with the shortest larva-adult period were FLA-025V, GLA-10, and DF-FLA-029V. The *M. esculenta* clones were included in the same group during all stages of the mite life cycle.

The viability of the immature stages of the cassava green mite varied among the *Manihot* species. The survival rate ranged from 76.67 to 100%, with the lowest viability being observed on accession GLA-19-DF (76.67%) of *M. carthaginensis* ssp. *glaziovii*. The viability was greater than 80% on the accessions of *M. esculenta* ssp. *flabellifolia*, and the survival rate was greater than 90% on GLA-03-DF and GLA-10-DF. Mortality was observed during the immature mite stage on the *M. peruviana* accessions and Cigana Preta and Sacaí clones (Table 1).

Table 1. Duration (X ± SE) and total viability (%) of stages of the life cycle (larva-adult period) of *M. tanajoa* as a function of the *Manihot* accession at a temperature of 25 ± 1°C, relative humidity of 70 ± 10%, and photoperiod of 12 hours. Cruz das Almas, Bahia State, Brazil, 2011.

	Duration (days) of cycle stage							Total viability (%)
Accession	L	PC	PN	DC	DN	TC	LA	
FLA-025V	1.13 ± 0.01 a	1.62 ± 0.14 a	1.03 ± 0.09 a	0.98 ± 0.10 b	1.04 ± 0.08 b	0.93 ± 0.08 a	5.53 ± 0.23 b	83.33
FLA-026V	1.26 ± 0.09 a	0.98 ± 0.04 b	1.23 ± 0.13 a	1.12 ± 0.06 b	1.27 ± 0.09 a	1.08 ± 0.07 a	6.43 ± 0.23 a	83.33
FLA-029V	1.13 ± 0.08 a	1.14 ± 0.07 b	1.08 ± 0.05 a	1.28 ± 0.11 a	1.13 ± 0.08 a	0.94 ± 0.08 a	6.07 ± 0.21 b	86.67
GLA-03-DF	1.39 ± 0.16 a	1.18 ± 0.09 b	0.98 ± 0.08 a	1.02 ± 0.06 b	1.08 ± 0.09 a	0.77 ± 0.08 b	6.32 ± 0.20 a	96.67
GLA-10-DF	1.39 ± 0.14 a	1.09 ± 0.06 b	0.96 ± 0.07 a	0.93 ± 0.03 b	0.93 ± 0.04 b	0.64 ± 0.06 b	5.98 ± 0.20 b	93.33
GLA-19-DF	1.40 ± 0.13 a	1.00 ± 0.00 b	1.29 ± 0.11 a	1.26 ± 0.10 a	1.16 ± 0.14 a	1.09 ± 0.12 a	7.01 ± 0.26 a	76.67
PER-006V	1.53 ± 0.16 a	1.41 ± 0.10 a	1.07 ± 0.06 a	1.12 ± 0.07 b	0.90 ± 0.02 b	0.68 ± 0.06 b	6.71 ± 0.12 a	100
PER-007V	1.33 ± 0.12 a	1.49 ± 0.09 a	1.04 ± 0.05 a	0.99 ± 0. 00 b	0.95 ± 0.03 b	0.69 ± 0.06 b	6.51 ± 0.08 a	100
PER-010V	1.43 ± 0.13 a	1.56 ± 0.09 a	1.14 ± 0.07 a	1.06 ± 0.05 b	0.94 ± 0.01 b	0.78 ± 0.06 b	6.92 ± 0.09 a	100
Cigana Preta	1.20 ± 0.09 a	1.20 ± 0.07 b	1.08 ± 0.05 a	1.04 ± 0.08 b	1.09 ± 0.06 a	1.01 ± 0.07 a	6.62 ± 0.20 a	100
Sacaí	1.07 ± 0.05 a	1.07 ± 0.05 b	1.23 ± 0.08 a	1.06 ± 0.05 b	1.19 ± 0.14 a	0.99 ± 0.06 a	6.62 ± 0.17 a	100
CV (%)	49.17	35.99	39.18	35.77	42.28	46.49	8.86	

Means followed by same letter belong to the same group by the Scott-Knott test at 5% probability. Stages: larva (L), protochrysalis (PC), protonymph (PN), deutochrysalis (DC), deutonymph (DN), teleiochrysalis (TC), larva-adult (LA). X ± SE = standard mean error.

These data show that there is genetic dissimilarity between the tested accessions, as there were significant differences in the green mite biological cycle among different genotypes of the same species.

Argolo et al. (2005) evaluated the resistance of cassava varieties (BGM 0876-Pretinha, BGM 0080-Engana Ladrão, BGM 0537-Do Céu, and BGM 0384-Sacai) selected as tolerant to *M. tanajoa* for the semiarid conditions of northeastern Brazil for varieties and found that the larva-adult period ranged from 7.39 to 9.43 days. The mite life cycle for clone Sacaí was 9.28 days, which was higher than that obtained in the present work for the same clone. However, there was a difference among the individual stages, with the protonymph, deutochrysalis, and deutonymph stages being longer in our study. These data showed that the duration of each stage does not interfere with the duration of the next stage or, consequently, in the development of cassava green mites.

The present study indicates that the characteristics having the greatest influence on mite development are related to the genetics of host species because the bioassays were performed under conditions of identical temperature, relative humidity, and availability of water and nutrients. It also indicates different types of antibiosis of resistance levels of these accessions, which should be considered as an indicator of pest resistance.

Burbano et al. (2007) evaluated and compared two Brazilian wild *Manihot* species (*M. flabellifolia* and *M. peruviana*) to commercial *M. esculenta* genotypes to determine their resistance potential to *M. tanajoa* and found variable degrees of mite damage. The *M. flabellifolia* genotypes were significantly different from *M. esculenta* but similar to the *M. peruviana* genotypes, which showed lower levels of damage.

Figure 1 shows the dissimilarity dendrogram constructed based on seven quantitative descriptors: number of days for the larval, protochrysalis, protonymph, deutochrysalis, deutonymph, teleiochrysalis, and larva-adult stages).

The average group matrix that defined the number of groups was 2.6. The groups formed by the UPGMA method, as based on the average Euclidean distance matrix, allowed the division of the eleven genotypes into six groups: Group 1 (FLA-025V), Group 2 (GLA-03-DF, GLA-10-DF), Group 3 (PER-006V, PER-010V, and PER-007V), Group 4 (FLA-026V, Cigana Preta, and Sacaí), Group 5 (FLA-029V), and Group 6 (GLA-19-DF) (Figure 1).

The largest distance was observed between FLA-025V and GLA-19-FD, i.e., genetically, the greatest dissimilarity was between the genotype of Group 6 and the genotype of Group 1.

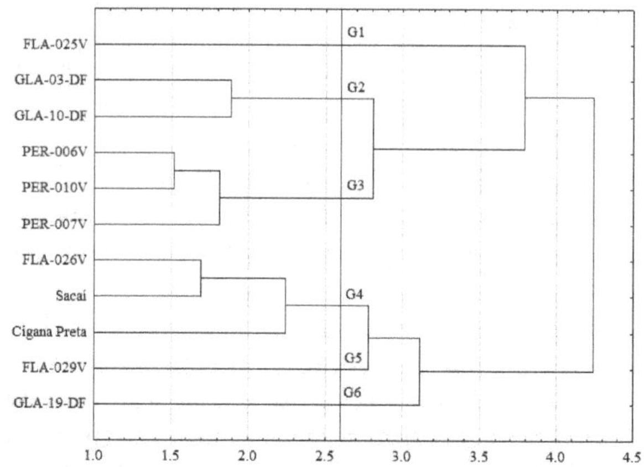

Figure 1. Dissimilarity dendrogram based on the mean Euclidean distance and UPGMA clustering method of the *M. tanajoa* life cycle (larva-adult) on 11 *Manihot* accessions.

The *M. flabellifolia* genotypes showed different behavior for the duration of the mite life cycle, and each accession was classified into a different group. However, there was similarity between the accessions in each group, indicating genotypic and/or phenotypic proximity among them, whereas dissimilarity was observed between the groups.

The feature that most contributed to the assessment of the genetic diversity according to the Singh (1981) criterion was the larva-adult period, at 57.30%.

Alves et al. (2011) assessed 14 accessions of six wild *Manihot* species with regard to biological aspects of *M. tanajoa* and found that the life cycle (larva-adult period) ranged from 5.76 to 8.18 days. The accessions of wild *M. anomala* species formed the group showing the longest mite development period, whereas *M. peruviana* and *M. flabellifolia* accessions showed shorter larva-adult periods.

The first two principal components explained 77.50% of the total accumulated variation. This value is considered satisfactory because the principal components explained approximately 80% of the variance found in the set of characters analyzed, enabling the formation of groups among the accessions and the construction of a scatter plot based on the diversity observed (Figure 1 and 2).

The first principal component explained 52.50% of the total variance, and the phases that most contributed to explaining this variability were the teleiochrysalis, deutonymph, and protonymph stages, as these variables showed higher weighting coefficients. The second principal component explained 25.00%, with the larval-adult and larva stages contributing the most (Figure 2).

The species distribution in the scatter plot (Figure 2) confirmed the greatest distance observed in the dendrogram (Figure 1) between FLA-025V and the other accessions. This result allowed the identification of the greatest genetic dissimilarity among individuals, in addition to grouping two *M. flabellifolia* accessions with *M. esculenta* and confirming the proximity of these accessions.

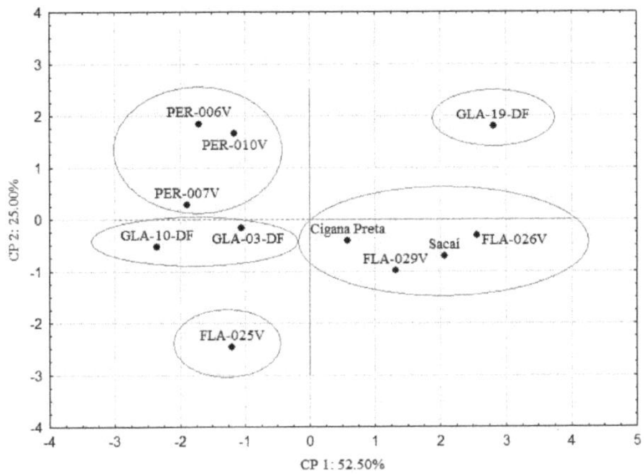

Figure 2. Graphic dispersion of the scores of 11 *Manihot* accessions in relation to principal components 1 and 2.

To verify the cassava ancestry, Carvalho (2005) studied 27 *Manihot esculenta* ssp. *flabellifolia* populations using two types of genetic markers (the glyceraldehyde 3-phosphate dehydrogenase [G3PDH] gene and five microsatellite loci) and confirmed that cassava was originally domesticated from *M. esculenta* ssp. *flabellifolia*.

There is complementarity between the two multivariate techniques used. In the cluster analysis (Figure 1) and principal component analysis (Figure 2), the GLA-19-DF and FLA-025V accessions showed no similarity with the other accessions, as they were allocated to different groups. A similar result was found for the *M. peruviana* and *M. carthaginensis* ssp. *glaziovii* accessions (GLA-10-DF and GLA-03-DF).

Conclusion

The time in days of a biological cycle phase (larva-adult period) has little influence on the duration of subsequent stages.

The longest larval to adult duration and lowest viability of the immature stage of *M. tanajoa* suggest that the GLA-19-DF accession is less susceptible to mite development when compared to the other genotypes tested.

The different behavior of the mite life cycle on accessions of the same species indicates that genetic differences exist between accessions of the same *Manihot* species.

The accessions of *M. esculenta* ssp. *peruviana* species are genetically close.

There is genetic variability among wild and domesticated *Manihot* species with regard to resistance to *Mononychellus tanajoa*.

Acknowledgements

The authors thank Capes for the scholarship granted to the first author by the Graduate Program in Plant Genetic Resources at UEFS and the Embrapa Cassava and Fruits.

References

AKINBO, O.; LABUSCHAGNE, M. T.; MARÍN, J.; OSPINA, C.; SANTOS, L.; BARRERA, E.; GUTIÉRREZ, J.; EWA, F.; OKOGBENIN, E.; FREGENE, M. QTL Analysis for root protein in a backcross family of cassava derived from *Manihot esculenta* ssp *flabellifolia*. **Tropical Plant Biology**, v. 5, n. 2, p. 161-172, 2012a.

AKINBO, O.; LABUSCHAGNE, M.; FREGENE, M. Introgression of whitefly (*Aleurotrachelus socialis*) resistance gene from F1 inter-specific hybrids into commercial cassava. **Euphytica**, v. 183, n. 1, p. 19-26, 2012b.

ALVES, A. A. C.; DITA, M. A.; NORONHA, A. C. S.; MENDES, R. A. Pré-melhoramento da mandioca-utilização de espécies silvestres como fonte de resistência a fatores bióticos e abióticos. In: LOPES, M. A.; FÁVERO, A. P.; FERREIRA, M. A. J.; FALEIRO, F. G.; FOLLE, S. M.; GUIMARÃES, E. P. (Ed.). **Pré-melhoramento de plantas - estado da arte e experiências de sucesso**. Brasília: Embrapa Informação Tecnológica, 2011. p. 527-548.

ARGOLO, P. S.; NORONHA, A. C. S.; OLIVEIRA,V. S.; FUKUDA, W. M. G. Aspectos da biologia e preferência para alimentação e oviposição de *Mononychellus tanajoa* (Bondar, 1938) em quatro variedades de mandioca. **Magistra**, v.17, n. 1, p. 23-27, 2005.

BURBANO, M.; CARABALÍ, A.; MONTOYA, J.; BELLOTTI, A. Resistencia natural de espécies silvestres de *Manihot* (Euphorbiaceae) a *Mononychellus tanajoa*, (Acariformes), *Aleurotrachelus socialis*, y *Phenacoccus herreni* (Homoptera). **Revista Colombiana de Entomologia**, v. 33, n. 2, p. 110-115, 2007.

CAMARGO, M. T. A. Estudo etnobotânico da mandioca (*Manihot esculenta* Crantz - Euphorbiaceae), na diáspora africana. **Fundação Gilberto Freire**, v. 24, n. 1, p. 22-30, 2003.

CARABALI, A.; BELLOTTI, A. C.; MONTOYA-LERMA, J.; FREGENE, M. *Manihot flabellifolia* Pohl, wild source of resistance to the whitefly *Aleurotrachelus socialis* Bondar (Hemiptera: Aleyrodidae). **Crop Protection**, v. 29, n. 1, p. 34-38, 2010.

CARVALHO, L. J. C. B. **Biodiversidade e biotecnologia em mandioca (*Manihot esculenta* Crantz)**. Brasília: Embrapa Recursos Genéticos e Biotecnologia, 2005.

CIAT-International Center for Tropical Agriculture. **Improved Cassava for the Developing World**. Annual Report Project IP3. Cali: CIAT, 2003. p. 8-65.

CRUZ, C. D. **Programa Genes**: versão Windows, aplicativo computacional em genética e melhoramento. Viçosa: UFV, 2001.

FERREIRA, D. F. Sisvar: a computer statistical analysis system. **Ciência e Agrotecnologia**, v. 35, n. 6, p. 1039-1042, 2011.

FUKUDA, C. Doenças e seu controle. In: SOUZA, L. S.; FARIAS, A. R. N.; MATTOS, P. L. P.; FUKUDA, W. M. G. (Ed.). **Aspectos socioeconômicos e agronômicos da mandioca**. Cruz das Almas: Embrapa Mandioca e Fruticultura Tropical, 2006. p. 433-454.

FUKUDA, W. M. G.; SILVA, S. O. E. Melhoramento de mandioca Brasil. In: CEREDA, M. P. (Org.). **Agricultura**: Tuberosas amilaceas latino americanas. São Paulo: Fundação Cargil, 2002. p. 242-257.

MORAES, G. J.; FLECHTMANN, C. H. W. **Manual de acarologia**: acarologia básica e ácaros de plantas cultivadas no Brasil. Ribeirão Preto: Holos, 2008.

NORONHA, A. C. S.; MORAES, G. J.; CIOCIOLA, A. I. Biologia de *Mononychellus tanajoa* (Bondar) (Acari: Tetranychidae) em variedades de mandioca. **Anais da Sociedade Entomológica do Brasil**, v. 24, p. 489-494, 1995.

SCOTT, A. J.; KNOTT, M. A. A cluster analysis method for grouping means in the analysis of variance. **Biometrics**, v. 30, n. 3, p. 507-512, 1974.

SINGH, D. The relative importance of characters affecting genetic divergence. **The Indian Journal of Genetics and Plant Breeding**, v. 41, n. 1, p. 237-245, 1981.

SNEATH, P. H.; SOKAL, R. R. **Numerical taxonomy**: The principles and practice of numerical classification. San Francisco: W.H. Freeman, 1973.

STATSOFT. **Statistica for Windows (data analysis software system)**, version 7.1. Statsoft. Tulsa: Statsoft Inc., 2005.

5

Genetic effects for controlling stripe rust (*Puccinia striiformis f. sp. tritici*) resistance in wheat through joint segregation analysis

Kalim Ullah[1], Naqib Ullah Khan[2*], Rozina Gul[2], Samrin Gul[2], Mohammad Irfaq Khan[3] and Imdad Ullah Khan[4]

*[1]PCCC Cotton Research Station, Dera Ismail Khan, Pakistan. [2]Department of Plant Breeding and Genetics, The University of Agriculture, Peshawar 25130, Pakistan. [3]Nuclear Institute for Food and Agriculture, Peshawar, Pakistan. [4]Department of Soil and Environmental Sciences, Faculty of Agriculture, Gomal University, Dera Ismail Khan, Pakistan. *Author for correspondence. E-mail: nukmarwat@yahoo.com; nukmarwat@aup.edu.pk*

ABSTRACT. Mixed inheritance analysis using joint segregation analysis (JSA) for stripe rust (*Puccinia striiformis f. sp. tritici*) resistance was carried out in six basic populations (P_1, F_1, P_2, BC_1, BC_2 and F_2) of four wheat crosses (Hashim-08 × LU-26, Farid-06 × Shafaq, Parula × Blue Silver, TD-1 × D-97603) during crop season 2009 to 2012. Genes controlling stripe rust resistance were assessed by using area under disease progress curve (AUDPC). The AUDPC was controlled by mixed two additive-dominant-epistatic major genes plus additive-dominant-epistasis of polygenes in cross Hashim-08 × LU-26 (model E), while in Farid-06 × Shafaq, it was controlled by mixed two major additive-dominant genes plus additive-dominant polygenes (model E-2). In cross Parula × Blue Silver, the AUDPC was managed by additive, dominance and epistasis of two major genes (model B-1), however, it was controlled by mixed one major gene and additive dominant polygenes in cross TD-1 × D-97603 (model D-1). Genetic variation and heritability was higher in major genes than polygene for all the crosses showing that AUDPC was mainly controlled by major genes. The genetic behavior of the AUDPC revealed that stripe rust resistance was controlled by mixed interaction of one to two major genes plus polygenes.

Keywords: area under disease progress curve (AUDPC), major gene and polygene inheritance, segregating populations, stripe rust (*Puccinia striiformis f. sp. tritici*), *Triticum aestivum* L.

Introduction

Rusts are the most important constraints in wheat production throughout the world. In our continent, leaf and stripe rust affects the production on approximately 60 (63%) and 43 (46%) million hectares, respectively, if susceptible cultivars are grown (Singh et al., 2014). Due to the mutagenic nature of rust pathogens, new resistant cultivars are rendered susceptible after sometime and usually become ineffective (Ghaledozdani, 2005). Use of resistance sources, however, is the most stable and economical method. Thus from the last few decades, rust management has been focused mainly through genetic resistance. In the beginning, rust was controlled through the integration of major resistance genes. However, due to the non-durability, later on the trend changed and hence growing the cultivars with adequate levels of durable resistance (additive effect of partial resistance of adult plant resistance genes) seems to be the best control strategy.

Stripe rust, also called yellow rust (*Yr*), is a fungal disease caused by *P. striiformis*. f. sp. *tritici*. It is the most damaging disease in the wheat growing areas in the world (Zhang, Yang, Li, Jin, & Yang, 2001). The disease requires cool and humid climatic conditions (Mamluk, 1992) and lower optimum temperature for its development (Ma & Singh, 1996). Presently, the yellow rust is a major wheat production constraint in most of the wheat growing areas of Pakistan. Grain yield losses of 10 to 70% have been reported depending upon the cultivar grown and the environmental conditions during ear emergence (Imtiaz, Cromey, Hampton, & Ogbonnaya, 2005). Tremendous losses of wheat production have been associated with yellow rust, when epiphytotics occurred under favourable conditions (Yahyaoui et al., 2002).

Stripe rust being a foliar disease has characteristic infection of powdery masses (uredial pustules) of yellow spores which forms stripes on the leaf surface. The uredial pustules have a characteristic torn margin and can occur on both sides of the leaves, stems, and heads. The pustules can cover large areas of affected leaves on young plants, and on older plants, the pustules arranged in parallel lines. This arrangement gives affected leaves a characteristic striped appearance. Severe attacks quickly give rise to yellowing, chlorosis and later necrosis. When the pustules mature, they break open to release a yellow-orange mass of urediospores. On seedlings, the pustules are scattered across the leaves in no order. In many wheat cultivars, the spores (pustules) develop in long, narrow, yellow stripes on leaves, leaf sheaths and heads. Once infected plants mature or become stressed, the tissue becomes brown and dry, and plants have a scorched and droughty appearance (Ma & Singh, 1996).

Most of the wheat cultivars released so far have race specific resistant with no resistance against the virulence of all pathogenic races. The gene *Yr9* controlling race specific resistance to stripe rust, after being widely used in early 1980s, was rendered ineffective by new stripe rust fungus races resulting in larger wheat yield losses. Another type of resistance introduced as durable resistance which remains effective in a cultivar is high temperature adult plant (HTAP) resistance which is non-race specific (Suenaga, Singh, Haurta-Espino, & William, 2003). In addition, other terms such as slow rusting and horizontal resistance are used to describe such type of resistance pattern. This type of resistance may be conditioned by group of minor genes and may not be easily and rapidly overcome. Wheat genotypes having slow rusting genes are often susceptible at seedling stage, but in adult stage, they exhibit moderate to high resistance to all types of the pathogens (Singh, Nelson, & Sorrells, 2000). Transgressive segregation has been observed in wheat for HTAP resistance and slow rusting resistance to stripe rust, and currently three stripe rust resistance genes named *Yr18, Yr29* and *Yr30* confer slow rusting (Singh, 1992; Suenaga, et al., 2003; William, Singh, Huerta-Espino, Ortiz, & Hoisington, 2003).

From the above information, regarding genetic control of stripe rust in wheat, it could be conferred that race specific or vertical resistance is of no more importance due to the occurrence of new pathotypes of fungus. At present, our ultimate aim should be to emphasize on non-race specific adult plant resistance and slow rusting which suggests that the stripe rust is not controlled by single major gene, and is inherited polygenically (Imtiaz, Cromey, Hampton, & Ahmad, 2003; Imtiaz et al., 2005). Therefore, the present study was designed to evaluate the genetic behavior and inheritance pattern of the genes controlling the area under disease progress curve (AUDPC) and to select the desirable genotypes for future breeding strategies.

Material and methods

Plant materials and procedure

Eight genetically diverse parental genotypes selected from wheat germplasm, were crossed in four combinations i.e. Hashim-08 × LU-26, Farid-06 × Shafaq, Parula × Blue Silver and TD-1 × D-97603. Six basic populations i.e. Parent-1 (P_1: pollen recipient), first filial generation (F_1), Parent-2 (P_2: pollen donor), Back cross-1 (BC_1: F_1 (pollen recipient) × P_1 (pollen donor), Back cross-2 (BC_2: F_1 (pollen recipient) × P_2 (pollen donor) and second filial generation (F_2) of each cross were developed during 2009-10 and 2010-11. The populations of each cross combination were planted in a randomized complete block (RCB) design with two replications during crop season 2011-12 at Faculty of Agriculture, Gomal University, Dera Ismail Khan, Pakistan (latitude 31°72′N, longitude 70°83′E and altitude of 174 m above mean sea level). As per minimum requirement of joint segregation analysis regarding number of individual observation in different populations, two rows (having 80 plants) were planted for parents (P_1, P_2) and F_1 populations, four rows (160 plants) for back cross populations (BC_1, BC_2) and six rows (240 plants) for F_2 populations in each replication. Row length, plants and rows spacing were kept at four meters, 10 and 30 cm, respectively. All other cultural practices and inputs were applied same to all the entries to minimize the environmental variations.

Each accession of the yellow rust screening nursery was bordered with a susceptible check of 'Morocco' as a spreader of yellow rust disease. Nursery was inoculated at tillering stage as to provide artificial yellow rust diseased condition in the field by making uniform spray of yellow rust urediospores (*Puccinia striiformis*) suspension at the end of February (Zadoks, Chang, & Konzak, 1974). The inoculum was contained mixture of urediospores of different yellow rust races prevalent in Pakistan. The urediospores were collected every year from different wheat growing regions in the country under Crop Disease Research Program (CDRP) at National Agricultural Research Centre (NARC), Islamabad, Pakistan. The inoculum was obtained from Wheat Research Institute Faisalabad, Pakistan. Preparation of spore's suspension, spores concentration of the inoculum, spraying of the spore's suspension on the nursery material and making of conditions conducive for disease development in the nursery were adapted as used by Khan, Ajab, Khattak, Mohammad, and Shah (2009b).

Measurement of the area under disease progress curve (AUDPC)

The disease severity data on ten randomly selected plants were visually recorded three times after seven days interval between the consecutive readings as suggested by Imtiaz et al. (2003). The severity was recorded from 0 to 100% rust infection on selected plants according to method of modified cob scale (Peterson, Campbell, & Hannah, 1948), and the response of individual plants was recorded (Table 1). Coefficient of infection (CI) was calculated by multiplying the response value with the intensity of infection in percent (Table 3). Average coefficient of infection (ACI) was derived from the sum of CI values of each entry divided by the number of replications. The area under disease progress curve (AUDPC) was calculated for individual plants using the CI values from the original rust severity data by using the following formula as suggested by Arama, Parlevliet, and Van-Silfhout (2000).

$$AUDPC = \sum [(X_i + X_{i+1})/2] t_i$$

where X_i and X_{i+1} are severity on date *i* and date *i* + 1, respectively and t_i is the number of days between date *i* and date *i* + 1.

Table 1. Response values of major infection type classes for stripe rust.

Reactions	Observations	Response value
No disease	O	0.0
Resistant	R	0.2
Resistant to moderately resistant	R-MR	0.3
Moderately resistant	MR	0.4
Moderately resistant to moderately susceptible	MR-MS	0.6
Moderately susceptible	MS	0.8
Moderately susceptible to susceptible	MS-S	0.9
Susceptible	S	1.0

Statistical analysis

The data regarding AUDPC were subjected to five groups consisting of 24 different genetic models of Joint Segregation Analysis (JSA) designed for six basic populations (Gai & Wang, 1998; Gai, Zhang, & Wang, 2003; Gai, Wang, Wu, & Chen, 2007) (Table 2). Suitable genetic models for each cross combination were determined by using maximum log of likelihood estimates (Wang & Gai, 1997) and Akaike's information criterion (Akaike, 1977). Further selection of best fit model was made on the basis of all non-significant or least number of significant values of three chi-square statistics i.e. U_1^2, U_2^2 and U_3^2 (Table 5 and 6). Two other important completely distribution free tests i.e. Smirnov's statistics (nW^2) and Kolmogorove's statistics (Dn) were used as goodness of fit tests to determine whether the selected model sufficiently explains the data (Table 5 and 6). However, if for a particular genetic model, none of these five statistics were found significant, then the data were adequately fit to the model (Gai & Wang, 1998; Ullah et al., 2014a). The data were analyzed by using *sin.exe software* and the major gene-polygene mixed inheritance model to a joint analysis of multi-generations (Gai et al., 2003). In case of the best fit model, the values of second order genetic parameters as well as variation due to major genes (σ_{mg}^2) and polygene (σ_{pg}^2) for BC_1, BC_2 and F_2 were worked out with the help of proposed formulae (Gai et al., 2003; Zhang, Gai, & Yang, 2003). Under the second order genetic parameters, the phenotypic variation (σ_p^2) partitioned into genetic and environmental variation (σ_e^2) for each cross. Based on Mather and Jinks (1982), the values from μ1 to μ69 presented in Table 7 are representing different means of component distributions (Wang, Podlich, Cooper, & Delacy, 2001; Zhang et al., 2003; Ullah et al., 2014b) regarding six basic populations (P_1, F_1, P_2, BC_1, BC_2 and F_2) which are to put in the formulae as suggested by Gai et al. (2003) for calculating 1st and 2nd order genetic parameters (Table 8).

Results

Plant population distribution for AUDPC in various crosses

In frequency distribution of plant population in P_1, F_1, P_2, BC_1, BC_2 and F_2 with respect to AUDPC level revealed that in cross Hashim-08 × LU-26, the tendency of both F_1 and BC_1 was towards disease susceptible parent (P_1) and AUDPC values ranged from 250 to 450, 150 to 400 and 200 to 500 for P_1, F_1 and BC_1, respectively (Table 3). While the tendency of BC_2 was towards stripe rust resistant parent (P_2) and AUDPC was recorded from 0 to 250 and 0 to 400 for P_2 and BC_2, respectively (Table 3). For cross Farid-06 × Shafaq, in F_1 and BC_1 populations the trend was towards P_1 and their corresponding AUDPC values were ranging from 200 to 550, 150 to 650 and 150 to 600 for P_1, F_1 and BC_1, respectively. The AUDPC values for P_2 and BC_2, were ranging from 50-300 and 50-550, respectively indicating the tendency of BC_2 towards P_2. Similar response and tendency of F_1, BC_1 and BC_2 was also observed in cross Parula × Blue Silver, and the ranges were 50 to 350, 50 to 400 and 0 to 400 for P_1, F_1 and BC_1, respectively. Similarly, P_2 (150 to 500) and BC_2 (150 to 700) revealed the tendency of BC_2 towards the respective parent. In cross TD-1 × D-97603, the F_1, BC_1 and BC_2 were equally distributed between the parents for AUDPC level because of slight differences between the P_1 (100 to 400) and P_2 (100 to 450). The F_2 segregating populations among four crosses were almost equally distributed between the both parents. The genetic behavior of AUDPC revealed that in F_2 and some BC_1 and BC_2 generations, the resistance to stripe rust was controlled by mixed interaction of one to two major genes plus polygene.

Table 2. Akaike's information criterion (AIC) values for AUDPC under five groups of 24 genetic models estimated through the iterated expectation and conditional maximization (IECM) algorithm.

Model group, code, and implication of model type	First order genetic parameters		Cross combination and AIC values			
	Major gene(s)	Polygene	Hashim-08 × LU-26	Farid × Shafaq	Parula × Blue silver	TD-1 × D-97603
Group 1: One major gene						
A-1: Additive-dominant	m, d, h	-	8606.43	9042.50	9071.51	8967.40
A-2: Additive	$m, d, (h=0)$	-	8692.75	9072.10	9093.51	8967.41
A-3: Completely dominant	$m, d \ (h = d)$	-	8612.37	9040.53	9068.94	8972.57
A-4: Completely negative dominant	$m, d \ (h = -d)$	-	8616.65	9145.16	9099.75	8961.11
Group 2: Two major genes						
B-1: Additive-dominance-epistasis	$m, d_a, d_b, h_a, h_b, i, j_{ab}, j_{ba}, l$	-	8543.06	9022.54	8963.88	8879.23
B-2: Additive-dominant	$m, d_a, d_b, h_a, h_b, i, j_{ab}, j_{ba}, l$	-	8600.58	9022.33	9003.56	8883.74
B-3: Additive	$m, d_a, d_b \ (h_a=h_b= 0)$	-	8769.38	9168.80	9128.21	8884.40
B-4: Equally additive	$m, d \ (d_a= d_b, \ h_a=h_b=0)$	-	8719.00	9082.75	9108.06	8969.47
B-5: Completely dominant	$m, d_a (= h_a), \ d_b (=h_b)$	-	8612.86	9039.94	9022.001	8913.64
B-6: Equally dominant	$m, d \ (= d_a= d_b= h_a= h_b)$	-	8641.34	9051.22	9100.02	8973.75
Group 3: Polygene						
C: Additive-dominant-epistasis	M	[d], [h], [*i*], [*j*], [l]	8592.27	9053.12	9102.21	8875.76
C-1: Additive-dominant	M	[d], [h]	8594.23	9054.43	9107.76	8884.04
Group 4: One major gene plus polygene						
D: Additive-dominant one major gene and additive-dominant-epistasis of polygene	m, d, h	[d], [h], [*i*], [*j*], [*l*]	8596.26	9032.26	8971.08	8879.76
D-1: Additive-dominant one major gene and additive-dominant polygene	m, d, h	[d], [h]	8559.86	9026.54	8968.51	8851.53
D-2: Additive one major gene and additive-dominant polygene	$m, d, (h = 0)$	[d], [h]	8555.71	9024.54	8966.51	8849.53
D-3: Completely dominant one major gene and additive-dominant polygene	$m, d \ (h = d)$	[d], [h]	8608.93	9025.67	8964.53	8872.04
D-4: Completely negative dominant one major gene and additive-dominant polygene	$m, d \ (h= -d)$	[d], [h]	8609.43	9022.87	8969.78	8874.54
Group 5: Two major genes plus polygene						
E: Additive-dominant-epistatic of two major genes and additive-dominant-epistasis of polygene.	$m_1 \sim m_6, d_a, d_b, h_a, h_b, i, j_{ab}, j_{ba}, l$	[d], [h], [*i*], [*j*], [*l*]	8555.80	9023.52	8964.10	8865.33
E-1: Additive-dominant epistasis of two major genes and additive-dominant polygene	$m, d_a, d_b, h_a, h_b, i, j_{ab}, j_{ba}, l$	[d], [h]	8553.61	9027.31	8969.07	8860.09
E-2: Additive-dominant two major genes and additive-dominant polygene	$m, d_a, d_b, h_a, h_b, i= j_{ab}= j_{ba}, l$	[d], [h]	8599.42	9023.16	9040.75	8876.83
E-3: Additive two major genes and additive-dominant polygene	$m, d_a, d_b, h_a = h_b = 0$	[d], [h]	8570.69	9019.87	8984.92	8869.63
E-4: Equally additive two major genes and additive-dominant polygene	$m, d \ (= d_a = d_b, \ (h_a = h_b = 0)$	[d], [h]	8626.89	9025.44	9091.28	8886.009
E-5: Completely dominant two major genes and additive-dominant polygene	$m, d_a = h_a, d_b = h_b,$	[d], [h]	8611.41	9028.92	9009.43	8882.19
E-6: Equally dominant two major genes and additive-dominant polygene	$m, d = d_a = d_b = h_a = h_b$	[d], [h]	8645.71	9027.98	9024.75	8882.65

m: Population mean. *d*, [d]: Additive effect due to major gene(s) and polygenes, respectively. *h*, [h]: Dominant component due to major gene(s) and polygenes, respectively. *i*, [*i*]: Additive × Additive component due to major gene(s) and polygene, respectively. j_{ab}: $d_a \times h_b$. First major gene with additive × second major gene with dominant effect. j_{ba}: $d_b \times h_a$. Second major gene with additive × first major gene with dominant effect. [*j*]: Additive-dominance epistasis. Source of different model groups and model types (Gai & Wang 1998; Gai et al., 2003; Zhang et al., 2003).

Table 3. Frequency distribution of plant population of area under disease progress curve (AUDPC) in P_1, F_1, P_2, BC_1, BC_2 and F_2 of four bread wheat crosses.

Crosses	Generations	Range of area under disease progress curve (AUDPC)															Size	Mean	Variance
		0	50	100	150	200	250	300	350	400	450	500	550	600	650	700			
Hashim -08 × LU-26	P_1	-	-	-	-	-	6	12	15	17	6	-	-	-	-	-	60	87.23	3337.03
	F_1	-	-	-	25	15	10	16	19	5	-	-	-	-	-	-	90	110.02	5136.00
	P_2	10	8	12	13	10	11	-	-	-	-	-	-	-	-	-	60	309.01	6659.57
	BC_1	-	-	-	-	23	22	21	12	21	26	25	-	-	-	-	150	141.58	8102.01
	BC_2	12	11	8	13	19	31	19	16	14	-	-	-	-	-	-	150	261.29	14641.75
	F_2	12	14	24	17	23	20	18	22	16	19	15	-	-	-	-	200	225.37	20299.61
Farid-06 × Shafaq	P_1	-	-	-	-	7	11	6	8	12	6	6	4	-	-	-	60	339.73	11891.83
	F_1	-	-	-	7	9	12	12	9	8	13	8	6	4	2	-	90	340.72	18271.91
	P_2	-	8	10	10	12	13	7	-	-	-	-	-	-	-	-	60	155.25	6651.03
	BC_1	-	-	-	12	13	9	16	22	28	17	13	15	5	-	-	150	347.53	15471.98
	BC_2	-	7	12	11	21	19	23	17	8	11	11	10	-	-	-	150	272.16	19126.62
	F_2	12	9	14	10	22	23	15	20	14	11	16	17	7	7	3	200	300.31	33193.92
Parula × Blue Silver	P_1	-	7	9	8	10	14	7	5	-	-	-	-	-	-	-	60	172.93	8805.38
	F_1	-	9	11	12	13	17	13	7	8	-	-	-	-	-	-	90	197.72	10909.15
	P_2	-	-	-	8	6	10	7	9	6	9	5	-	-	-	-	60	295.66	12292.97
	BC_1	8	12	13	17	25	19	16	31	9	-	-	-	-	-	-	150	209.38	14740.10
	BC_2	-	-	-	4	6	12	13	15	19	21	12	15	13	9	11	150	421.96	21620.88
	F_2	9	8	13	15	19	16	21	17	22	23	16	6	6	7	2	200	301.38	28682.10
TD-1 × D-97603	P_1	-	-	8	7	10	12	13	7	3	-	-	-	-	-	-	60	220.61	7314.41
	F_1	-	-	8	12	14	9	17	13	11	6	-	-	-	-	-	90	252.11	10519.45
	P_2	-	-	6	10	8	7	8	9	7	5	-	-	-	-	-	60	246.36	12271.25
	BC_1	-	7	13	17	15	18	25	31	14	10	-	-	-	-	-	150	242.78	12938.37
	BC_2	-	-	11	15	14	16	20	30	26	13	7	-	-	-	-	150	282.85	12093.24
	F_2	8	12	20	17	13	23	17	15	14	21	12	9	13	6	-	200	289.55	32906.28

Genetic analysis of stripe rust in various wheat crosses

According to Akaike's information criterion (Table 2), maximum log of likelihood values (Table 4), and goodness of fit tests (Table 5), the model E was selected as the best fit genetic model for cross Hashim-08 × LU-26. The model E revealed that AUDPC was controlled by mixed two additive-dominant-epistatic major genes plus additive-dominant-epistasis of polygene. Regarding AUDPC, the negative and equal additive (da, db) effect (-53.68) was exhibited by two major genes (A & B). While dominant effects (ha, hb) were exhibited by two major genes (A & B) regarding AUDPC with values of 175.16 and 108.22, respectively. Dominant ratios (ha/da & hb/db) of major genes A and B were recorded as -3.26 and -2.02, respectively. Additive × additive effect (i) of the major genes plus polygene was found to be 44.08. The additive × dominant effect of gene A over gene B (Jab) and that of B over A (Jba) were -114.91 and -47.96, respectively. Negative dominant × dominant type of non-allelic interaction (l) was recorded as -206.43.

The second order genetic parameters in cross Hashim-08 × LU-26, represents the phenotypic variation for AUDPC in segregating generations BC_1, BC_2 and F_2 (Table 8). The genetic variation was further divided into variation due to major genes and polygene. According to the best-fit model E, as the resistance was controlled by two major genes plus polygene, therefore higher phenotypic variation (σ_p^2) was observed in BC_1, BC_2 and F_2. The major-gene heritability (h_{mg}^2) values which are among the most important second order parameters were recorded as 58, 49.88 and 76.74% for BC_1, BC_2, and F_2, respectively. The polygene heritability (h_{pg}^2) values were estimated as 6.98, 23.06 and 3.74% for BC_1, BC_2 and F_2, respectively. Maximum variation and heritability were estimated in major genes than polygene, however, the polygene heritability revealed its contribution in controlling the said trait.

For cross Farid-06 × Shafaq, the Model E-2 was selected as the best fit model on the basis of Akaike's information criterion (Table 2), maximum log of likelihood values (Table 4) and goodness of fit test (Table 5). Genetic model E-2 represents that AUDPC was controlled by mixed two major additive-dominant genes plus additive-dominant polygene. The population mean for the cross Farid-06 × Shafaq was 247.52, while the additive components were estimated to be 113.43 and 0.18, respectively. The dominant components for the said cross were recorded as 3.28 and 0.16, respectively. Adverse additive effect was negative (-102.07) while dominance effect was positive (92.04) produced by of polygene. The second order genetic parameters for the said cross indicates that the phenotypic variation among BC_1, BC_2 and F_2 was mainly due to major genes and environment (Table 8). The major gene heritability values for BC_1, BC_2 and F_2 were 27.61, 41.44 and 66.26%, respectively. The zero values for polygene variation as well as polygene heritability with respect to BC_1 and BC_2 indicate that very few minor genes (polygene) were involved in controlling the AUDPC by contributing very small effects towards controlling of the trait.

Table 4. Maximum log of likelihood estimates for area under disease progress curve (AUDPC) under various genetic models estimated through IECM algorithm.

Model	Hashim-08 × LU-26	Farid × Shafaq	Parula × Blue silver	TD-1 × D-97603
A-1	-4300.21	-4517.25	-4531.75	-4479.7
A-2	-4343.37	-4533.05	-4543.75	-4480.7
A-3	-4303.18	-4517.26	-4531.47	-4483.28
A-4	-4405.32	-4569.58	-4546.87	-4477.55
B-1	-4261.53	-4501.27	-4471.94	-4429.61
B-2	-4294.29	-4505.16	-4495.78	-4435.87
B-3	-4380.69	-4580.40	-4560.11	-4438.4
B-4	-4356.50	-4538.37	-4551.03	-4481.73
B-5	-4302.43	-4515.97	-4507.00	-4452.82
B-6	-4317.67	-4522.61	-4547.01	-4483.87
C	-4286.13	-4521.62	-4547.45	-4427.88
C-1	-4285.12	-4522.32	-4543.66	-4435.02
D	-4286.13	-4504.13	-4473.54	-4427.88
D-1	-4270.93	-4504.27	-4475.25	-4416.76
D-2	-4269.85	-4504.27	-4475.25	-4416.76
D-3	-4296.43	-4504.45	-4473.52	-4428.02
D-4	-4299.67	-4502.23	-4479.23	-4428.04
E	-4259.90	-4493.76	-4464.05	-4414.66
E-1	-4260.80	-4498.65	-4469.53	-4415.04
E-2	-4288.71	-4500.58	-4509.37	-4427.41
E-3	-4276.34	-4500.93	-4483.46	-4425.81
E-4	-4305.44	-4504.72	-4537.64	-4435.004
E-5	-4296.70	-4505.46	-4495.71	-4432.09
E-6	-4314.85	-4505.99	-4504.37	-4433.32

For cross Parula × Blue Silver, the suitable model was B-1 which selected on the basis of smallest AIC values (Table 3), likelihood estimates (Table 4) and goodness of fit tests (Table 6). However, the model B-1 represents that AUDPC was controlled by additive, dominance and epistasis of two major genes. The population mean for the AUDPC was recorded as 248.70. The additive effect due to the first major gene (A) was negative (-109.77) while that was positive (41.05) due to the 2nd major gene (B). Positive dominant effects were exhibited both by first and second major genes with values of 191.53 and 34.94, respectively. Under the epistasis of the two major genes, the additive × additive and dominant × dominant effects were negative i.e. -7.44 and -261.018, respectively. The additive × dominant epistatic effects were found negative due to first major gene and second major gene. The second order genetic parameters of the cross revealed the phenotypic variations in the segregating generations BC_1, BC_2 and F_2 which were controlled by genetic and environmental components (Table 8). Heritability values of major genes were for BC_1 (66.03%), BC_2 (95.63%) and F_2 (75.50%). However, polygene variation and heritability values were zero because no minor genes (polygene) were involved for expression of AUDPC in cross Parula × Blue Silver.

For cross TD-1 × D-97603, on the basis of likelihood and smallest AIC values (Tables 3 and 4) and goodness of fit tests (Table 6), the best fit genetic model was D-1 representing the effect of mixed one major gene and additive dominant polygene for AUDPC. The population mean of the cross was found to be 249.10. Due to major genes, negative additive effect (-189.57) was recorded while polygene showed positive additive and dominant effects with values of 167.54 and 42.98, respectively. As evident from the 2nd order genetic parameters (Table 8), the variance and broad sense heritability of the major genes were higher than those of the polygene in cross TD-1 × D-97603 for AUDPC. The values with respect to polygenic variance as well as polygenic heritability were equal to zero for BC_1, BC_2 and F_2 generations. The lower values of polygenic variation and heritability might be due to epistatic effects between the major and minor genes.

Discussion

Joint segregation analysis (JSA) is used to analyze the segregating data of quantitatively controlled traits like the Mendelian procedure. Unlike quantitative trait locus (QTL) mapping, the JSA neither can identify many QTLs nor locate the position of the major genes on a particular chromosome (Gai et al., 2007). While designing breeding experiments for the improvement of quantitative traits, selecting parents for crosses, progeny selection and gene pyramiding, the JSA is strongly recommended as a simple and useful technique. The technique is helpful to know the number of major genes, their kinds of genetic effects, heritability values of polygene as well as genetic information on all kinds of genetic effects (Wang & Gai, 2001; Gai et al., 2007). In addition, JSA can also be used as a check for QTLs mapping and it should be conducted before QTL mapping is performed so that plant breeders can have a first impression on the genetic system of involved trait (Gai et al., 2007).

Because of durability of slow rusting resistance, plant breeders are interested to study the genetic nature of this type of resistance and its introgression into breeding materials (Zhang, et al., 2001; Zhang et al., 2003; Khan, Ajab, Hongxiang, & Khattak, 2009a; Khan et al., 2009b). In present studies, in cross Hashim-08 × LU-26, the AUDPC was controlled by mixed two additive dominant epistatic major genes plus additive-dominant epistasis of polygene which revealed that AUDPC was mainly governed by dominant effects of the major genes with epistatic nature with polygene. The additive ×

dominant negative effects also favor the epistasis of major genes and polygene. Past studies revealed that the AUDPC in wheat crosses was controlled by additive dominance epistatic effect of two major genes plus polygene (Khan et al., 2009a). In QTLs mapping for leaf and stripe rust in wheat, Singh et al. (2014) also identified several epistatic effects both for both rusts resistance QTL. In wheat cultivars, the AUDPC was controlled by three/four genes in cv. Sonoita 81, two/three genes in Tanager-S, three genes in Galvez 87, and cvs. Ures-81 and Moncho-S have two genes for slow rusting resistance to leaf rust (Das, Rajaram, Kronstad, Mundt, & Singh, 1993; Das, Rajaram, Mundt, & Kronstad, 2004). Leaf rust resistance was reported in 50 Mexican wheat cultivars and several genotypes were identified with adult plant resistance (Singh & Rajaram, 1991). For stripe rust resistance, the wheat cvs. Libellula and Xian Nong-4 showed two to three resistance genes, and cv. San Pastore appears to have two to four resistance genes (Zhang et al., 2001).

In cross Farid-06 × Shafaq, the resistance to stripe rust was controlled by mixed two major additive-dominant genes plus additive dominant polygene. Due to first and second major genes, positive additive effects were recorded. This revealed that AUDPC was under the control of additive effect of both major and minor genes. In previous studies, the AUDPC was under the control of two major genes with additive dominance epistatic effect plus additive dominant polygene in wheat crosses (Khan et al., 2009b). In inheritance of resistance to stripe rust in winter wheat cultivars, the dominant component was detected in some cases, and the resistance was conditioned by oligogenes (polygene) with heritability of the resistance ranging from 50 to 79% (Feng et al., 2007). Three and four QTL controlled the slow leaf rusting resistance in the wheat crosses (William et al., 2003).

Table 5. Test for goodness of fit regarding area under disease progress curve (AUDPC) of models C, D and E.

Cross-1: Hashim-08 × Lu-26						
Models	Generations	U_1^2	U_2^2	U_3^2	nW^2	Dn
E	P_1	0.04 (0.82)	0.02 (0.86)	0.03 (0.84)	0.1 (> 0.05)	0.09*
	F_1	0.01 (0.91)	0.51 (0.47)	10.78**	0.33 (> 0.05)	0.11 (> 0.05)
	P_2	0.01 (0.90)	0.66 (0.41)	14.01***	0.31 (> 0.05)	0.12 (> 0.05)
	BC_1	0.07 (0.78)	0.008 (0.92)	0.44 (0.50)	0.12 (> 0.05)	0.07*
	BC_2	0.26 (0.61)	0.21 (0.64)	0.01 (0.89)	0.16 (> 0.05)	0.07*
	F_2	0.54 (0.46)	0.86 (0.35)	0.74 (0.38)	0.13 (> 0.05)	0.08*
B-1	P_1	0.53 (0.46)	0.43 (0.50)	0.03 (0.86)	0.12**	0.1*
	F_1	0.30 (0.57)	1.64 (0.19)	8.92**	0.33 (> 0.05)	0.12 (> 0.05)
	P_2	0.06 (0.79)	1.24 (0.26)	11.92***	0.26 (> 0.05)	0.13 (> 0.05)
	BC_1	2.14 (0.14)	0.79 (0.37)	4.45*	0.43 (> 0.05)	0.12 (> 0.05)
	BC_2	0.10 (0.74)	0.41 (0.52)	1.65 (0.19)	0.18 (> 0.05)	0.07**
	F_2	1.05 (0.30)	2.43 (0.11)	5.15*	0.3 (> 0.05)	0.11*
E-1	P_1	0.04 (0.82)	0.02 (0.86)	0.03 (0.84)	0.1*	0.09*
	F_1	0.01 (0.91)	0.51 (0.47)	10.78**	0.33 (> 0.05)	0.11 (>0.05)
	P_2	0.01 (0.90)	0.66 (0.41)	14.01***	0.31 (> 0.05)	0.12 (>0.05)
	BC_1	0.07 (0.78)	0.008 (0.92)	0.44 (0.50)	0.12*	0.07*
	BC_2	0.26 (0.61)	0.21 (0.64)	0.01 (0.89)	0.16 (> 0.05)	0.07*
	F_2	0.54 (0.46)	0.86 (0.35)	0.74 (0.38)	0.13*	0.08*
Cross-2: Farid-06 × Shafaq						
Models	Generations	U_1^2	U_2^2	U_3^2	nW^2	Dn
E	P_1	0.006 (0.93)	0.05 (0.81)	1.50 (0.21)	0.08*	0.09*
	F_1	0.09 (0.75)	0.63 (0.42)	19.31***	0.47 (> 0.05)	0.12 (> 0.05)
	P_2	0.003 (0.95)	0.16 (0.68)	3.38 (0.06)	0.09*	0.09*
	BC_1	0.01 (0.89)	0.00 (0.99)	0.23 (0.62)	0.04*	0.05*
	BC_2	0.10 (0.74)	0.26 (0.60)	0.67 (0.41)	0.06*	0.06*
	F_2	0.06 (0.80)	0.03 (0.85)	0.06 (0.79)	0.06*	0.04*
E-1	P_1	0.20 (0.64)	0.65 (0.41)	2.13 (0.14)	0.11 (> 0.05)	0.08*
	F_1	0.04 (0.83)	0.94 (0.33)	21.83***	0.51 (> 0.05)	0.12 (> 0.05)
	P_2	0.13 (0.70)	0.59 (0.44)	2.69 (0.10)	0.09*	0.09*
	BC_1	2.25 (0.13)	2.86 (0.09)	0.90 (0.34)	0.29 (> 0.05)	0.1 (> 0.05)
	BC_2	0.03 (0.84)	0.05 (0.81)	0.02 (0.86)	0.08*	0.06*
	F_2	0.08 (0.77)	0.56 (0.45)	3.66 (0.055)	0.16 (> 0.05)	0.07*
E-2	P_1	0.02 (0.88)	0.04 (0.83)	1.96 (0.16)	0.09*	0.1 (> 0.05)
	F_1	0.90 (0.34)	0.06 (0.79)	22.18***	0.6 (> 0.05)	0.15 (> 0.05)
	P_2	0.85 (0.35)	1.72 (0.18)	2.76 (0.09)	0.15 (> 0.05)	0.11 (> 0.05)
	BC_1	0.07 (0.77)	0.01 (0.90)	2.49 (0.11)	0.12 (> 0.05)	0.06*
	BC_2	0.84 (0.35)	0.71 (0.39)	0.03 (0.85)	0.13 (> 0.05)	0.07*
	F_2	0.13 (0.71)	0.08 (0.77)	6.59*	0.26 (> 0.05)	0.08*

Table 6. Test for goodness of fit regarding area under disease progress curve (AUDPC) of models C, D, E, E-1 and B-1.

Cross-3: Parula × Blue silver						
Models	Generations	U_1^2	U_2^2	U_3^2	nW^2	Dn
E	P_1	0.004 (0.94)	0.05 (0.81)	0.46 (0.49)	0.05*	0.08*
	F_1	0.001 (0.97)	0.28 (0.59)	5.15*	0.14 (> 0.05)	0.08*
	P_2	0.001 (0.98)	0.52 (0.46)	8.99**	0.23 (> 0.05)	0.11 (> 0.05)
	BC_1	0.003 (0.95)	0.00 (0.98)	0.07 (0.78)	0.07*	0.07*
	BC_2	0.04 (0.83)	0.009 (0.92)	0.17 (0.67)	0.04*	0.04*
	F_2	0.02 (0.87)	0.01 (0.91)	1.10 (0.29)	0.07 (> 0.05)	0.04*
E-1	P_1	0.007 (0.93)	0.002 (0.96)	0.24 (0.62)	0.04*	0.08*
	F_1	1.21 (0.27)	0.29 (0.58)	4.34*	0.22 (> 0.05)	0.1 (> 0.05)
	P_2	1.24 (0.26)	0.12 (0.72)	8.36**	0.33 (> 0.05)	0.15 (> 0.05)
	BC_1	0.14 (0.70)	0.16 (0.68)	0.03 (0.85)	0.09*	0.07*
	BC_2	6.38*	4.19*	2.53 (0.11)	0.71 (> 0.05)	0.11 (> 0.05)
	F_2	0.01 (0.92)	0.05 (0.81)	0.33 (0.56)	0.05*	0.04*
B-1	P_1	0.008 (0.92)	0.03 (0.84)	0.17 (0.67)	0.04*	0.07*
	F_1	2.19 (0.13)	0.86 (0.35)	4.05*	0.3 (> 0.05)	0.11 (> 0.05)
	P_2	0.97 (0.32)	0.06 (0.79)	7.69**	0.29 (> 0.05)	0.14 (> 0.05)
	BC_1	0.09 (0.76)	0.28 (0.59)	0.95 (0.32)	0.14 (> 0.05)	0.08*
	BC_2	8.66**	6.79**	0.94 (0.33)	0.84 (> 0.05)	0.12 (> 0.05)
	F_2	0.30 (0.58)	0.21 (0.64)	0.07 (0.78)	0.13 (> 0.05)	0.06*
Cross-4: TD-1 × D-97603						
Models	Generations	U_1^2	U_2^2	U_3^2	nW^2	Dn
E	P_1	0.01 (0.91)	0.04 (0.82)	0.19 (0.65)	0.07*	0.09*
	F_1	0.002 (0.96)	0.85 (0.35)	12.52***	0.34 (> 0.05)	0.12 (> 0.05)
	P_2	0.009 (0.92)	0.84 (0.35)	16.30***	0.38 (> 0.05)	0.15 (> 0.05)
	BC_1	0.50 (0.47)	0.35 (0.54)	0.12 (0.72)	0.24 (> 0.05)	0.12 (> 0.05)
	BC_2	0.003 (0.95)	0.02 (0.86)	0.22 (0.63)	0.08*	0.05*
	F_2	0.61 (0.43)	0.24 (0.61)	1.08 (0.29)	0.12 (>0.05)	0.05*
E-1	P_1	0.00 (0.99)	0.01 (0.91)	0.18 (0.66)	0.06*	0.09*
	F_1	0.01 (0.89)	1.02 (0.31)	12.50***	0.34 (>0.05)	0.11 (>0.05)
	P_2	0.008 (0.92)	1.19 (0.27)	16.22***	0.38 (>0.05)	0.14 (>0.05)
	BC_1	0.94 (0.33)	1.26 (0.26)	0.53 (0.46)	0.33 (>0.05)	0.13 (>0.05)
	BC_2	0.04 (0.82)	0.08 (0.77)	0.08 (0.77)	0.09*	0.05*
	F_2	0.88 (0.34)	0.33 (0.56)	1.74 (0.18)	0.17 (>0.05)	0.06*
D-1	P_1	0.006 (0.93)	0.01 (0.89)	0.73 (0.39)	0.06*	0.08*
	F_1	0.001 (0.97)	0.21 (0.64)	3.07 (0.07)	0.15 (>0.05)	0.09*
	P_2	0.007 (0.93)	0.30 (0.58)	6.36*	0.2 (>0.05)	0.12 (>0.05)
	BC_1	0.13 (0.71)	0.64 (0.42)	3.27 (0.07)	0.37 (>0.05)	0.12 (>0.05)
	BC_2	0.28 (0.59)	0.77 (0.37)	2.14 (0.14)	0.3 (>0.05)	0.1 (>0.05)
	F_2	0.08 (0.76)	0.02 (0.86)	3.26 (0.07)	0.2 (>0.05)	0.06*

In above crosses, the maximum variation and heritability were shown by major genes than polygene. The heritability due to major genes was higher than polygene in various wheat populations (Khan et al., 2009a). The slow leaf rusting resistance heritability has been reported to be moderate to high in wheat (Bjarko & Line, 1988). In most of the wheat crosses, the broad and narrow-sense heritability were high while some crosses revealed moderate narrow-sense heritability (Zhang et al., 2001). Moderate to high heritability was reported for slow rusting resistance, and suggested selection of tolerant genotypes in early segregating generations against leaf rusting disease (Das, Rajaram, Mundt, & Kronstad, 1992; 2004). Positive values of dominant effects and additive × additive effect of the major genes plus polygene, and negative values of dominant ratios of major genes exhibited that contribution of major genes in controlling the AUDPC was maximum, however, was of epistatic nature with polygene. In various wheat crosses, the negative and positive signs of the additive as well as dominant effect due to major genes and polygene may occur due to varied genetic background of the parental lines involved (Wang & Gai, 2001). However, additive × dominant negative effects also favor the epistasis of major genes and polygene. Higher phenotypic variation was observed in BC1, BC2 and F2 populations because the resistance was controlled by two major genes plus polygene. Leaf rust resistant genes were reported in 70 wheat cultivars and suggested 11 cultivars with adult-plant resistance genes (Singh et al., 2000). In adult-plant resistance of leaf rust in wheat, the leaf rust resistance gene Lr34 and two to three other effective genes were reported (Singh & Rajaram, 1993).

In cross Parula × Blue Silver, the AUDPC was controlled by additive, dominance and epistasis of two major genes and no polygene were involved in controlling the said trait. Low estimates regarding polygenic variation and heritability might be due to epistatic effects between the major and minor genes among the segregating generations (Wang & Gai, 2001). Additive × additive and dominant × dominant effects were negative which proves that no epistasis was involved due to major genes. Positive dominant effects of both the major genes were noted which revealed that AUDPC was under the influence of combined dominant effects of the major genes. The polygenic variance and heritability values for BC_1, BC_2 and F_2 were equal to zero because no minor genes (polygene) were involved in controlling the said trait. Additive dominant as well as dominant × dominant epistasis for leaf rust have been reported in wheat crosses (Bjarko & Line, 1988). In segregating generations for stripe rust resistance, the additive / modifying action of two genes was reported in a cross between susceptible and resistant cultivars of wheat (Imtiaz et al., 2003).

The two genes with additive effect were found to be responsible for stripe rust resistance in a cross between resistant and susceptible wheat parental lines (Ma, Singh, & Abdalla, 1997).

The AUDPC was controlled by mixed one major gene and additive dominant polygene in cross TD-1 × D-97603. A higher and positive value of additive component of polygene indicates that leaf rust resistance was mainly controlled by the favorable interaction of polygene found in both parents. Non-race specific resistance and slow rusting suggests that the stripe rust resistance was not controlled by a single major gene, but is inherited polygenically (Imtiaz et al., 2005). Additive effect with respect to stripe rust resistance has also been reported in wheat crosses (Ma et al., 1997). The involvement of epistasis with significant additive x additive component in controlling AUDPC for adult plant resistance has been reported in parental cultivars, F_1, F_2 and backcross populations in wheat (Chen, Soria, Yan, Sun, & Dubcovsky, 2003). Using generation mean analysis, Das et al. (2004) reported digenic epistasis with predominant additive gene effect for stripe rust resistance in different wheat crosses. Across the environments, eight QTL were detected for leaf rust resistance and ten QTL for the quantitative expression of leaf tip necrosis in various wheat cultivars, and at these loci, the alleles of cv. Forno increased the level of resistance as well as the extent of leaf tip necrosis, indicating pleiotropy (Messmer et al., 2000). In a genetic analysis for latent period of *Puccinia recondita* in wheat, four genes with unequal and epistatic effects controlled the latent period, whereas some findings revealed that at least five genes were found to be involved in the prolongation of the latent period (Shaner, Buechley, & Nyquist, 1997).

In segregating populations of all the crosses, the major gene heritability was higher than polygene showing that the leaf rust resistance was mainly under the control of major genes with a very small contribution of polygenes. Higher environmental variation was also observed in all the crosses indicating that the resistance to stripe rust was under greater influence of environmental fluctuations. Similar fluctuations due to environment for stripe rust resistance at adult plant stage were also reported (Das et al., 2004). The contradictions between the present and the previous findings might be because of the different statistical approaches i.e. diallel or generation mean analysis used to measure the genetic effect as the polygenic system and have no power to determine the effect of the individual major genes and aggregate effect of the polygene.

Table 7. Maximum likelihood estimates of component parameters regarding area under disease progress curve (AUDPC) in four wheat crosses in their respective best fit model.

Variables	Hashim-08 × Lu-26	Farid-06 × Shafaq	Parula × Blue Silver	TD-1 × D-97603
	Model Type: E	Model Type : E-2	Model Type: B-1	Model Type: D-1
μ_1	87.23	340.63	172.54	222.05
μ_2	110.02	349.68	214.16	253.28
μ_3	309.01	167.8	309.99	246.68
μ_{41}	73.16	437.68	172.54	160.84
μ_{42}	76.08	437.66	112.48	314.49
μ_{43}	209.96	252.65	331.54	-
μ_{44}	213.41	252.63	214.16	-
μ_{51}	145.97	354.7	214.16	210.54
μ_{52}	345.82	354.37	454.81	347.66
μ_{53}	345.82	163.12	548.93	-
μ_{54}	220.46	162.78	309.99	-
μ_{61}	63.12	488.72	172.54	90
μ_{62}	66.03	488.7	112.48	288.21
μ_{63}	82.32	488.37	105.32	513.31
μ_{64}	199.92	303.69	331.54	-
μ_{65}	203.35	303.66	214.16	-
μ_{66}	403.21	303.33	454.8	-
μ_{67}	82.32	112.1	406.96	-
μ_{68}	403.21	112.08	548.93	-
μ_{69}	277.84	111.75	309.99	-
Σ^2	3961.21	11200.07	9417.97	8243.68
Σ_4^2	3961.21	11200.07	9417.97	8243.68
Σ_5^2	3961.21	11200.07	9417.97	8243.68
Σ_6^2	3961.21	11200.07	9417.97	8243.6

Table 8. Estimates of first and second order genetic parameters for area under disease progress curve (AUDPC) in four wheat crosses.

First order parameter				2nd order parameter			
Cross-1: Hashim-08 × LU-26 (model E)							
					BC_1	BC_2	F_2
M_1	150.51	h_a	175.16	σ_p^2	8102.01	14641.8	20299.6
M_2	33.06	h_b	108.22	σ_{mg}^2	4699.10	7303.91	15578.80
M_3	157.57	h_a/d_a	-3.26	σ_e^2	3961.21	3961.21	3961.21
M_4	136.45	h_b/d_b	-2.02	σ_{pg}^2	558.28	3376.63	759.60
M_5	69.01	I	44.08	h_{mg}^2(%)	58.00	49.88	76.74
M_6	98.91	j_{ab}	-114.91	h_{pg}^2(%)	6.89	23.06	3.74
d_a	-53.68	j_{ba}	-47.96	-	-	-	-
d_b	-53.68	L	-206.43	-	-	-	-
Cross-2: Farid-06 × Shafaq (model E-2)							
					BC_1	BC_2	F_2
M	247.52	-	-	σ_p^2	15471.98	19126.62	33193.92
d_a	113.43	-	-	σ_{mg}^2	4271.9	7926.55	21993.85
d_b	0.18	-	-	σ_e^2	11200.07	11200.07	11200.07
h_a	3.28	-	-	σ_{pg}^2	0.00	0.00	0.00
h_b	0.16	-	-	h_{mg}^2(%)	27.61	41.44	66.26
[d]	-102.07	-	-	h_{pg}^2(%)	0.00	0.00	0.00
[h]	92.04	-	-	-	-	-	-
Cross-3: Parula × Blue Silver (Model B-1)							
					BC_1	BC_2	F_2
M	248.70	h_a	191.53	σ_p^2	14740.1	21620.9	28682.1
	-	h_b	34.94	σ_{mg}^2	9658.4	20698.29	21655.67
	-	h_a/d_a	-1.74	σ_e^2	9417.97	9417.97	9417.97
	-	h_b/d_b	0.85	σ_{pg}^2	0.00	0.00	0.00
	-	I	-7.44	h_{mg}^2(%)	65.52	95.73	75.50
	-	j_{ab}	-61.39	h_{pg}^2(%)	0.00	0.00	0.00
d_a	-109.77	j_{ba}	-149.74	-	-	-	-
d_b	41.05	L	-261.08	-	-	-	-
Cross-4: TD-1 × D-97603 (Model D-1)							
	-	-	-		BC_1	BC_2	F_2
M	249.10	-	-	σ_p^2	12939.37	12093.24	32906.28
D	-189.57	-	-	σ_{mg}^2	4695.69	3849.56	24662.68
H	-9.328	-	-	σ_e^2	8243.68	8243.68	8243.69
[d]	167.54	-	-	σ_{pg}^2	0.00	0.00	0.00
[h]	42.98	-	-	h_{mg}^2(%)	36.29	31.83	74.95
	-	-	-	h_{pg}^2(%)	0.00	0.00	0.00

Conclusion

In most of the crosses, mixed two additive-dominant-epistatic major genes with polygene controlled the AUDPC. A significantly larger portion of genetic variation and heritability could be attributed to genes with major effects rather than the polygene. However, the environmental variations could not be ignored as the stripe rust resistance was under influence of environmental fluctuations. The genetic behavior of AUDPC revealed that in all crosses the resistance to stripe rust was controlled by mixed interaction of one to two major genes plus polygene, and suggested the selection of resistant genotypes in early generation i.e. $F_2 \sim F_3$.

References

Akaike, H. (1977). On the Entropy Maximum Principle. In P. R. Krishnaiah (Ed.), *Applications of Statistics* (p. 27-41). Amsterdam, NL: North-Holland Publishing Company.

Arama, P. F., Parlevliet, J. E., & Van-Silfhout, C. H. (2000). Trangressive segregation for resistance in wheat to Septoria tritici blotch. *African Crop Science Journal, 8*(3), 213-222.

Bjarko, M. E., & Line, R. F. (1988). Heritability and number of genes controlling leaf rust resistance in four cultivars of wheat. *Phytopathology, 78*(4), 457-461.

Chen, X., Soria, M. A., Yan, G.; Sun, J., & Dubcovsky, J. (2003). Development of sequence tagged site and cleaved amplified polymorphic sequence markers for wheat stripe rust resistance gene *Yr5*. *Crop Science, 43*(6), 2058-2064.

Das, M. K., Rajaram, S., Kronstad, W. E., Mundt, C. C., & Singh, R. P. (1993). Associations and genetics of three components of slow rusting resistance in leaf rust of wheat. *Euphytica, 68*(1), 99-109.

Das, M. K., Rajaram, S., Mundt, C. C., & Kronstad, W. E. (1992). Inheritance of slow rusting resistance to leaf rust in wheat. *Crop Science, 32*(6), 1452-1456.

Das, M. K., Rajaram, S., Mundt, C. C., & Kronstad, W. E. (2004). Number of genes controlling slow rusting resistance to leaf rust in five spring wheat cultivars. *Annals of Applied Biology, 145*(1), 91-94.

Feng, J., Zhang, Z., Li, G., Zhou, Y., Wang, H., Guo, Q., & Sun, J. (2007). Inheritance of resistance to stripe rust in winter wheat cultivars Aquileja and Xian Nong 4. *Journal of Applied Genetics, 48*(1), 43-46.

Gai, J., Wang, Y., Wu, X., & Chen, S. (2007). A comparative study on segregation analysis and QTL mapping of quantitative traits in plants with a case in soybean. *Frontiers of Agriculture in China, 1*(1), 1-7.

Gai, J. Y., & Wang, J. K. (1998). Identification and estimation of QTL model and its effects. *Theoretical and Applied Genetics, 97*(7), 1162-1168.

Gai, J. Y., Zhang, Y., & Wang, J. (2003). Mixed genetic model for two major genes plus multi-genes in genetic system of quantitative traits in plants (p. 277-285). Beijing, YAN: Science press.

Ghaledozdani, H. (2005). Variability of *Puccinia striiformis* in Iran. In *Procedings of the 2nd Asian Conference on Plant Pathology* (p. 64). Singapore: National University of Singapore.

Imtiaz, M., Cromey, M. G., Hampton, J. G., & Ahmad, M. (2003). Inheritance of durable adult plant resistance to stripe rust (*P. striiformis* f. sp. tritici) in 'Otane' wheat (*T. aestivum* L.). *New Zealand Journal of Crop and Horticultural Science, 31*(1), 23-31.

Imtiaz, M., Cromey, M. G., Hampton, J. G., & Ogbonnaya, F. C. (2005). Genetics of stripe rust resistance in 'Karamu' wheat. *Australian Journal Agricultural Research, 56*(6), 619-624.

Khan, M. I., Ajab, M., Hongxiang, M., & Khattak, G. S. S. (2009a). Assessment of genes controlling area under disease progress curve (AUDPC) for stripe rust (*P. striiformis* f. sp. tritici) in two wheat (*T. aestivum* L.) crosses. *Cytology and Genetics, 43*(4), 25-38.

Khan, M. I., Ajab, M., Khattak, G. S. S., Mohammad, T., & Shah, S. J. A. (2009b). Genetic behavior of controlling area under disease progress curve for stripe rust (*Puccinia striiformis* f. sp. *tritici*) in two wheat (*T. aestivum* L.) crosses. *Phytopathology, 99*(11), 1265-1272.

Ma, H., & Singh, R. P. (1996). Expression of adult resistance to stripe rust at different growth stages of wheat. *Plant Disease, 80*(4), 375-379.

Ma, H., Singh, R. P., & Abdalla, O. (1997). Resistance to stripe rust in five durum wheat cultivars. *Plant Disease, 81*(1), 27-30.

Mamluk, O. F. (1992). Durum wheat diseases in West Asia and North Africa (WANA). In S. Rajaram, E. E. Sarri, G. P. Hettel (Ed.), *Durum wheats: challenges and opportunities* (9th ed., p. 89-106). Mexico: CIMMYT Wheat Special Report.

Mather, K.; Jinks, J. L. (1982). *Biometrical Genetics* (2nd ed.). London, UK: Chapman and Hall Ltd.

Messmer, M. M., Seyfarth, R., Keller, M., Schachermayr, G., Winzeler, M., Zanetti, S., ... Keller, B. (2000). Genetic analysis of durable leaf rust resistance in winter wheat. *Theoretical and Applied Genetics, 100*(3), 419-431.

Peterson, R. F., Campbell, A. B., & Hannah, A. E. (1948). A diagnostic scale for estimating rust intensity on leaves and stems of cereals. *Canadian Journal of Research, 26*(5), 496-500.

Shaner, G., Buechley, G., & Nyquist, W. E. (1997). Inheritance of latent period of *Puccinia recondita* in wheat. *Crop Science, 37*(3), 748-756.

Singh, A., Knox, R. F., Depauw, R. M., Singh, R. K., Cuthbert, R. D., Campbell, H. L., ... Bhavani, S. (2014). Stripe rust and leaf rust resistance QTL mapping, epistatic interactions, and co-localization with stem rust resistance loci in spring wheat evaluated over three continents. *Theoretical and Applied Genetics, 127*(11), 2465-2477.

Singh, R. P. (1992). Genetic association of leaf rust resistance gene *Lr34* with adult plant resistance to stripe rust in bread wheat. *Phytopathology, 82*(8), 835-838.

Singh, R. P., Nelson, J. C., & Sorrells, M. E. (2000). Mapping *Yr28* and other genes for resistance to stripe rust in wheat. *Crop Science, 40*(4), 1148-1155.

Singh, R. P., & Rajaram, S. (1991). Resistance to *Puccinia recondita* f. sp. *tritici* in 50 Mexican bread wheat cultivars. *Crop Science, 31*(6), 1472-1479.

Singh, R. P., & Rajaram, S. (1993). Genetics of adult plant resistance to stripe rust in ten spring bread wheats. *Euphytica, 72*(1), 1-7.

Suenaga, K., Singh, R. P., Haurta-Espino, J., & William, H. M. (2003). Microsatellite markers for genes *Lr34/Yr18* and other quantitative trait loci for leaf rust and stripe rust resistance in bread wheat. *Phytopathology, 93*(7), 881-890.

Ullah, K., Khan, N. U., Khan, S. J., Khan, I. U., Khan, M. I., Khan, R. U., & Gul, S. (2014a). Genetic analysis for grain filling duration in wheat using joint segregation analysis. *Turkish Journal of Agriculture and Forestry, 38*(6), 807-819.

Ullah, K., Khan, N. U., Khan, S. J., Khan, M. I., Khan, I. U., Gul, S., ... Khan, R. U. (2014b). Cell membrane thermo-stability studies through joint segregation analysis in various wheat populations. *Pakistan Journal of Botany, 46*(6), 1243-1252.

Wang, J., & Gai, J. (1997). Identification of major gene and polygene mixed inheritance model and estimation of genetic parameters of a quantitative trait from F_2 progeny. *Acta Genetica Sinica, 24*(5), 432-440.

Wang, J., & Gai, J. (2001). Mixed inheritance model for resistance to agromyzid beanfly (*Melanagromyza sojae* Zehntner) in soybean. *Euphytica, 122*(1), 9-18.

Wang, J., Podlich, D. W., Cooper, M., & Delacy, I. H. (2001). Power of the joint segregation analysis method for testing mixed major-gene and polygene inheritance models of quantitative traits. *Theoretical and Applied Genetics, 103*(5), 804-816.

William, M., Singh, R. P., Huerta-Espino, J., & Ortiz, I. S., Hoisington, D. (2003). Molecular marker mapping of leaf rust resistance gene Lr46 and its association with stripe rust resistance gene *Yr29* in wheat. *Phytopathology, 93*(2), 153-159.

Yahyaoui, A., Hakim, M. S., Al Naimi, M. & Rbeiz, N. (2002). Evolution of physiologic races and virulence of

Puccinia striiformis on wheat in Syria and Lebanon. . *Plant Disease, 86*(5), 499-504.

Zadoks, J. C., Chang, T. T., & Konzak, C. F. (1974). A decimal code for the growth stages of cereals. *Weed Research, 14*(6), 415-421.

Zhang, Y. M., Gai, J. Y., & Yang, Y. H. (2003). The EIM algorithm in the joint segregation analysis of quantitative traits. *Genetics Research, 81*(2), 157-163.

Zhang, Z. J., Yang, G. H., Li, G. H., Jin, S. L., & Yang, X. B. (2001). Transgressive segregation, heritability, and number of genes controlling durable resistance to stripe rust in one Chinese and two Italian wheat cultivars. *Phytopathology, 91*(7), 680-686.

6

Immature tassels as alternative explants in somatic embryogenesis and plant regeneration in south Brazilian maize genotypes

Magali Ferrari Grando[1*], Madelaine Lurdes Varnier[1], Marilia Rodrigues de Silva[1], Beatriz Marti Emydio[2], Luiz Ricardo Pereira[2] and Marilei Suzin[1]

*[1]Laboratório de Biotecnologia Vegetal, Faculdade de Agronomia e Medicina Veterinária, Universidade de Passo Fundo, BR-285, Km 171, 99052-900, Campus I, São José, Passo Fundo, Rio Grande do Sul, Brazil. [2]Empresa Brasileira de Pesquisa Agropecuária, Embrapa Trigo, Passo Fundo, Rio Grande do Sul, Brazil. *Author for correspondence. E-mail: magali@upf.br*

ABSTRACT. Somatic embryogenesis and *in vitro* plant regeneration are fundamental processes in the obtainment of transgenic maize plants. Explant, genotype and culture medium are determining factors in these processes. Immature embryo explants and the American Hi-II genotype have been widely employed to acquire genetically modified plants in this species. However, the use of more readily available explants is desired as well as the development of genetic transformation protocols for productive genotypes adapted to local conditions. This study provides an evaluation of immature tassel explants in relation to embryogenic callus production and plant regeneration in South Brazilian maize genotypes for their use in genetic transformation experiments. Immature tassels from 5 hybrids were cultivated in different callus-induction media. The frequency and the fresh mass of embryogenic calli were evaluated. The frequency was influenced by genotype, and the fresh mass was influenced by genotype and culture medium. In plant regeneration, shoots, complete seedlings and acclimatized and fertile plants were quantified. Treatments producing long term embryogenic calli from immature tassels of South Brazilian genotypes with the capacity to regenerate were identified.

Keywords: embryogenic callus, tissue culture, explants, *Zea mays*.

Introduction

Maize is a liliopsida of great agricultural importance for the South Region of Brazil, and it is increasingly requested due to the expansion of activities associated with industrialization and animal exploitation.

Products molecularly improved from this species are already being cultivated in Brazilian farms. In addition, genetic engineering has permitted the introduction of genes into maize that improve its nutritional value (CARNEIRO et al., 2000; GRANDO et al., 2005; NAQVIA et al., 2009; ZHU et al., 2007) and tolerance to important abiotic stresses due to drought or cold temperature (AL-ABED et al., 2007; SHOU et al., 2004; ZHANG et al., 2005).

The first stage in the obtainment of transgenic plants involves the development of tissue culture protocols because the gene is introduced at the level of topipotent cells, and the plants can be regenerated from these cells. Indirect somatic embryogenesis has been the morphogenetic process of preference in the regeneration of transgenic plants (FRAME et al., 2002, 2006, 2011; HUANG; WEI, 2005; ISHIDA et al., 2007; VEGA et al., 2008; ZHAO et al., 2001).

Nevertheless, maize is considered a recalcitrant species *in vitro*; therefore, responses have only been obtained through the use of very young explants containing embryogenic or meristematic cells (SHARMA et al., 2005). Highly competent explants produce long term embryogenic calli that survive through a long selection period, an obligatory stage during the process of genetic transformation, and maintain their plant regeneration potential. Plant regeneration from these calluses occurs via germination of somatic embryos with well-defined root/shoot axes (ARMSTRONG; GREEN, 1985). Plant regeneration using maize tissue culture was first reported by Green and Phillips (1975) who used immature embryos as explants. Plants regenerated from somatic embryos present higher genetic integrity when compared with plants regenerated through the process of organogenesis.

The success of regeneration protocols is subject to limitations associated with the genotype, somaclonal variation, low frequencies of callus induction and the very process of *in vitro* plant regeneration (TOMES; SMITH, 1985). Generally, these factors limit the effective production of transgenic plants (SAIRAM et al., 2003).

The first stage of plant regeneration from maize somatic embryos is related to somatic embryo maturation and shoot induction development via cultivation in a culture medium with a high sucrose concentration (60 g L^{-1}), and this regeneration initially occurs in the absence of light and without the addition of growth regulators. The second stage of the *in vitro* plant regeneration of maize is related to the development of shoots and roots, which are more frequently obtained through a reduction of the sucrose concentration (FRAME et al., 2002, 2006; LEE et al., 2007; VEGA et al., 2008).

Two types of embryogenic calli have been described in maize culture. Type I is hard, compact, nodular, yellow-white colored and may proliferate as a mixture of complex tissues showing somatic embryos (ARMSTRONG; GREEN, 1985). Furthermore, these authors characterized these Type I calli by slow growth or difficulty of being cultivated throughout a long period of time. Type II calli have been observed in few maize genotypes and are characterized as being highly friable, soft and fast at growth. In addition, they have a high embryogenic activity rate, which permits its cultivation for a long period of time.

Embryogenic callus induction is dependent on various factors such as culture medium, genotype and the explant type employed to start the *in vitro* process. In relation to the composition of the medium for callus induction, the most widely used nutritious media for embryogenic callus induction in grasses is the MS basic medium (MURASHIGE; SKOOG, 1962). This medium has been used for the production of genetically modified plants of specific lines by Frame et al. (2006, 2011). Alternatively, N6 basic medium (CHU et al., 1975) has been successfully employed in the induction of the type II friable callus in model genotypes such as A188 and B73, the Hi-II hybrid and other genotypes (ZHAO et al., 2001; FRAME et al., 2002, 2011; HUANG; WEI, 2005; ISHIDA et al., 2007; VEGA et al., 2008). Gorji et al. (2011) observed that the N6 medium, which contains lower nitrogen levels than the MS medium, promoted a higher induction and maintenance of embryogenic calluses in elite maize lines.

The most widely used synthetic auxin for embryogenic callus induction in grasses is 2,4-D (2,4-Dichlorophenoxyacetic Acid). However, the addition of citokynin in the presence of 2,4-D may increase the embryogenic response in some species of grasses (GRANDO et al., 2002).

Genotype is another factor that has a fundamental role during embryogenesis induction in maize (FRAME et al., 2006; HUANG; WEI, 2005; ISHIDA et al., 2007; LEE et al., 2007, WANG et al., 2007). The embryogenic potential of Brazilian maize genotypes was reported by Santos-Serejo and Aguiar-Perecin (2000), Carneiro et al. (2000), Petrillo et al. (2008) and Fernandes et al. (2008). However, no reports exist regarding transgenic plants obtained from South Brazilian maize genotypes.

The most widely used explant for establishing the embryogenic callus culture for genetic transformation in maize is the immature zygotic embryo (FRAME et al., 2002, 2006; HUANG; WEI, 2005; ISHIDA et al., 2007; VEGA et al., 2008). However, the production of immature embryos is a demanding process, which requires time and laborious activities, such as planting, pollination, excision of the immature embryo, and dependence on plant growth in the greenhouse. These explants are seasonally available, and a narrow time range exists for the explant to be in its best condition for in vitro cultivation (FRAME et al., 2011; GORJI et al., 2011). The use of more readily available explants is one of the greatest challenges for rendering the genetic transformation process faster and more effective (CHENG et al., 2004). Therefore, the use of explants that do not rely on the immature embryo aims at accelerating the process of obtaining in vitro plants.

In maize, the immature tassel was first used by Songstad et al. (1992) in a cultivation system based on N6 medium supplemented with 2,4-D, L-proline and $AgNO_3$ for the obtainment of type II embryogenic calli in A188 and B73 model lines as well as its Hi-II hybrid. The advantage of using the immature tassel is its rapid obtainment of plants within 6-7 weeks, with no need for growing them until the reproductive stage, pollination, fertilization, or development of the immature embryo. Primordia of young and non-emerged flowers with pre-meiotic flower buds are considered excellent explants for starting embryogenic cultures in grasses. In these explants, embryogenic calluses are primarily formed through the proliferation of the young floral primordium.

To accelerate the obtainment process of *in vitro* plants for genetic manipulation, this study evaluated the immature tassel response of 5 South Brazilian maize genotypes in 4 different culture media in relation to the capacity of embryogenic callus production and growth as well as the regeneration of plants from these calli.

Material and methods

The experiment was developed in the Plant Biotechnology Laboratory of the University of Passo Fundo. Maize genotypes were grown in the greenhouse at Embrapa Trigo in Passo Fundo, Rio Grande do Sul State.

Callus induction from maize immature tassels

Genotypes of simple maize hybrids Hs1 (PF96317/PF963244), Hs2 (PF963027/PF963244), Hs3 (PF972334/PF963004), Hs4 (PF972291/PF963106) and Hs5 (PF972218/PF963173) belong to the Embrapa Trigo maize improvement program and were seeded in vases in greenhouses. To acquire immature tassels, 6-7-week-old plants (Figure 1a) were collected according to Songstad et al. (1992). Immature tassels measuring 2-3 cm (Figure 1b) were segmented (2-3 mm) and inoculated in petri dishes containing different callus-induction media.

Four types of culture media were used, including N6 (CHU et al., 1975) and MS (MURASHIGE; SKOOG, 1962) supplemented with different components: M1= N6 basic medium, 100 mg L^{-1} hydrolyzed casein, 2.88 g L^{-1} L-proline, Erickson vitamins, 10 μM $AgNO_3$, 100 mg L^{-1} myo-inositol, 20 g L^{-1} sucrose, 2 g L^{-1} phytagel and 1.0 mg L^{-1} 2.4-D (SONGSTAD et al., 1992); M2= equal to M1 medium without $AgNO_3$; M3= MS basic medium, 100 mg L^{-1} hydrolyzed casein, 2.88 g L^{-1} L-proline, Erickson vitamins, 10 μM $AgNO_3$, 100 mg L^{-1} myo-inositol, 20 g L^{-1} sucrose, 2 g L^{-1} phytagel and 1.0 mg L^{-1} 2.4-D; and M4= equal to M3 medium + 1 mg L^{-1} BAP (6-Benzylaminopurine). Erickson vitamins contain 0.5 mg L^{-1} thiamine HCl, 2 mg L^{-1} glycine, 0.5 mg L^{-1} pyridoxine HCL and 0.5 mg L^{-1} nicotinic acid. The pH of the media was adjusted to 5.8 with KOH (potassium hydroxide) before autoclaving (121°C for 20 min.). Next, $AgNO_3$ was sterilized via filtering and added to the culture medium.

A completely randomized experimental design was employed with 5 repetitions per treatment. The experimental unit was composed of one petri dish containing 10 segments of immature tassel. The cultures were maintained in the dark at 28±2°C for six weeks and subcultivated with the same fresh medium every 3 weeks.

After 42 days, embryogenic calluses were evaluated, weighed and transferred to maintenance medium for fresh mass evaluation. The maintenance medium had a similar composition to the induction medium, differing only in the absence of $AgNO_3$ and the increase in 2.4-D concentration to 2 mg L^{-1}. Cultures were maintained in the dark at 28±2°C and subcultivated every 21 days in maintenance medium to multiply embryogenic calli. The frequency and the fresh mass of embryogenic calli obtained at 42 days of cultivation were evaluated. The fresh mass was obtained from the average of calli obtained from 10 segments of immature tassels cultivated in a petri dish.

Calli were maintained *in vitro* for 8 subcultivations (21 weeks) to evaluate their longevity. Frequency data of embryogenic calli were transformed according to the arcsin formula ASEN [root((x+0.5)/100))*57.2958], submitted to variance analysis and the difference between the averages was compared using the Tukey test at 5%. Alternatively, data related to the fresh mass of embryogenic callus were compared using the average ± 1 standard deviation.

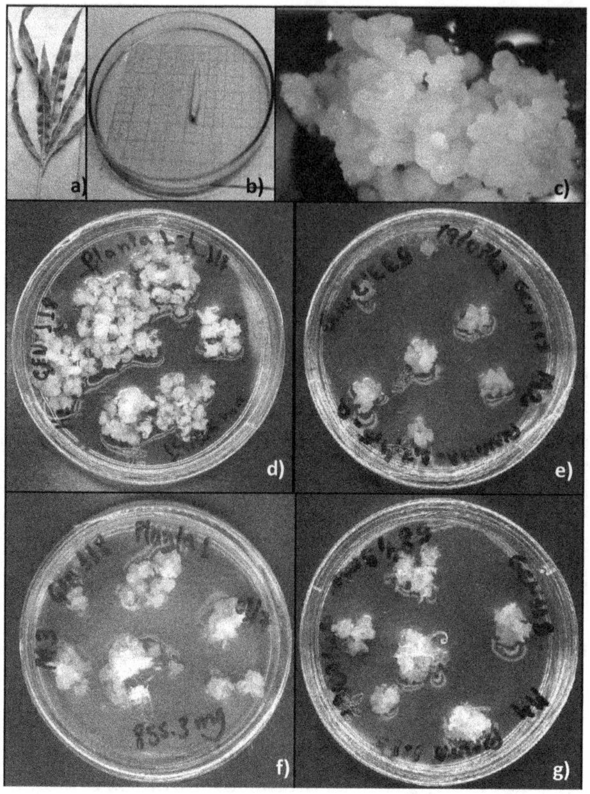

Figure 1. a) Maize plant at the immature tassel obtainment stage (6-7 weeks of cultivation); b) immature tassel used as an explant for callus induction; c) embryogenic callus of the Hs1 genotype; d) embryogenic calluses of the Hs1 genotype of maize during the maintenance stage originating from different callus induction media (M1 (d), M2 (e), M3 (f) and M4 (g)) after 7 subcultivations.

Maize plant regeneration from embryogenic calluses

Plant regeneration began after 21 weeks of induction of embryogenic calli, aiming at assessing the regenerative capacity of calluses maintained for a long time in cultivation. The regeneration system used was described by Sellmer et al. (1994). Embryogenic calluses were cultivated in regeneration medium (R1) composed of MS basic medium supplemented with modified White vitamins (0.2 g L^{-1} glycine; 0.5 mg L^{-1} nicotinic acid; 0.5 mg L^{-1} pyridoxine HCl; 0.5 mg L^{-1} thiamine HCl), 100 mg L^{-1} myo-inositol, 60 g L^{-1} sucrose, 2 g L^{-1} phytagel, pH 5.8. The cultures were grown for one week in the dark at 28±2°C and then transferred to a growth chamber with a photoperiod of 16 hours light: 8 hours dark for 3 more weeks with ± 25 mmol m^{-2} s^{-1} lux illuminance. The calli were transferred to fresh medium of equal composition until the formation of shoots. The calluses/structures that presented shoots were transferred for rooting (R2) in medium similar to the regeneration medium but with sucrose reduced to 20 g L^{-1}.

Seedlings were removed from the medium and transferred to 500 ml pots containing Eucatex sterilized substrate buttonhole and vermiculite type substrate (1:1), and they were maintained for 4 days in a chamber with high humidity that was gradually decreased. After 7-10 days, seedlings were transferred to the greenhouse, initially protected from direct sunlight for 2-3 days and gradually exposed to the sun. The developed plants were transferred to 10 L vases with a Eucatex buttonhole and a vermiculite type substrate until they reached maturity to assess fertility and their morphological aspects.

To evaluate the effectiveness of plant regeneration among the different genotypes, the number of shoots, complete seedlings (2-5 including root) and the number of acclimatized plants per gram of embryogenic callus transferred for regeneration medium were quantified. For example, one gram of fresh callus was homogenously distributed in the surface of a petri dish containing 25 ml of regeneration medium with 5 repetitions. Data were submitted to t-test at the 5% level of significance.

Results and discussion

Embryogenic calli induction

Of the 5 South Brazilian maize hybrids evaluated, 4 produced type I embryogenic calluses from immature tassel segments. The genotype influenced the frequency of the calli obtained ($p > 0.0007$); however, the culture media utilized did not influence the frequency of embryogenic calluses ($p > 0.3655$) (data not shown).

In total, 12.7 and 12% of calluses were produced by Hs1 and Hs2 genotypes, respectively, and these percentages were higher than the production rates of Hs3 and Hs4 genotypes (Figure 2). Figure 1c shows the characteristics of the embryogenic calluses that were classified as type I.

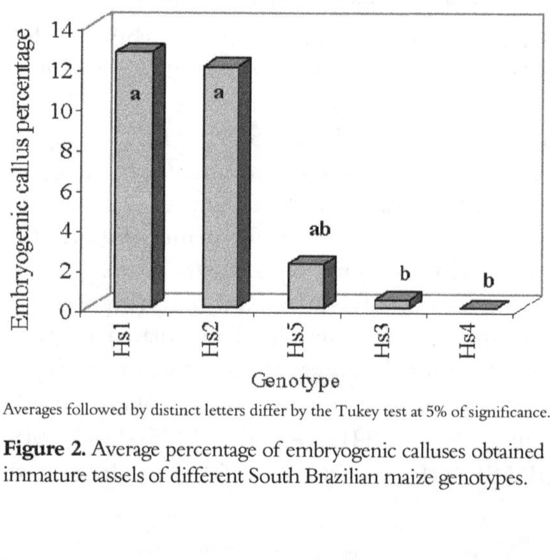

Averages followed by distinct letters differ by the Tukey test at 5% of significance.

Figure 2. Average percentage of embryogenic calluses obtained from immature tassels of different South Brazilian maize genotypes.

Genotype plays an important role in the formation of embryogenic calluses in different cultures, including maize. Genotypic differences regarding the formation of embryogenic maize cultures from other maize explants have been extensively reported in the literature (FERNANDES et al., 2008; FRAME et al., 2006; ISHIDA et al., 2007; LEE et al., 2007; TOMES; SMITH, 1985).

The expression pattern during somatic embryogenesis in different maize genotypes is controlled by a gene or a block of genes (WILLMAN et al., 1989). Hodges et al. (1986) suggest that nuclear genes exhibiting dominance are important for the formation of somatic embryos or plant regeneration in maize. Alternatively, Tomes and Smith (1985) reported that the genetic heritage control in maize involved in embryogenic callus formation is the additive type, with a significant maternal effect and heterosis. Armstrong et al. (1992) found RFLP markers associated with chromosomes 1, 2, and 9, and the most critical region for callus formation is in the long arm of maize chromosome 9. These authors proposed the existence of a primary gene (or genes) in this region that promotes embryogenic callus initiation and plant regeneration. Recently, this group of researchers (LOWE et al., 2006) developed a molecular marker-assisted selection methodology to assess embryogenic callus initiation in maize.

According to Krakowsky et al. (2006), the genetic basis for the production of maize regenerative calluses is not completely elucidated. These authors discovered the existence of 11 QTLs in 8 chromosomes, 8 QTLs showed a main effect and 3 presented an epistatic interaction.

The obtainment of genetically modified plants greatly depends on the capacity of the *in vitro* response of the genotype used. Nevertheless, some maize genotypes, especially elite material, present a poor tissue culture response, which limits the number of genotypes that were effectively transformed (FRAME et al., 2006; ISHIDA et al., 2007; LEE et al., 2007). The Hi-II hybrid, due to its high *in vitro* response, has been used for genetic transformation; however, it has poor agronomic characteristics. Thus, the application of this technology in agriculturally relevant genotypes, such as locally developed hybrids by genetic breeding programs, is highly desirable (FRAME et al., 2006; GORJI et al., 2011; HUANG; WEI, 2005; ISHIDA et al., 2007; LEE et al., 2007), which justifies the selection of *in vitro* responsive genotypes.

The purpose of this study is to evaluate the *in vitro* response of Brazilian genotypes and alternative explants that can be used in tissue culture and maize genetic transformation.

In maize culture, the immature zygotic embryo is the most widely used explant for the establishment of embryogenic callus cultures and genetic transformation (FRAME et al., 2002, 2006, 2011; ISHIDA et al., 1996, 2007; LEE et al., 2007; VEGA et al., 2008) because the immature embryo presents cells that are competent for somatic embryogenesis, unlike other tissues that are more differentiated. However, the production of immature embryos is a demanding process that requires time and painstaking activities for the growth of plants and pollination. Thus, the use of more readily available alternative explants, such as immature tassels, can accelerate the obtainment of *in vitro* plants.

Maize immature tassels were used by Songstad et al. (1992) to acquire embryogenic cultures in model genotypes such as A188 and B73 lines and its Hi-II hybrid. Embryogenic calli obtained from immature tassels have been used to introduce genes into this hybrid via bio-ballistics (HE et al., 2003; GRANDO et al., 2005). Rhodes et al. (1986) evaluated 13 North American maize genotypes in relation to their ability to start embryogenic calluses from immature tassels, which determines the genotypic effect for this characteristic. They observed that not all genotypes expressed the capacity of regenerating *in vitro* plants and that, within the genotypes, *in vitro* responses of immature embryos and tassels were qualitatively and quantitatively similar.

Fresh mass of embryogenic calli

The fresh mass of the embryogenic calluses was evaluated at 42 days of culture. On average, Hs1 genotype calli produced a greater amount of fresh mass (670 mg) when compared to other genotypes (10 and 70 mg) (Table 1).

Table 1. Fresh mass of embryogenic calli acquired from immature tassels of 4 maize hybrids cultivated for 42 days in 4 culture media.

Genotype	Media	Fresh mass (mg)*
Hs1	M1	670 S
	M2	500
	M3	860 S
	M4	640 S
Average		670
Hs2	M1	10
	M2	20
	M3	190
	M4	60
Average		70
Hs3	M1	40
	M2	-
	M3	-
	M4	-
Average		10
Hs5	M1	-
	M2	20
	M3	160
	M4	20
Average		70
	General Average	270
	Standard deviation	310

Average + standard deviation = superior (S); Average – standard deviation = inferior (I).

Although the culture media did not influence the percentage of embryogenic calluses, the media did influence the fresh mass of induced calli. Regarding the Hs1 genotype, the calli cultivated in M1, M3, and M4 (containing silver nitrate) media were considered superior to other treatments.

The media M3 (MS basic medium) and M4 (MS+BAP basic medium) promoted the growth of embryogenic calli of the Hs1, Hs2 and Hs5 genotypes; however, the calli maintained in these media through 8 subcultivations (21 weeks) lost their embryogenicity and presented differentiated structures in the form of leaves and shoots (Figure 1g and f). These observations are in agreement with Vasil (1987), who reported that embryogenic calluses cultivated in MS medium may become organogenic or germinate early.

However, the calli induced in M1 and M2 media (N6 basic medium) maintained their embryogenic features for approximately 5 months of culturing (Figure 1d and e). Normally, the embryogenic calli of cereal species are characterized by slow growth because they are unable to be cultivated during a long period of time (NABORS et al., 1983). This experiment shows that the use of N6 basic medium allows embryogenic calli to be kept for a longer period compared to MS medium. Similarly, Gorji et al. (2011) observed that maize embryogenic calli induced in N6 medium were of better quality than calli obtained in MS medium.

MS medium has been more widely used for the induction of type I maize embryogenic calli (FRAME et al., 2006), and N6 medium has been used for the induction of type II embryogenic calli, primarily the A188 genotype and its hybrids, including Hi-II (FRAME et al., 2002, 2011). These findings support the need to evaluate the behavior of the interaction between genotype and the culture medium. In this experiment, even using alternative genotypes, N6 basic medium fostered the formation of type I embryogenic calli of best quality.

The various responses observed during the production of somatic embryos and maize calluses in MS and N6 media can be attributed to its different nitrogenous compounds (ARMSTRONG; GREEN, 1985). MS medium contains 60 mM nitrogen (20.6 mM NH_4^+ and 29.4 mM NO_3^-) compared with N6 medium, which contains 35 mM nitrogen (7 mM NH_4^+ and 28 mM NO_3^-). The ammonia: nitrate ratio is approximately 1:2 in MS medium; however, it is 1:4 in N6 medium. The form of nitrogen may have a fundamental role in the expression of genes that control the embryogenic process because this difference is the greatest between N6 and MS media (HODGES et al., 1986).

In this study, N6 basic medium fostered a greater proliferation of embryogenic calluses throughout successive subcultivations when silver nitrate was added. Figure 1d and e shows the greater proliferation of embryogenic calli cultivated in M1 medium (N6+$AgNO_3$) compared to calli cultivated in M2 medium (N6 without $AgNO_3$) after 8 subcultivations.

Silver nitrate has been used to increase the production of type II calli and the regeneration of maize plants. The addition of 10 μM $AgNO_3$ in culture medium increased the production of embryogenic calli in 59% of Hi-II hybrid maize immature tassels (SONGSTAD et al., 1992). The effect of $AgNO_3$ is related to its action as an ethylene antagonist, which prevents the formation of endogenous ethylene that is generated by mechanical damage caused to the explants by the process of in vitro cultivation and affects the embryogenic response.

The importance of evaluating the effects of genotypes and culture media throughout subcultivations is justified by the need of the calluses passing through a long period in selection medium during the genetic transformation process. Thus, the embryogenicity of these calluses should be maintained during this period to ensure the regeneration of genetically modified plants.

Maize Plant regeneration from embryogenic calluses

Only Hs1 calli induced in M1 and M2 media that remained embryogenic throughout subcultivations were evaluated for regeneration potential. The culture medium used in callus induction did not influence plant regeneration. This genotype produced approximately 11.7 shoots, 4.9 seedlings, and 3.3 acclimatized plants per gram of callus transferred to the regeneration medium (Table 2).

Table 2. Average shoot production, 2-5 cm rooted seedlings, and acclimatized plants per gram of embryogenic callus of the Hs1 genotype of maize cultivated in regeneration medium.

CIM	NS	NSd	NAP	NRP
M1	8.9 ns*	4.0 ns	3.2 ns	29
M2	14.5	5.8	3.3	13
Average/total	11.7	4.9	3.3	42

CIM = callus induction media; NS = number of shoots per gram of embryogenic callus; NSd = number of seedlings per gram of embryogenic callus; NAP = number of acclimatized plants per gram of embryogenic callus; NRP = number total of regenerated plants; *ns= not significant according to t-test in 5% level of significance.

Approximately, 42% of the shoots produced roots that formed seedlings, and 69% of these were acclimatized. The plants reached maturity and became fertile. Figure 3 displays the stages of the process of plant regeneration and acclimatization of the Hs1 maize hybrid.

Figure 3. Stages of *in vitro* plant regeneration of Hs1 maize. a) somatic embryo germination in R1 medium; b) maize seedling development in R2 medium; c) seedling acclimatization in vase with substrate in high humidity environment; d) plant development in greenhouse; e) ear of corn showing the fertility of *in vitro* regenerated plants.

In this study, *in vitro* plants of a South Brazilian maize genotype were regenerated from embryogenic calli induced from immature tassel explants. Plant regeneration from indirect somatic embryogenesis is a prerequisite for the production of genetically modified plants, mainly in cereals (FRAME et al., 2002, 2006; ISHIDA et al., 2007). The identification of productive genotypes that are adapted to environmental conditions and that also present the capacity of responding *in vitro* allows genetic transformation to be performed with genotypes alternative to the Hi-II model genotype.

The Hi-II genotype has been internationally used to produce maize transgenic plants; however, this genotype presents low agricultural value, which increases the time needed as well as the costs for transgene transference to locally adapted lines by backcrossing (LUPOTTO et al., 2004). Furthermore, backcrossing is hindered by the incompatibility of heterotic groups and low combination capacity (O'KENNEDY et al., 2001).

The use of the immature tassels accelerates the obtainment of *in vitro* plants due to its higher speed of development in comparison to the immature embryo. The tassel becomes available 6-7 weeks after seeding, but the immature embryo can only be collected 14 weeks after seeding. In addition, artificial pollination remains necessary.

Conclusion

The immature tassel has proven to be an effective alternative explant in the induction of embryogenic calli and the regeneration of *in vitro* maize plants.

A South Brazilian maize genotype (HS1) and a culture medium (N6 basal medium + $AgNO_3$) that permit induction, proliferation and longevity of embryogenic calli were identified, enabling the regeneration of complete and fertile maize plants. These parameters can be used in experiments of genetic manipulation aiming at the obtainment of genetically modified plants.

References

AL-ABED, D.; MADASAMY, P.; TALLA, R.; GOLDMAN, S.; RUDRABHATLA, S. Genetic engineering of maize with the *Arabidopsis* DREB1A/CBF3 gene using split-seed explants. **Crop Science**, v. 47, n. 6, p. 2390-2402, 2007.

ARMSTRONG, C. L.; GREEN, C. E. Establishment and maintenance of friable embryogenic maize callus and the involvement of L-proline. **Planta**, v. 164, n. 2, p. 207-214, 1985.

ARMSTRONG, C. L.; ROMERO-SEVERSON, J.; HODGES, T. K. Improved tissue culture response of an elite maize inbred through backcross breeding, and identification of chromosomal regions important for regeneration by RFLP analysis. **Theoretical and Applied Genetics**, v. 84, n. 5-6, p. 755-762, 1992.

CARNEIRO, A. A.; CARNEIRO, N. P.; CARVALHO, C. H. S.; VASCONSELOS, M. J. V.; PAIVA, E.; LOPES, M. A. Milho transgênico: melhoria da qualidade nutricional de grão. **Biotecnologia, Ciência e Desenvolvimento**, v. 15, n. 3, p. 42-46, 2000.

CHENG, M.; LOWE, B. A.; SPENCER, T. M.; YE, X.; ARMSTRONG, C. L. Factors influencing Agrobacterium-mediated transformation of monocotyledonous species. ***In vitro* Cell and Developmental Biology-Plant**, v. 40, n. 1, p. 31-45, 2004.

CHU, C. C.; WANG, C. C.; SUN, C. S.; HSU, C.; YIN, K. C.; CHU, C. Y. Establishment of an efficient medium for rice anther culture through comparative experiments on the nitrogen source. **Scientia Sinica**, v. 18, n. 5, p. 659-668, 1975.

FERNANDES, E. H.; PRIOLI, A. J.; SCAPIM, C. A.; SCHUSTER, I.; VIEIRA, E. S. N.; AMARAL JR., A. T.; MOTERLE, L. M. Embriogênese somática a partir de embriões imaturos em genótipos de milho. **Ciência Rural**, v. 38, n. 9, p. 2604-2607, 2008.

FRAME, B. R.; MCMURRAY, J. M.; FONGER, T. M.; MAIN, M. L.; TAYLOR, K. W.; TORNEY, F. J.; PAZ, M. M.; WANG, K. Improved Agrobacterium-mediated transformation of three maize inbred lines using MS salts. **Plant Cell Reports, Genetic Transformation and Hybridization**, v. 25, n. 4, p. 1024-1034, 2006.

FRAME, B. R.; SHOU, H.; CHIKWAMBA, R. K.; ZHANG, Z.; XIANG, C.; FONGER, T. M.; PEGG, S. E. K.; LI, B.; NETTLETON, D. S.; PEI, D.; WANG, K.

Agrobacterium tumefaciens-mediated transformation of maize embryos using a standard binary vector system breakthrough technologies. **Plant Physiology**, v. 129, n. 1, p. 13-22, 2002.

FRAME, B.; MAIN, M.; SCHICK, R.; WANG, K. Genetic transformation using maize immature zygotic embryos. In: THORPE, A.; YEUNG, E. C. (Ed.). **Plant embryo culture**: methods and protocols. Methods in Molecular Biology. New York: Springer Science, 2011. p. 327-341. (v. 710)

GORJI, A. H.; ZOLNOORI, M.; JAMASBI, A.; ZOLNOORI, Z. *In vitro* plant generation of tropical maize genotypes. **International Proceedings of Chemical, Biological and Environmental Engineering**, v. 16, p. 52-59, 2011.

GRANDO, M. F.; FRANKLIN, C. I.; SHATTERS JUNIOR, R. G. Optimizing embryogenic callus production and plant regeneration from "Tifton 9" bahiagrass (*Paspalum notatum* Flügge) seed explant for genetic manipulation. **Plant Cell Tissue and Organ Culture**, v. 17, n. 3, p. 213-222, 2002.

GRANDO, M. F.; SMITH, R. L.; MOREIRA, C.; SCULLY, B. T.; SHATTERS, R. Developmental changes in abundance of the VSPb protein following nuclear transformation of maize with the Soybean vspb cDNA. **BMC Plant Biology**, v. 5, n. 3, p. 1-9, 2005.

GREEN, C. E.; PHILLIPS, R. J. Plant regeneration from tissue cultures of maize. **Crop Science**, v. 15, n. 3, p. 417-421, 1975.

HE, X.; HALL, M. B.; GALLO-MEAGHER, M.; SMITH, R. L. Improvement of forage quality by downregulation of maize O-Methyltransferase. **Crop Science**, v. 43, n.6, p. 2240-2251, 2003.

HODGES, T. K.; KAMO, K. K.; IMBRIE, C. W.; BECWAR, M. R.; SCHROLL, S. Genotype specificity of somatic embryogenesis and regeneration in maize. **Bio/Technology**, v. 4, n. 3, p. 219-223, 1986.

HUANG, X. Q.; WEI, Z. M. Successful *Agrobacterium*-mediated genetic transformation of maize elite inbred lines. **Plant Cell, Tissue and Organ Culture**, v. 83, n. 2, p. 187-200, 2005.

ISHIDA, Y.; HIEI, Y.; KOMARI, T. *Agrobacterium* mediated transformation of maize. **Nature Protocols**, v. 2, n. 7, p. 1614-1621, 2007.

ISHIDA, Y.; SAITO H.; OHTA, S.; HIEI, Y.; KOMARI, T.; KUMASHIRO, T. High efficiency transformation of maize *(Zea mays L.)* mediated by *Agrobacterium tumefaciens*. **Nature Biotechnology**, v. 14, n. 6, p. 745-750, 1996.

KRAKOWSKY, M. D.; LEE, M.; GARAY, L.; WOODMAN-CLIKEMAN, W.; LONG, M. J.; SHAROPOVA, N.; FRAME, B.; WANG, K. Quantitative trait loci for callus initiation and totipotency in maize (*Zea mays* L.). **Theoretical and Applied Genetics**, v. 113, n. 5, p. 821-830, 2006.

LEE, B. L.; KENNON, A. R.; CHEN, X.; JUNG, T. W.; AHN, B. O.; LEE, J. Y.; ZHANG, Z. J. Recovery of transgenic events from two highly recalcitrant maize (*Zea mays* L.) genotypes using *Agrobacterium*-mediated standart-binary-vector transformation. **Maydica**, v. 52, n. 4, p. 457-469, 2007.

LOWE, B. A.; WAY, M. M.; KUMPF, J. M.; ROUT, J.; WARNER, D.; JOHNSON, R.; ARMSTRONG, C. L.; SPENCER, M. T.; CHOMET, P. S. Marker assisted breeding for transformability in maize. **Molecular Breeding**, v. 18, n. 3, p. 229-239, 2006.

LUPOTTO, E.; CONTI, E.; REALI, A.; LANZANOVA, C.; BALDONI, E.; ALLEGRI, L. Improving *in vitro* culture and regeneration conditions for *Agrobacterium*-mediated maize transformation. **Maydica**, v. 49, n. 1, p. 21-29, 2004.

MURASHIGE, T.; SKOOG, F. A revised medium for rapid growth and bioassays with tobacco tissue cultures. **Plant Physiology**, v. 15, n. 3, p. 473-497, 1962.

NABORS, M. W.; HEYSER, J. W.; DYKES, T. A.; DEMONTT, K. J. Long-duration, high frequency plant regeneration from cereal tissue culture. **Planta**, v. 157, n. 3, p. 385-391, 1983.

NAQVIA, S.; ZHUA, C.; FARREA, G.; RAMESSARA, K.; BASSIEA, L.; BREITENBACHB, J.; CONESAC, D. P.; ROSC, G.; SANDMANNB, G.; CAPELLA, T.; CHRISTOUA, P. Transgenic multivitamin corn thorough biofortification of endosperm with three vitamins representing three distinct metabolic pathways. **PNAS**, v. 106, n. 19, p. 7762-776, 2009.

O'KENNEDY, M. M.; BURGER, J. T.; BERGER, D. K. Transformation of elite white maize using the particle inflow gun and detailed analysis of a low-copy integration. **Plant Cell Reports**, v. 20, n. 8, p. 721-730, 2001.

PETRILLO, C. P.; CARNEIRO, N. P.; PURCINO, A. A. C.; CARVALHO, C. H. S.; ALVES, J. D.; CARNEIRO, A. A. Optimization of particle bombardment parameters for the genetic transformation of brazilian maize inbred lines. **Pesquisa Agropecuária Brasileira**, v. 43, n. 3, p. 371-378, 2008.

RHODES, C. A.; GREEN, C. E.; PHILLIPS, R. L. Factors affecting tissue culture initiation from maize tassels. **Plant Science**, v. 46, n. 3, p. 225-232, 1986.

SAIRAM, R. V.; PARANI, M.; FRANKLIN, G. Z.; SMITH, L. B.; MACDOUGALL, J.; WILBER, C.; SHEIKHI, H.; KASHIKAR, N.; MEEKER, K.; AL-ABED, D.; BERRY, K.; VIERLING, R.; GOLDMAN, S. L. Shoot meristem: an ideal explant for *Zea mays* L. transformation. **Genome**, v. 46, n. 2, p. 323-329, 2003.

SANTOS-SEREJO, J. A.; AGUIAR-PERECIN, M. L. R. Genótipos de milho com alta capacidade para embriogênese somática e regeneração de plantas obtidas a partir de calos. **Scientia Agricola**, v. 57, n. 4, p. 717-722, 2000.

SELLMER, J. C.; RITCHIE, S. W.; KIM, I. S.; HODGES, T. K. Initiation, maintenance and plant regeneration of type II callus and suspension cells. In: FREELING, M.; WALLBOT, V. (Ed.). **The maize handbook**. New York: Springer Verlag, 1994. p. 671-677.

SHARMA, V. K.; HANSCH, R.; MENDEL, R. R.; SCHULZE, J. Mature embryo axis-based high frequency somatic embryogenesis and plant regeneration from multiple cultivars of barley (*Hordeum vulgare* L.). **Journal of Experimental Botany**, v. 56, n. 417, p. 1913-1922, 2005.

SHOU, H.; FRAME, B. R.; WHITHAM, S. A.; WANG, K. Assessment of transgenic maize events produced by particle bombardment or *Agrobacterium*-mediated transformation. **Molecular Breeding**, v. 13, n. 2, p. 201-208, 2004.

SONGSTAD, D. D.; PETERSEN, W. L.; ARMSTRONG, C. L. Establishment of friable embryogenic (type II) callus from immature tassels of *Zea mays* (Poaceae). **American Journal of Botany**, v. 76, n. 7, p. 761-764, 1992.

TOMES, D. T.; SMITH, O. S. The effect of parental genotypes on initiation of embryogenic callus from elite maize (*Zea mays* L.) germoplasm. **Theoretical and Applied Genetics**, v. 70, n. 5, p. 505-509, 1985.

VASIL, I. K. Developing cell and tissue culture systems for the improvement of cereal and grass crops. **Journal of Plant Physiology**, v. 128, n. 3, p. 193-218, 1987.

VEGA, J. M.; YU, W.; KENNON, A.; CHEN, X.; ZHANG, Z. J. Improvement of *Agrobacterium*-mediated transformation in Hi-II maize (*Zea mays*) using standard binary vectors. **Plant Cell Reports**, v. 27, n. 2, p. 297-305, 2008.

WANG, Y.; FU, S.; WEN, Y.; ZHANG, Z.; XIA, Y.; LIU, Y.; RONG, T.; PAN, G. Selection of maize inbred lines with high regeneration and susceptibility to *Agrobacterium tumefaciens*. **Journal of Genetics and Genomics**, v. 34, n. 8, p. 749-755, 2007.

WILLMAN, M. R.; SCHROLL, S. M.; HODGES, T. K. Inheritance of somatic embryogenesis and planta regeneration from primary (Type 1) callus in maize. ***In vitro* Cell Development Biology**, v. 25, n. 1, p. 95-100, 1989.

ZHANG, Y.; YIN, X.; YANG, A.; LI, G.; ZHANG, J. Stability of inheritance of transgenes in maize (*Zea mays* L.) lines produced using different transformation methods. **Euphytica**, v. 144, n. 1-2, p. 11-22, 2005.

ZHAO, Z.; GU, W.; CAI, T.; TAGLIANI, L.; HONDRED, D.; BOND, D.; SCHROEDER, S.; RUDERT, M.; PIERCE, D. High throughput genetic transformation mediated by *Agrobacterium tumefaciens* in maize. **Molecular Breeding**, v. 8, n. 4, p. 323-333, 2001.

ZHU, C.; NAQVI, S.; GOMEZ-GALERA, S.; PELACHO, A. M.; CAPELL, T.; CHRISTOUH, P. Transgenic strategies for the nutritional enhancement of plants. **Trends in Plant Sciences**, v. 12, n. 12, p. 548-555, 2007.

7

QTLs identification for characteristics of the root system in upland rice through DNA microarray

Thiago Gledson Rios Terra[1], Haroldo Silva Rodrigues[2], Paulo Hideo Nakano Rangel[3], Rafael Simões Tomaz[4], Cosme Damião Cruz[5] and Aluízio Borém[6*]

*[1]Universidade Federal do Tocantins, Gurupi, Tocantins, Brazil. [2]Departamento de Fitotecnia, Universidade Federal de Viçosa, Viçosa, Minas Gerais, Brazil. [3]Embrapa Arroz e Feijão, Santo Antônio de Goiás, Goiás, Brazil. [4] Universidade Estadual Paulista, Dracena, São Paulo, Brazil. [5]Departamento de Biologia Geral, Universidade Federal de Viçosa, Viçosa, Minas Gerais, Brazil. [6]Departamento de Fitotecnia, Universidade Federal de Viçosa, Viçosa, Minas Gerais, Brazil. *Author for correspondence. E-mail: borem@ufv.br*

ABSTRACT. The aim of this work is the construction of a genetic map and identification of quantitative trait loci (QTLs) that control characteristics of the root system of rice. We evaluated a $F_{2:3}$ population composed of 150 families from the cross between the varieties IAC 165 × BRS Primavera. Genotyping was performed in the F_2 population using 3,742 SNP (Single Nucleotide Polymorphism) markers. The evaluation of the root system in the F_3 population was performed through a large-scale phenotyping method based on image generation with a CI-600 root scanner and on quantification through the WinRhizo® software. The experiment was arranged in a randomized block design with three replications performed under greenhouse. The variables analyzed were root length, root surface area and root volume at the depths of 5 to 25 cm and 25 to 45 cm. The SNP markers analysis allowed the construction of the genetic map with a full length of 1424 cM. The linkage group with the largest coverage area was number 3 with 270 cM (100 SNPs), followed by linkage group 1 with 249 cM (170 SNPs) and linkage group 2 with 163 cM (99 SNPs). The genetic analysis allowed the detection of QTLs for all the characteristics.

Keywords: genomic, *Oryza sativa* L., DNA chip, QTL mapping, abiotic stress.

Introduction

Upland rice is responsible for approximately 21% of Brazilian agricultural production. However, the lack of water throughout its production process has a major negative impact on the culture. According to Toescher, Righes, and Carlesso (1997) and Stone (1985), drought occurrence impacts mainly nutrient absorption by the root system, negatively affecting the crop productivity. The understanding of the genetic mechanisms associated with the rice root system development aiming selection could help with the optimization of water absorption, and this strategy seems to be essential for breeding of the culture.

In this context, DNA markers have supported plant breeding in the orientation of crosses, formation of base populations, development of divergent populations (Li-Na, Gui-Lan, & Long-

Zhi, 2013), construction of genetic maps (Jun-Zhi, Xiao, Chao, Ke, & Yan-Jun, 2009), identification of quantitative trait loci (QTL) (Uga, Okuno, & Yano, 2011) and identification of functional markers (Andersen & Lübberstedt, 2003), as well as in marker assisted selection (MAS) and genomic selection (GS) (Steele, Virk, Kumar, Prasad, & Witcombe, 2007; Uga et al., 2013).

For such purposes, single nucleotide polymorphisms (SNPs) have been highly used in genetic analysis because they are stable, widely distributed in the genome and allow for automation, which facilitates high-throughput genotyping and QTL detection. Since 1995, QTL mapping has been conducted in rice cultivation for both temporary and permanent populations (Sabouri, Sabouri, Jafarzadeh, & Mollashahi, 2011; Liang, 2013; Liang, Zhen-Hua, & Jie-Yun, 2013).

Drought tolerance is very dependent on the root system. This dependence on the root system in turn influences other important agronomic traits in upland rice, such as efficiency in nutrient absorption and resistance to pests and root diseases (Oliveira, Freitas, Fiúza, Menezes, & Dotto, 2010; Henry, Gowda, Torres, Mcnally, & Serraj, 2011; Uga et al., 2013). For such reasons, the root system is frequently studied given the potential for the stabilization or increase of productivity under adverse conditions (Topp et al., 2013; Gamuyao et al., 2012).

DNA microarrays are a very useful tool for QTL identification, Marker Assisted Selection (MAS) and genomic selection (GS) studies because they provide greater genome coverage. Related to MAS, it is particularly important to have an adequate linkage map saturation to detect more accurate QTL positioning and its effect estimation, as it is then easier to associate it with a causative gene. Considering the importance of the root system for rice cultivation, this work sought to evaluate a segregating $F_{2:3}$ population of upland rice to construct a linkage map aiming QTL identification for root system traits by means of DNA microarrays technology.

Material and methods

Genetic material

The genetic material evaluated in this experiment was constituted by a segregating population ($F_{2:3}$) derived from a single F_1 plant obtained from the cross between the varieties of upland rice (tropical japonica subspecies) IAC 165 and BRS Primavera, composed by 148 families and its two genitors.

IAC 165 is a commercial variety of upland rice developed by the Agronomic Institute of Campinas from the cross between "Dourado Precoce" and IAC 1245. It has a moderate resistance to lodging, strong root development, tolerance to water deficit, relatively short cycle and long and thick grain. BRS Primavera is a commercial variety released for cultivation in highlands and was developed by Embrapa Rice and Beans. It is derived from the cross between IRAT 10 and LS 85-158 cultivars, has intermediate size (average of 100 cm), medium tillering, low lodging resistance and early cycle (average of 100 days). It presents average productivity, susceptibility to drought, long and thin high quality grains, and shows a very narrow harvest stage.

Root phenotyping

The root phenotyping was performed in the experimental field of Fazenda Palmital, of EMBRAPA Rice and Beans, in the municipality of Goianira, Goiás State, Brazil, from October 2012 to June 2013. The experiment was installed in a randomized block outlining with three repetitions and was performed under controlled conditions in a greenhouse. The experimental unit was constituted by a polyethylene pipe with an internal diameter of 30 cm and a height of 80 cm and was filled with a clayey red latosol that presented constitution of: 19.0 g dm^{-3} M.O., pH 5.4, 0.5 g dm^{-3} P(Mel.); 0.24 $cmol_c$ dm^{-3} K, 2.0 $cmol_c$ dm^{-3} Ca, 0.8 $cmol_c$ dm^{-3} Mg, 0.0 $cmol_c$ dm^{-3} Al, 2.5 $cmol_c$ dm^{-3} H+Al; 590.0 g kg^{-1} sand, 120.0 g kg^{-1} silt, 290.0 g kg^{-1} clay, with an added 250 grams of 5-25-15 fertilizer. Installed inside of the experimental unit was an acrylic tube of 6 cm of internal diameter with rubber covers for protection against the entry of soil, water or organic waste.

Around the acrylic tube, three rice plants were transplanted with one week of emergence. Two weeks after transplantation, the generation of the images was initiated at the depth of 5-25 cm, and three weeks after transplantation, imaging at the depth of 25-45 cm was performed. The images were digitalized using a *scanner* (CI – 600 *Cano Scan*) inserted in the acrylic tube, which scanned the surface around the tube. The evaluations were performed weekly for until the plans were eight weeks of age.

The generated images were analyzed using the image quantification software *WinRhizo*, version 2008, which makes available, among other variables, the total length of the roots (COM) measured in centimeters (cm), the estimated contact area of the roots (ACR) measured in square centimeters (cm^2) and the total volume of the roots (VOL) measured in cubic centimeters (cm^3).

DNA Microarray Genotyping

For the genotyping of population $F_{2:3}$, a microarray composed of SNP-type markers was used. For the microarray construction, genetic sequencing data from eight varieties of *upland* rice (*japonica* subspecies) were employed whose characteristics related to the root system are described in Table 1.

A total of 3,742 SNP markers were developed and distributed along the 12 rice chromosomes. These markers were submitted to the DNA microarray technique (Schena, Shalon, Davis, & Brown, 1995). Any SNPs that did not present a polymorphism, had not been evaluated in at least 80% of the population, or that presented segregation distortion by the chi-square test at 5% probability level were eliminated on the basis of the genetic map's construction and the QTL identification.

Data analysis

The phenotyping process was performed through the *WinRhizo*™ software (Arsenault, Pouleur, Messier, & Guay, 1995). The areas under the growth curve (AUC) were estimated for each root variable over the weeks of evaluation using the trapezoidal method. The data from the AUC were submitted to variance homoscedastic testing followed by joint variance analysis. Considering the two depths, the AUC was used to estimate a value closer to the real growth rate of the root system (Boyer, 1996) rather than using only an average calculated from the data collected along the eight weeks of this study. The variance was estimated from data of each genotype AUC related to the genotype's averages. The analysis aimed to verify significant differences within the population for the analyzed characteristics of the root system. The GENES software (version 2013) was used in this step (Cruz, 2013).

The construction of the genetic map was performed through the *IcIMapping* QTL software (version 2012 3.2), considering an LOD *Score* of 3.0 for QTL identification (Collard, Jahufer, Brouwer, & Pang, 2005). For the identification of QTLs related to the root characteristics, the simple interval mapping method was performed using the Windows QTL *Cartographer* software (version 2.5_011, 2012) (Basten, Weir, & Zeng, 2004).

Results and discussion

The phenotypic information showed variance homoscedasticity and the joint variance analysis for the two depths ($p < 0.01$), which is presented in Table 2. The results demonstrated great variability as well as significant differences for all variables and all considered factors (genotypes, depth and interaction).

Table 1. Upland rice varieties (*japonica* subspecies) and their characteristics related to the root system.

Variety	Characteristics related to the root system
Azucena	Variety from the Philippines, known in the international literature as a genotype of good root development and tolerance to the stress of drought and toxic aluminum.
Moroberekan	Variety from West Africa, also known in the international literature as a genotype of good root development, good root penetration capacity on soil, tolerant to drought and toxic aluminum.
Chorinho	Traditional variety collected in the state of Minas Gerais, Brazil, with intermediate tolerance to stress by water deficiency in the field, with tall plants.
Puteca	Traditional rice variety collected in the state of Goiás, Brazil, susceptible to stress from water deficiency, with tall plants.
IAC 165	Commercial rice variety developed by the Campinas Agronomic Institute, with high rusticity, good root development and tolerance to the stress of drought.
BRS Primavera	Commercial variety developed by Embrapa Rice and Beans, presenting susceptibility to drought stress, whose plants are of intermediate size (average of 100 cm).
Catetão	Traditional variety collected in the state of MatoGrosso, Brazil, tolerant to drought stress, with tall plants.
Ligeiro	Traditional rice variety collected in the state of Maranhão, Brazil, tolerant to drought stress, presenting a short cycle and low productivity.

Table 2. Variance analysis for the variables of the root system: length (COM), contact area (ACR) and volume (VOL) of the roots in the depths of 5-25 cm and 25-45 cm.

		Mean square					
F. V.	GL	COM		ACR		VOL	
Block	2	7943175.93		227320.49		42.24	
Genotypes(G)	149	645107.39	**	17818.13	**	3.56	**
Depths(P)	1	44593404.61	**	2080284.10	**	550.40	**
G x P	149	596237.47	**	16523.02	**	3.18	**
Residue	598	179663.06		5420.12		1.23	
Average		1657.23		275.73		3.69	
Average depth 1		1879.83		323.80		4.48	
Average depth 2		1434.64		227.65		2.91	
CV (%)		25.57		26.70		30.07	

**Significant at the level of 1% probability by the F test.

Genetic map based on the F_2 population from the cross between varieties IAC 165 and BRS Primavera

Having identified significant differences in the root system characteristics, genomic analysis was then performed by constructing a genetic map and QTLs detection for such characteristics. SNPs that did not cover 80% of the segregating population, were not polymorphic, or that presented segregation distortion evaluated through the chi-square test ($p < 0.05$) were eliminated from the analysis. Then, 1,061 polymorphic markers were considered for genetic analysis. The genetic maps were constructed based on these 1,061 polymorphic markers, considering an LOD *score* of 3.0 and a maximum recombination frequency of 0.3. The mapping process allowed the construction of 12 linkage groups (Figure 1), reconstituting the number of chromosomes of rice species considered, according to Cheng, Buell, Wing, Gu, and Jiang (2001).

The markers on the genetic map presented the same ordering presented in the physical map from which the SNPs were selected, based on the database available by the Gramene portal (Jaiswal et al., 2006). Linkage groups 1 and 7 presented the highest number of markers (170 SNPs on each chromosome), followed by linkage groups 11 (130 SNPs) and 3 (100 SNPs). The smaller linkage group was group 9, with only 18 SNPs markers. Many markers presented were completely connected, having been omitted from the linkage group (Figure 1). Overall, the markers presented good distribution along the chromosomes, with the exception of chromosome 9, which had a total length of 6 cM. This result is corroborated by Sabouri et al. (2011), in whose work on *Oryza sativa* L. also reported the shorter length of this chromosome in relation to others. The longest length was found in chromosome 3, with 270 cM, followed by chromosome 1 with 249 cM and by chromosome 2 with 163 cM. The total length of the genetic map constructed was 1,424 cM. The size of the genome obtained is similar to that reported by other authors in several works for the species *O. sativa* L. (Liang, 2013).

According to Uga et al. (2011), the characteristic COM is governed by a smaller quantity of genes when compared to other characteristics of the root system, such as growth in depth and the angle of growth of the roots. For the characteristic COM, QTLs with a significant effect associated with the two depths evaluated were detected in 5 to 25 cm (Figure 2) and 25 to 45 cm (Figure 3), as these are positioned at the linkage groups 3 and 1, respectively. For the depth of 5 to 25 cm, the QTL was positioned in the distal fraction of the linkage group 3, at the position 258-260 cM. This QTL was identified between the markers OsC1r3_18199829 and OsC1r3_18195592 with a LOD of 3.22. For the depth of 25 to 45 cm, the QTL was positioned in the proximal fraction of the linkage group 1, positioned between the markers OsC1r1_142021996 and OsC1r1_141152310 with a LOD of 3.37.

The characteristic ACR, in turn, is intimately associated with the capacity to absorb water and soil nutrients, a quality of great importance for the growth and development of plants. For this characteristic, at the most superficial depth (5 to 25 cm,), one QTL was detected, as well as in the distal fraction of the linkage group associated with chromosome 3, positioned at 256-260 cM, between the markers OsC1r3_18199829 and OsC1r3_18195592 with LOD = 3.02 (Figure 4). It is possible that this QTL has a pleiotropic effect for root characteristics because it is located in the same region of the QTL detected for COM characteristics. The same happens for the second QTL detected for this characteristic in the depth of 25 to 45 cm (Figure 5), detected at the beginning of the linkage group 1, at a distance of 5-8 cM between the markers OsC1r1_142021996 and OsC1r1_141152310. The LOD associated with this QTL was of 3.06.

For the characteristic VOL a QTL in linkage group 3 was detected at the depth of 5 to 25 (Figure 6). This QTL is located at the position 96-104 cM, between the markers OsC1r3_129673798 and OsC1r3_132585678, with a LOD of 2.73. Two QTLs for the VOL characteristic were located at the depth of 25 to 45 cm in linkage group 1 (Figure 7). The first QTL at the proximal fraction of the linkage group was detected at the position of 5-9 cM between the markers OsC1r1_142021996 and OsC1r1_141152310 with LOD of 2.99. The second QTL was detected at the position 67-76 cM between the markers OsC1r1_128642534 and OsC1r1_124653796 with a LOD of 3.05.

The identification of two regions controlling a quantitative characteristic suggests an inheritance of polygenic control for the VOL characteristic. This finding allows us to infer information about the existence of more genes involved in the VOL control that would allow the plant to reach the deepest regions on the soil (Uga et al., 2011). This author highlights that such genes would allow the plant to stimulate root growth in a vertical way through, for example, elongation of the root cells. Other works conducted with rice (*O. sativa* L.) also identified QTLs for characteristics, such as COM, ACR and VOL, in these same chromosomes, corroborating with the results found in the present research. Courtois, Shen, Petalcorin, Carandang, and Mauleon (2003) studied a population of endogamic lineages from the cross between the varieties IAC 165 (*O. sativa* L. subesp. *japonica*) and Co39 (*O. sativa* L. subesp. *indica*) and found the existence of a main QTL on chromosomes 1 and 4 for root system characteristics, results similar to those obtained by us. This is possibly due to the genetic pool used because the IAC 165 genetic material is a common genitor in both studies.

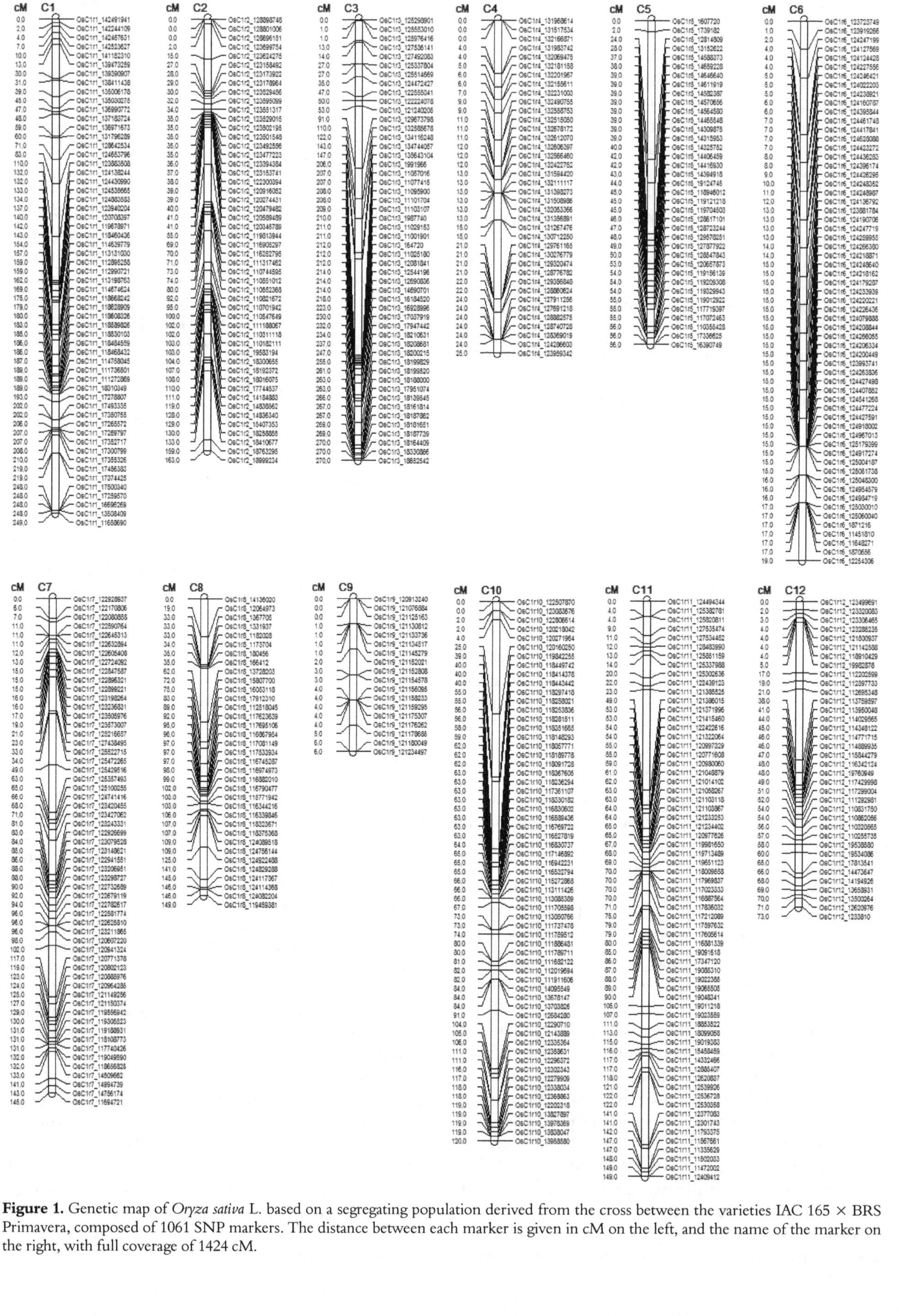

Figure 1. Genetic map of *Oryza sativa* L. based on a segregating population derived from the cross between the varieties IAC 165 × BRS Primavera, composed of 1061 SNP markers. The distance between each marker is given in cM on the left, and the name of the marker on the right, with full coverage of 1424 cM.

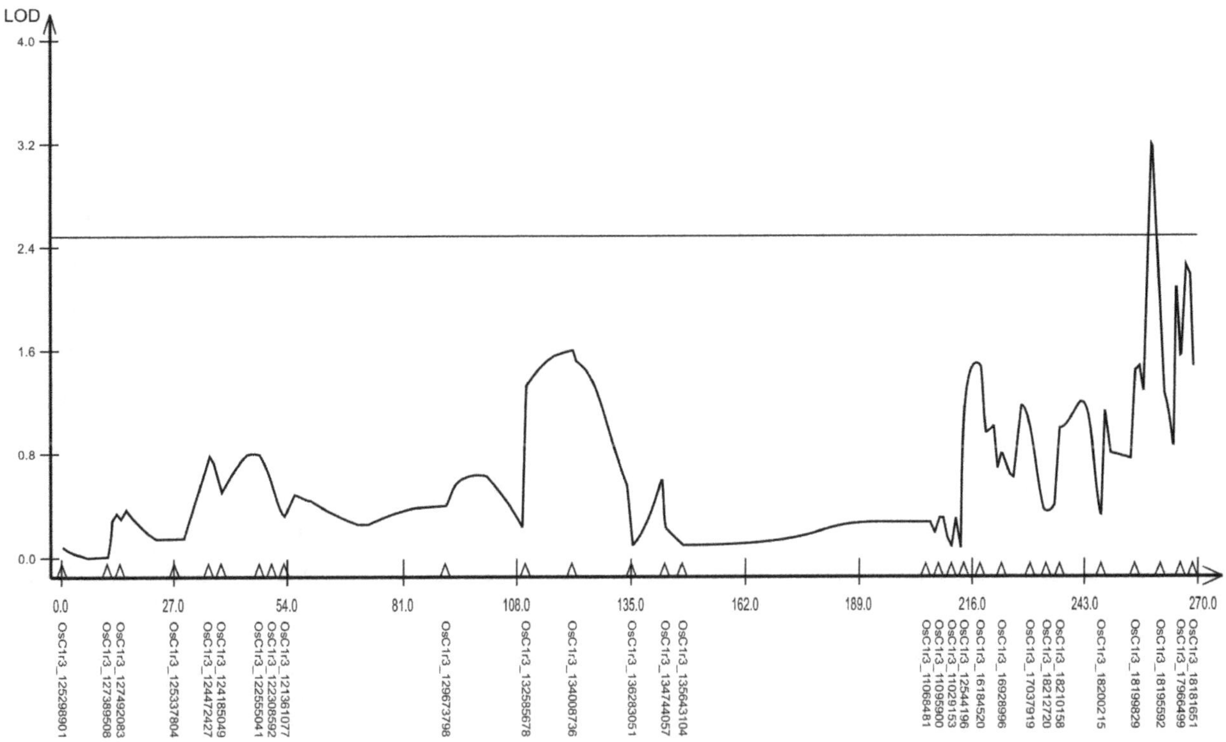

Figure 2. QTL associated to the characteristic COM at the depth of 5 to 25 cm in a $F_{2:3}$ population of rice on the chromosome 3 by the Composed Interval Mapping analysis. The critical value of LOD of 2.5 is represented by the horizontal line.

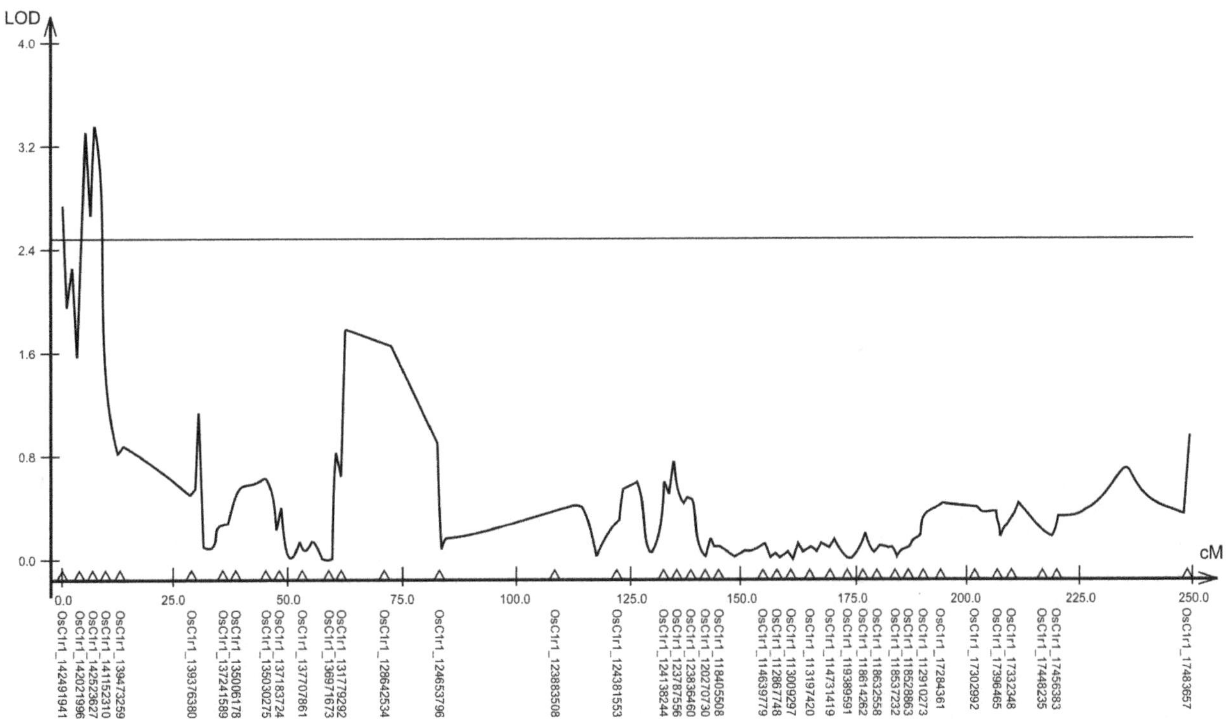

Figure 3. QTL associated to the characteristic COM at the depth of 25 to 45 cm in a $F_{2:3}$ population of rice on the chromosome 1 by the Composed Interval Mapping analysis. The critical value of LOD of 2.5 is represented by the horizontal line.

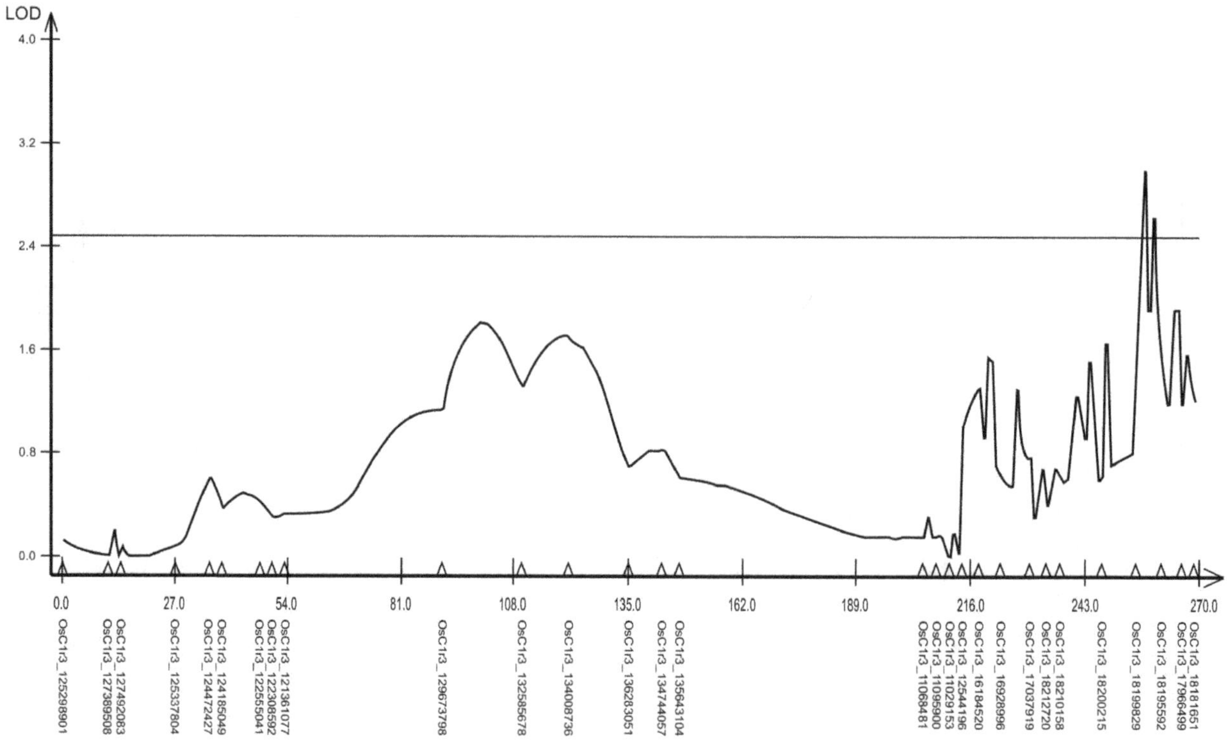

Figure 4. QTL associated to the characteristic root contact area (ACR) at a depth of 5 to 25 cm in a $F_{2:3}$ population of rice on the chromosome 3 by the Composed Interval Mapping analysis. The critical value of LOD of 2.5 is represented by the horizontal line.

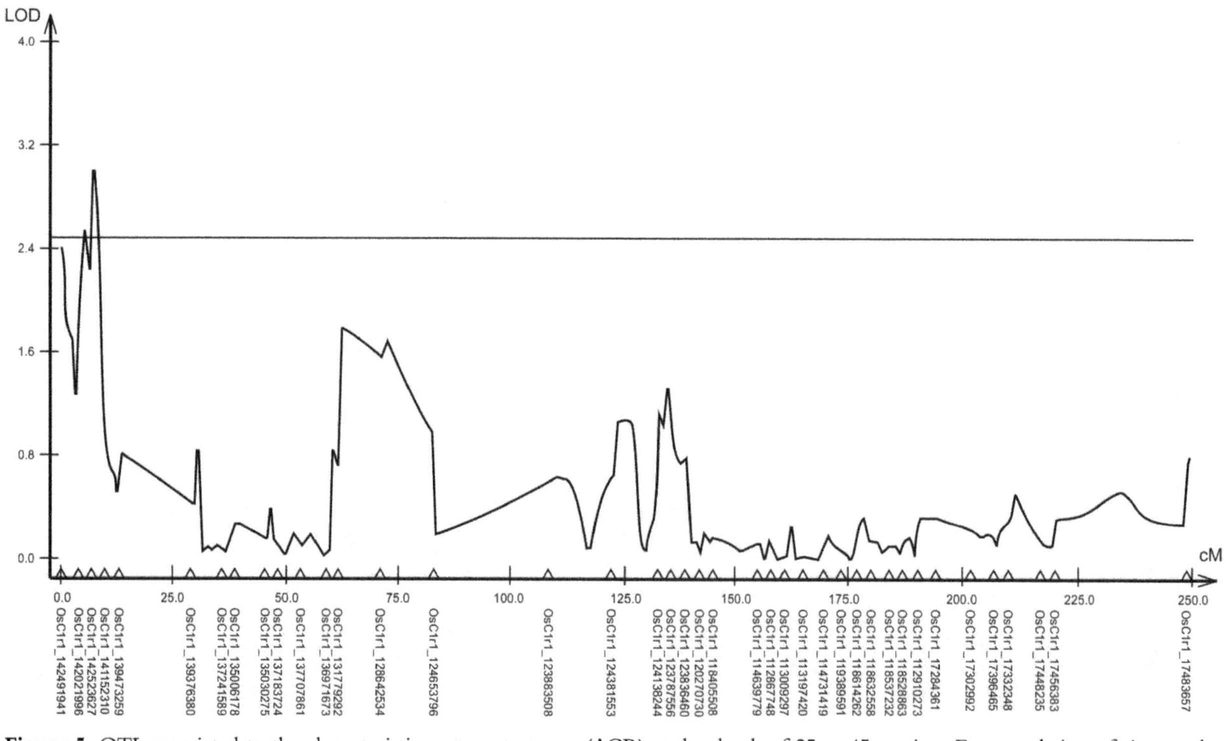

Figure 5. QTL associated to the characteristic root contact area (ACR) at the depth of 25 to 45 cm in a $F_{2:3}$ population of rice on the chromosome 1 by the Composed Interval Mapping analysis. The critical value of LOD of 2.5 is represented by the horizontal line.

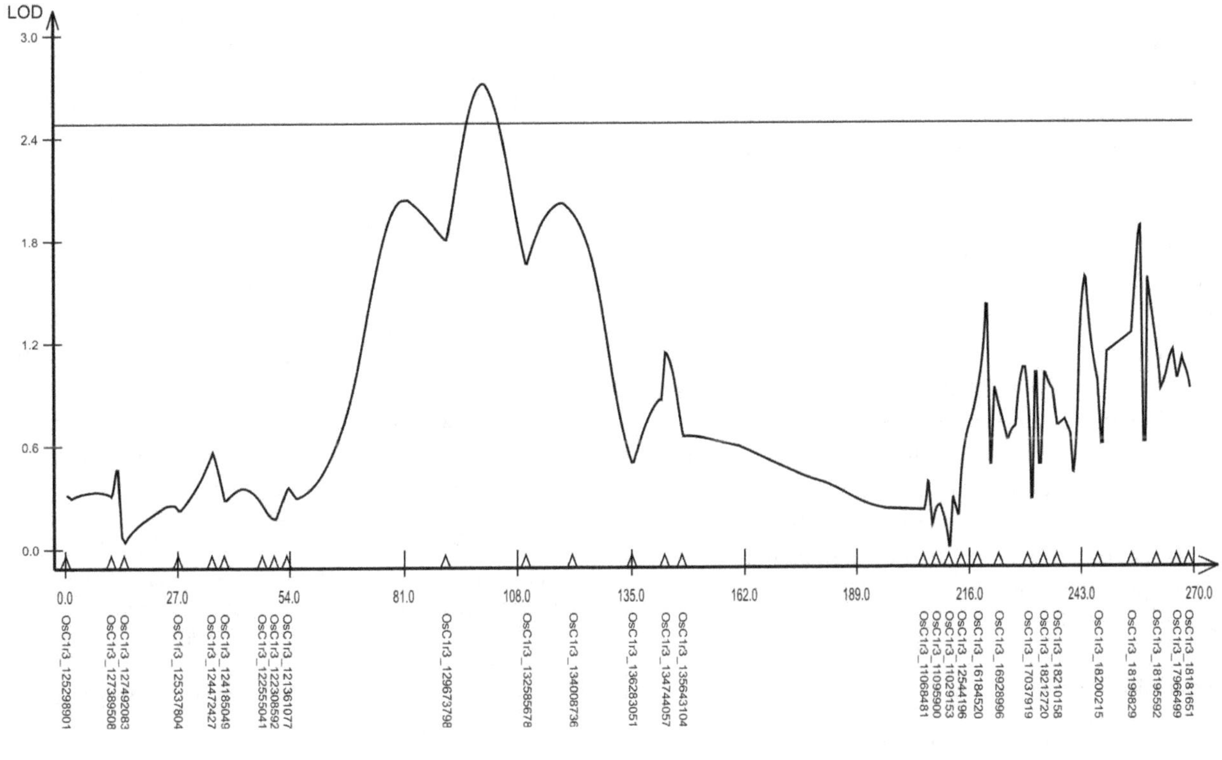

Figure 6. QTL associated to Volume of the roots at the depth of 5 to 25 cm in a $F_{2:3}$ population of rice in the liaison group 3 by the Composed Interval Mapping analysis. The critical value of LOD of 2.5 is represented by the horizontal line.

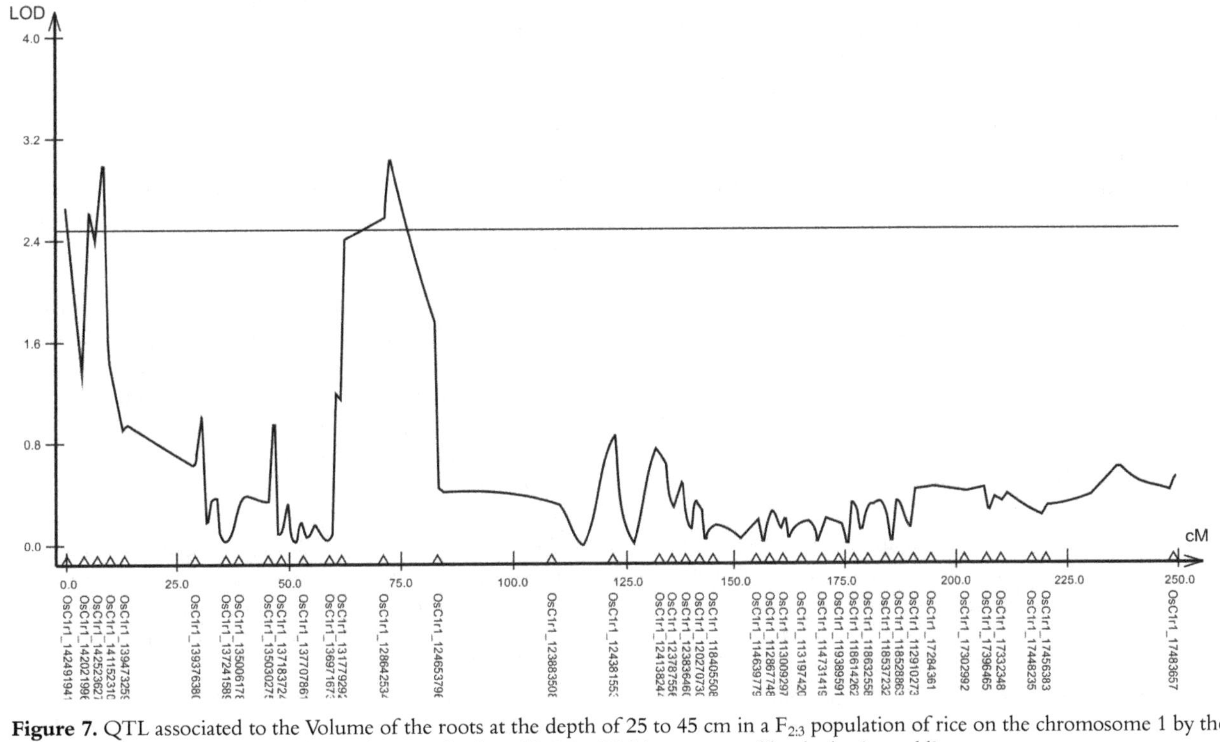

Figure 7. QTL associated to the Volume of the roots at the depth of 25 to 45 cm in a $F_{2:3}$ population of rice on the chromosome 1 by the Composed Interval Mapping analysis. The critical value of LOD of 2.5 is represented by the horizontal line.

The limited progress for drought tolerance in rice culture has been attributed, in part, to the large quantity of genes involved and the large environmental influence associated with this characteristic expression. Uga et al. (2013), however, were able to develop a rice variety that was productive even when submitted to water stress. The lineage was developed through backcrossing, using MAS to monitor the gene *dro*1. This gene, located on chromosome 9, is reported to be responsible for the elongation of root cells.

In rice culture, many of the QTLs identification work related to root system characteristics have been performed through crossings between subspecies of genotypes *O. sativa* indica with *O. sativa* japonica (Steele et al., 2007; Uga et al., 2011; Courtois et al., 2003; Ping et al., 2003). In this type of crossing, there is a huge segregation, enabling the study and detection of QTL for certain characteristics, given the knowledge of contrasting genitors. The crossing of such genetic materials results in the segregation of smaller amounts of characteristics, which enables both the most refined molecular study and cultivar development. This study, in turn, allows for the identification of genes and their subsequent use in SAM operational procedures, such as those used by Uga et al. (2013).

Conclusion

The resulting segregating population from the cross between the varieties IAC 165 and BRS Primavera presented wide variability for the root characteristics studied.

The DNA microarray technique allowed for the construction of an upland rice genetic map with broad genome coverage and a total length of 1,424 cM, covering all chromosomes of the species.

The linkage groups 1 and 3 presented QTLs related to the characteristics COM, ACR and VOL of the root system.

References

Andersen, J. R., & Lübberstedt, T. (2003). Functional markers in plants. *Trends in Plant Science*, *8*(11), 554-560.

Arsenault, J. L., Pouleur, S., Messier, C., & Guay, R. (1995). WinRHIZO™, a root-measuring system with a unique overlap correction method. *HortScience*, *30*(4), 906-906.

Basten, C. J., Weir, B. S., & Zeng, Z. B. (2004). *QTL Cartographer*, version 1.17. Raleigh, NC: Department of Statistics, North Carolina State University.

Boyer, C. B. (1996). *Introdução à história da matemática*. Campinas, SP: Editora Edgard Blucher.

Cheng, Z., Buell, C. R., Wing, R. A., Gu, M., & Jiang, J. (2001). Toward a Cytological Characterization of the Rice Genome. *Genome Research*, *11*(12), 2133-2141.

Collard, B. C. Y., Jahufer, M. Z. Z., Brouwer, J. B., & Pang, E. C. K. (2005). An introduction to markers, quantitative trait loci (QTL) mapping and marker-assisted selection for crop improvement: the basic concepts. *Euphytica*, *142*(1-2), 169-196.

Courtois, B., Shen, L., Petalcorin, W., Carandang, S., & Mauleon, R., & Li, Z. (2003). Locating QTLs controlling constitutive root traits in the rice population IAC 165 × Co39. *Euphytica*, *134*(3), 335-345.

Cruz, C. D. (2013). GENES - a software package for analysis in experimental statistics and quantitative genetics. *Acta Scientiarum. Agronomy*, *35*(3), 271-276.

Gamuyao, R., Chin, J. H., Pariasca-Tanaka, J., Pesaresi, P., Catausan, S., Dalid, C., ... Heuer, S. (2012). The protein kinase Pstol1 from traditional rice confers tolerance of phosphorus deficiency. *Nature*, *488*(7412), 535-539.

Henry, A., Gowda, V. R. P., Torres, R. O., Mcnally, K. L., & Serraj, R. (2011). Variation in root system architecture and drought response in rice (*Oryza sativa*): Phenotyping of the Oryza SNP panel in rainfed lowland fields. *Field Crops Research*, *120*(2), 205-214.

Jaiswal, P., Ni, J., Yap, I., Ware, D., Spooner, W., Youens-Clark, K., ... Mccouch, S. (2006). Gramene: a bird's eye view of cereal genomes. *Nucleic Acids Research*, *34*(suppl 1), D717-D723.

Jun-Zhi, Z., Xiao, L., Chao, L., Ke, X., & Yan-Jun, D. (2009). Genetic analysis and molecular mapping of light-sensitive red-root mutant in rice. *Rice Science*, *16*(1), 27-32.

Liang, G., Zhen-Hua, Z., & Jie-Yun, Z. (2013). Quantitative trait loci for heading date and their relationship with genetic control of yield traits in rice (*Oryza sativa*). *Rice Science*, *20*(1), 1-12.

Liang, Y. S. (2013). Locating QTLs controlling several adult root traits in an elite Chinese hybrid rice. *Gene*, *526*(2), 331-335.

Li-Na, Z., Gui-Lan, C., & Long-Zhi, H. (2013). Genetic Diversity of Rice Landraces from Lowland and Upland Accessions of China. *RiceScience*, *20*(4), 259-266.

Mammadov, J., Aggarwal, R., Buyyarapu, R., & Kumpatla, S. (2012). SNP Markers and their impact on plant breeding. *International Journal of Plant Genomics*, *2012*(2012) 1-11.

Oliveira, J. V., Freitas, T. F. S., Fiúza, L. M., Menezes, G., & Dotto, G. (2010). *Manual de insetos associados à cultura do arroz irrigado*. Cachoeirinha, RS: Instituto Rio Grandense do Arroz.

Ping, M., Zichao, L., Chunping, L., Hongliang, Z., Changming, W., Chen, L., & Xiangkun, W. (2003). QTL mapping of the root traits and their correlation analysis with drought resistance using DH lines from paddy and upland rice cross. *Chinese Science Bulletin*, *48*(24), 2718-2724.

Sabouri, H., Sabouri, A., Jafarzadeh, M. R., & Mollashahi, M. (2011). Detection of QTLs controlling field blast resistance in rice (*Oryza sativa* L.). *Plant Omics Journal*, *4*(1), 1-5.

Schena, M., Shalon, D., Davis, R. W., & Brown, P. O. (1995). Quantitative monitoring of gene expression patterns with a complementary DNA microarray. *Science*, *270*(5235), 467-470.

Steele, K. A., Virk, D. S., Kumar, R., Prasad, S. C., & Witcombe, J. R. (2007). Field evaluation of upland rice lines selected for QTLs controlling root traits. *Field Crops Research, 101*(2), 180-186.

Stone, L. F. (1985). Absorção de P, K, Mg, Ca e S por arroz influenciada pela deficiência hídrica, vermiculita e cultivar. *Pesquisa Agropecuária Brasileira, 20*(11), 1251-1258.

Toescher, C. F., Righes, A. R., & Carlesso, R. (1997). Volume de água aplicada e produtividade do arroz sob diferentes métodos de irrigação. *Revista da FZVA, 4*(1), 49-57.

Topp, C. N., Iyer-Pascuzzi, A. S., Anderson, J. T., Lee, C., Zurek, P. R., Symonoa, E., ... Benfey, P. N. (2013). 3D phenotyping and quantitative trait locus mapping identify core regions of the rice genome controlling root architecture. *Proceedings of the National Academy of Sciences of the United States of America, 110*(18), 1695-1704.

Uga, Y., Okuno, K., & Yano, M. (2011). Dro1, a major QTL involved in deep rooting of rice under upland field conditions. *Journal of Experimental Botany, 62*(8), 2485-2494.

Uga, Y., Sugimoto, K., Ogawa, S., Rane, J., Ishitani, M., Hara, N., ... Inoue, H. (2013). Control of root system architecture by DEEPER ROOTING 1 increases rice yield under drought conditions. *Nature Genetics, 45*(9), 1097-1102.

8

Genetic parameters and productivity of quinoa in western Paraná State, Brazil

Edmar Soares de Vasconcelos*, Lívia Maria Lemos Hoepers, Rafaela Goularte Amaral, Vanessa Aline Egewarth and Andressa Strenske

*Departamento de Agronomia, Centro de Ciências Agrárias, Universidade Estadual do Oeste do Paraná, Rua Pernambuco, 1777, 85960-000, Marechal Cândido Rondon, Paraná, Brazil. *Author for correspondence. E-mail: edmar.vasconcelos@unioeste.br*

ABSTRACT. Quinoa has been gaining attention because of its nutritional quality, low cholesterol and lack of gluten; in Brazil, the cultivation efforts in the different regions are mainly related to breeding. This study aimed to determine the genetic parameters and evaluate the productivity of the different genotypes of quinoa for detecting genotypes amenable to selection. The experiment was conducted in crop years 2010/11 and 2011/12 in environment 1 and environment 2, respectively. In environment 1, the evaluation of 61 genotypes was performed, and in environment 2, 31 genotypes were evaluated. The experimental design was a randomized block with two replications; each plot measured 2.0 x 5.0 m (10 m^2) and consisted of four rows spaced at 0.45 m. Data collected on the productivity, plant height at maturation and growth cycle were analyzed using an analysis of variance, average tests and estimates of the genetic parameters. The genotypes N24 and N08 were the only genotypes more productive than the other 46 genotypes, with values of productivity of 1,446.23 and 1,428.93 kg ha^{-1} and with a growth cycle of 117 and 111 days, respectively. The heritability values determined demonstrate the possibility of genetic gain using joint selection that involves two environments.

Keywords: *Chenopodium quinoa*, genotypic selection, agronomic performance.

Introduction

Quinoa is an annual crop from the family Chenopodiaceae. It was introduced in Brazil during the 1990s as part of an effort to diversify the production system in the Cerrado. The first attempts to adapt it to the cultivation were made through selection in hybrid populations using descendant from Cambridge, England (Spehar, 2007).

For commercial production in Brazil, quinoa is desired that presents rapid growth, absence of lodging, photoperiod insensitivity, low ramification, perianth and seed indehiscence, an early cycle and uniform maturation. The genotypes must present various cycles, high yields of grains and biomass, seed protein quality, tolerance to saline conditions and 100-grain weight between 2.0 and 3.5 g (Adolf, Shabala, Andersen, Razzaghi, & Jacobsen, 2012; Gonzalez, Konishi, Bruno, Valoy, & Prado, 2012; Silva, Fuentes, Zamora, Jacobsen, & Schwember, 2014). Such characteristics have been the target of quinoa improvement programs, including for BRS

Piabiru (Spehar, 2007) and BRS Syetetuba (Spehar, Rocha, & Santos, 2011) Brazilian savannah and in southern Brazil (Vasconcelos, Egewarth, Oliveira, & Piano, 2013).

The estimation of genetic and phenotypic parameters, such as heritability, genetic and phenotypic correlation and gains with selection, is important for genetic improvement programs because these estimates facilitate the choice of methods and characters used in the initial and advanced phases of improvement programs, thereby allowing the study of mechanisms, genetic values and variability for one character (Cruz, Regazzi, & Carneiro, 2012; Vasconcelos, Reis, Sediyama, & Cruz, 2012). However, characteristics related to the quinoa cultivation have not been studied with regard to genetic values for quantity.

From a genetic point of view, the improvement and evolution of wild or cultivated plants have many common features; they require variability, some type of selection and some degree or form of insulation to preserve the characteristics of the types that are cultivated. Studies are aimed at determining the degree of heterogeneity among the possible genotypes for selection to ensure that only the best genotypes stay in a breeding program (Silva, Chaves, Arnhold, & Cruz, 2009), particularly for quinoa.

The variability among cultivars reflects the heterogeneity of the genetic material, improves food security threatened by climate change and offers the possibility of identifying promising material for use in a breeding plant program (Ruiz et al., 2014). In the case of quinoa, Gonzalez, Konishi, Bruno, Valoy, and Prado (2012) showed that variability related to the cultivation area caused variation in yield and seed quality.

In the South of Brazil, the study of quinoa behavior to evaluate productive characteristics and the possibility of selecting genotypes with potential for a new cultivar is necessary. As a result, quinoa will be able to contribute to crop rotation and expand the spectrum of possibilities in regional agriculture (Vasconcelos et al., 2013).

The present study aimed to determine the genetic parameters and analyze the productivity of 61 quinoa genotypes (*Chenopodium quinoa* Willd) during the crop year of 2010/11 and 31 genotypes during the crop year of 2011/12 in the region of Marechal Cândido Rondon, Paraná State, Brazil.

Material and methods

The experiments were conducted in the crop years of 2010/11 and 2011/12 in Marechal Cândido Rondon, Paraná State, Brazil, where the average annual temperature varies between 22 and 23°C, the total normal average annual precipitation for the region is between 1,600 and 1,800 mm, and the local soil is classified as Eutoferric Red Latosol - ERL, with clay texture.

In the crop year 2010/11, which constitutes the first experiment, 61 genotypes derived from selections performed in the year 2009/10 were used, whereas in the second experiment, crop year 2011/12, the 31 genotypes analyzed were derived from selections made in the year of 2010/11; 10 genotypes were sown in both the first and second experiments to serve as common checks. The genotypes were selected for plant height between 1.00 and 1.20 m, with growth cycles between 110 and 125 days and panicle size greater than 15 cm.

The experimental design adopted was a randomized block with two replications because of the low number of available seeds, with each plot measuring 2.0 x 5.0 m (10 m^2) and formed by four 5.0 m long rows spaced at 0.45 m.

The implementation of the experiments was made by manual direct seeding in an area that previously had been used to cultivate corn in the first year and soybean in the second. The fertilization was based on a dose of 250 kg ha^{-1} of the formula 08 - 20 - 20 during the seeding, which was provided 5 cm below the seeds. The seeding of the first environment was performed on 19 October 2010, whereas the second environment was sown on 15 February 2011.

'Engeopleno' (Thiamethoxam) was applied at a dose of 150 mL ha^{-1} during flowering for pest control of the 'vaquita de San Antonio' (*Diabrotica speciosa*) and stink bug. Weed control was performed whenever necessary through manual weeding. The grain harvest and threshing was also performed manually with sieves.

Due to seed inviability, 15 genotypes of the first experiment did not germinate, which made the analysis of those genotypes impractical. Because of these losses, the number of analyzed genotypes in the first environment was 35, in addition to the 10 genotypes that were used as common checks. For the second environment, 21 genotypes, plus the genotypes used for common checks, were analyzed, amounting to a combined analysis of 66 genotypes. The traits were initially corrected by the values obtained in the common checks, and these corrected values were then used in the analysis of variance (Cruz, 2006).

With the data on productivity, plant height and cycle, quantitative genetic analyses were performed with the aim of determining if, and how much, gain was accomplished with selection.

The variance analysis was performed for each of the experiments following the equation 1:

$$Y_{ijk} = m + G_i + T_j + GT_{ij} + B_k + \epsilon_{ijk} \quad (1)$$

where:
Y_{ijk} = the characteristic value,
m = the overall average of the experiments,
G_i = the effect of the ith genotype (i = 1, 2, ..., g),
T_i = the effect of the jth common checks (j = 1, 2, ..., r),
GT_{ij} = the effect of the interaction Genotype (i) x Common checks (j),
B_k = the effect of the kth block, and
ϵ_{ijk} = random error.

The statistical model adopted for the combined analysis was as follows equation 2:

$$Y_{ijkl} = m + {}^{G}/_{A_{il}} + T_j + A_l + TA_{jl} + {}^{(G/T)}/_{A_{ijl}} + {}^{B}/_{A_{kl}} + \epsilon_{ijkl} \quad (2)$$

where:
Y_{ijkl} = the characteristic value,
m = the overall average of the experiments,
G A_{il}^{-1} = the effect of the ith genotype in the lth environment,
T_j = the effect of the jth common checks,
A_l = the effect of the lth environment,
TA_{jl} = the effect of the interaction between the jth common checks and the lth environment, (G T^{-1}) A_{ijl}^{-1} = the effect of the G T^{-1} interaction in the lth environment,
B A_{kl}^{-1} = the effect of the kth block in the lth environment, and
ϵ_{ijkl} = the random error.

Based on the variance analysis of both experiments, we determined the following genetic parameters:

σ_g^2 = the genotypic variance, which was calculated based on the expectancy of the mean square genotypes (MSg), according as equation 3:

$$\sigma_g^2 = \frac{MS_g - MS_e}{b} \quad (3)$$

where:
$\sigma_e^2 = MS_e$ = the environmental variance;
σ_g^2 = the genetic variance;
b = the number of blocks, and
MS_e = the mean square residue.
h_r^2 = narrow sense heritability, which was calculated based on combined variance according as equation 4:

$$h_r^2 = \frac{\sigma_g^2}{\frac{2MS_g}{b}} \quad (4)$$

where:
CVg = the coefficient of genetic variation, which was calculated according as equation 5:

$$h_r^2 = \frac{\sigma_g^2}{\frac{2MS_g}{b}} \quad (5)$$

where:
m = the experiment general mean.
CVe = the coefficient of environmental variation, which was calculated according as equation 6:

$$CV_e = \frac{100 * \sqrt{MS_e}}{m} \quad (6)$$

The ratio CVg CVe^{-1} corresponds to the ratio between the coefficients of genetic and environmental variation.

The variance analysis, genetics parameters and the comparison averages (Tukey's test at a 5% probability) were obtained with the aid of the computer application GENES (Cruz, 2013).

Results and discussion

In the variance analysis (Table 1), a significant difference was verified among genotypes, among common checks and among genotypes versus common checks (test F at 5%), for all analyzed characteristics. This result determines the existence of variability in the population, a basic condition that enables selection among these materials. The combined variance analysis of the productivity and cycle characteristics showed the existence of variability among the subjects of different populations analyzed in different years.

The overall average values found for productivity (1,155.08 kg ha^{-1} in environment 1, 288.33 kg ha^{-1} in environment 2 and 801.97 kg ha^{-1} in the combined analysis – Table 1) compared to the mean value found by Vasconcelos, Vasconcelos, Balan, and Silvério (2012) during an experiment performed with quinoa in the city of Campo Mourão, Paraná State, which was 846 kg ha^{-1}, demonstrates the need to perform a selection for this characteristic in west Paraná. According to Spehar (2007), quinoa can reach productivity averages up to 3,000 kg ha^{-1} under favorable conditions.

Table 1. Resume of the analysis of individual variance and combined variance of the variables Productivity, Cycle and Plant Height, evaluated in Marechal Cândido Rondon, in the crop years 2010/2011 and 2011/2012.

		Mean Square			
Environment	F.V.	GL	Productivity	Cycle	Plant Height
1	Treatments	44	453959.87*	35.49*	224.44*
	Genotypes	34	402971.91*	23.66*	174.43*
	Common checks	9	474789.82*	67.80*	337.09*
	G X Witn.	1	2000081.15*	146.75*	911.03*
	Residue	44	17811.28	5.55	34.92
	Overall Average		1155,08	133.69	54.98
	CV%		11.55	1.76	10.75
2	Treatments	30	41214.74*	7.99*	82.16^{NS}
	Genotypes	20	43049.34*	7.72*	97.40^{NS}
	Common checks	9	22849.37^{NS}	9.47*	55.52^{NS}
	G x Witn.	1	169810.89*	0.24^{NS}	16.95^{NS}
	Residue	30	12632.87	0.61	50.70
	Overall Average		288,33	100.71	39.72
	CV%		38.98	0.78	17.93
Combined	Environment	1	27626408.65*	39927.41*	8550.90*
	Common checks	9	239672.30*	50.04*	252.01*
	Witn x Envir.	9	257966.90*	27.22*	140.60^{NS}
	Genot/Envir	54	269667.25*	17.75*	145.90*
	(W vs G)/Envir	2	1084946.02*	73.49*	463.99^{NS}
	Residue	74	15711.92	3.55	41.32
	Overall Average		801.97	120.24	48.75
	CV%		15.63	1.57	13.18

*Significant by test F at a 5% probability level; ^{NS}Nonsignificant by test F at a 5% probability level.

The variable cycle averaged 134, 101 and 120 days, for environment 1, environment 2 and combined analysis, respectively (Table 1). The results were found within the variable average of the crop, determined by Spehar (2007), which lies between 80 and 150 days in Brazilian conditions. The height averages found in the present experiment were 54.98 cm, 39.72 cm and 48.75 cm for environment 1, environment 2 and the combined analysis, respectively (Table 1), showing higher results in environment 1, which was sown on 19 October 2010 in the normal harvest period. The sowing date for which Vasconcelos, Vasconselos et al. (2012) obtained the greater plant height was 18 March, a period currently referred to as interim harvest, where an average plant height of 65 cm was observed in the city of Campo Mourão, Paraná State. In contrast to the great variation observed between environments 1 and 2, the averages obtained for the combined analysis presented results of mean productive performance, growth cycle and plant height closer to the results obtained for the most productive environment (environment 1), characterizing greater possibility of selection for the two environments combined, than if selection was performed in only one of the environments.

The coefficient of variation (CV) represents the residual standard deviation expressed as a percentage of the overall average. The distribution of the CV values of a response variable allows researchers to establish limits of values, which indicate the validity of study results, with low CV values suggesting higher precision. Pimentel-Gomes (2009), studying the coefficients of variation obtained in agricultural assays, classifies these CVs as follows: low (CV below 10%), medium (CV between 10 and 20%), high (CV between 20 and 30%), very high (CV values above 30%). Based on this classification, the coefficients of variation found for the productivity variable presented medium indexes (11.54 and 15.63%) for environment 1 and the combined analysis, respectively, and a very high index (38.98%) for environment 2 (Table 1). The CV values observed for plant height (10.75% in environment 1, 17.93% in environment 2 and 13.19% in the combined analysis - Table 1), coincide with the CV values obtained by Vasconcelos, Vasconselos et al. (2012), which are classified as medium values, showing the reliability of data, as well as the precision of the experiments.

The growth cycle character presented low coefficients for all environments: 1.76% in environment 1, 0.78% in environment 2 and 1.57% in the combined analysis. The highest coefficients of variation were found in environment 2 for productivity (38.98%), and for plant height (17.93%) (Table 1).

In the test of averages, the productivity averages were weighted by the common checks according to Cruz (2006), and genotypes N24 and N08 surpassed the averages of the 46 other genotypes, presenting productivity mean values of 1,446.2 and 1,428.9 kg ha^{-1}, respectively. The real productivity values are presented in Table 2, where it is possible to observe that the maximum productivity average (2,501.3 kg ha^{-1}) was achieved by genotype N08. Vasconcelos, Vasconselos et al. (2012), who analyzed the effect of different sowing periods in quinoa genotypes in the city of Campo Mourão, observed a superior maximum average value of 1,884.4 kg ha^{-1}.

The genotypes N13, N14, N25, P08 and Preta 06 presented cycle averages that surpassed 13 other genotypes, having a higher number of days when compared to the others (Table 2). Possibly, the averages of those genotypes that are superior to others exhibit the longest growth cycles. The plant height variable, on the other hand, presented a single genotype (genotype N19) that surpassed 22 other genotypes, and this genotype had an average height of 68.27 cm (Table 2).

Table 2. Mean values of weighted productivity, real productivity, growth cycle and plant height for quinoa genotypes cultivated in the crop years 2010/11 and 2011/12.

Genotypes	Productivity ($Kg\ ha^{-1}$)	Real productivity ($Kg\ ha^{-1}$)	Cycle (days)	Plants Height (cm)
N24	1,446.23 a	1,981.50	117.25 abcd	53.82 abc
N08	1,428.93 a	2,045.03	111.50 d	52.62 abc
N31	1,130.43 ab	1,665.70	117.75 abcd	54.97 abc
N33	1,115.53 abc	1,650.80	118.25 abcd	58.22 ab
P47	1,084.45 abc	549.18	117.75 abcd	56.49 abc
N03	1,053.79 abc	1,053.79	111.25 d	43.25 bcd
N10	1,053.48 abcd	1,588.75	111.25 d	48.12 abc
N19	1,035.97 abcd	1,035.97	121.25 abc	68.28 a
N13	1,024.23 abcd	1,559.50	124.25 a	33.92 bcd
P56	1,010.35 abcde	475.08	116.75 abcd	55.50 abc
N42	990.47 abcde	990.47	114.50 cd	46.56 abc
N23	978.23 abcde	1,513.50	122.75 ab	50.77 abc
P52	976.05 abcde	440.77	117.75 abcd	55.71 abc
N06	924.92 abcde	924.92	117.00 abcd	54.15 abc
P48	920.14 abcdef	384.87	116.75 abcd	52.43 abc
P32	907.29 abcdef	372.02	117.75 abcd	56.57 abc
P50	894.92 abcdef	359.65	116.25 abcd	43.35 bcd
P05	870.88 abcdef	335.60	118.25 abcd	50.89 abc
N20	843.40 abcdef	308.13	116.25 abcd	50.09 abc
P58	835.27 abcdef	300.00	116.25 abcd	45.90 abcd
N05	807.96 bcdef	807.96	115.25 bcd	42.91 bcd
P37	792.58 bcdef	257.31	116.75 abcd	53.38 abc
N32	790.98 bcdef	1,326.25	120.75 abc	45.32 abcd
P62	782.10 bcdef	246.82	112.75 cd	54.19 abc
N39	781.38 bcdef	1,316.65	117.75 abcd	45.92 abcd
N44	763.08 bcdef	1,298.35	119.75 abc	43.67 abcd
Real	757.00 bcdef	757.00	121.25 abc	43.45 bcd
P3	750.18 bcdef	1,285.45	119.75 abc	51.07 abc
Casa + aberta	747.93 bcdef	1,283.20	121.25 abc	46.62 abc
P61	746.41 bcdef	211.14	113.25 cd	53.24 abc
N29	744.83 bcdef	1,280.10	119.75 abc	46.07 abcd
Casa	743.78 bcdef	1,279.05	117.75 abcd	47.62 abc
Casa + fechada	743.05 bcdef	743.04	113.50 cd	47.43 abc
N34	739.08 bcdef	1,274.35	117.75 abcd	50.47 abc
P60	719.61 bcdef	184.33	111.25 d	58.66 ab
P54	715.26 bcdef	179.98	115.75 abcd	47.12 abc
N30	704.68 bcdef	1,239.95	119.75 abc	50.17 abc
P34	704.43 bcdef	169.15	116.75 abcd	38.27 bcd
P15	693.62 bcdef	158.35	112.75 cd	60.13 ab
N11	676.19 bcdef	676.19	117.75 abcd	50.91 abc
N36	675.28 bcdef	140.00	116.25 abcd	47.94 abc
N14	672.93 bcdef	1,208.20	124.25 a	48.92 abc
P21	654.43 bcdef	119.15	117.75 abcd	37.69 bcd
N 09	644.13 bcdef	1,179.40	121.25 abc	52.62 abc
N41	631.13 bcdef	1,166.40	119.75 abc	32.67 bcd
P57	606.52 bcdefg	71.25	116.75 abcd	43.83 abcd
N22	576.11 cdefg	576.11	117.25 abcd	57.48 ab
N 01	561.38 cdefgh	1,096.65	118.25 abcd	47.52 abc
P38	558.62 cdefgh	23.34	116.75 abcd	39.02 bcd
N35	553.18 cdefgh	1,088.45	119.75 abc	28.57 bcd
P14	545.57 cdefgh	10.29	113.75 cd	40.82 bcd
N15	533.53 cdefgh	1,068.80	110.75 d	46.17 abcd
N43	503.58 cdefgh	1,038.85	116.25 abcd	39.57 bcd
N 07	493.23 cdefgh	1,028.50	119.75 abc	38.97 bcd
N38	492.33 cdefgh	1,027.60	119.75 abc	30.62 bcd
N45	439.88 defgh	975.15	118.25 abcd	50.42 abc
N21	382.88 efghi	918.15	117.75 abcd	45.97 abcd
N25	382.63 efghi	917.90	124.25 a	41.92 bcd
N27	380.28 efghi	915.55	116.25 abcd	35.97 bcd
Guara 15	314.78 fghi	850.05	119.75 abc	40.87 bcd
Preta 06	-4.12 ghij	531.15	124.25 a	13.92 d
Orange 01	-51.52 hij	483.75	110.75 d	35.92 bcd
N12	-198.12 ij	337.15	119.75 abc	38.82 bcd
Orange 02 & 07	-372.87 j	162.40	121.25 abc	25.62 cd
P8	-445.42 j	89.85	124.25 a	35.92 bcd
P7	-497.52 j	37.75	117.25 abcd	48.92 abc

The same letters used in a column indicate that there is no significant difference by Tukey's Test at 5% significance.

The coefficient of genetic variation, which corresponds to the deviation of the genetic pattern and is expressed as a percentage of the average, is an indicator of the relative magnitude of change in a character, which can be obtained through selection in an improvement program. The relation CVg CVe^{-1} and the heritability coefficient in the narrow sense (h_r^2) are parameters that measure the reliability of the data and the success in the selection of superior genotypes. The estimations of high coefficients of heritability are associated with a greater genetic variability, greater selective accuracy (Cargnelutti Filho, Storck, & Ribeiro, 2009) and greater possibilities for success in selecting lineages with higher productivity of grain. The relations among the coefficient of genetic variation, the coefficient of environmental variation (CVg CVe^{-1}) and the heritability coefficient (h^2) are measures of environmental influence on the expression of phenotypes (Faluba et al., 2010).

For grain productivity, the estimation of the ratio CVg CVe^{-1} was 3.29 and 1.1, and the estimated heritability coefficient was 95.58 and 70.65%, for the individual analyses of environment 1 and environment 2, respectively. For the combined analysis, estimated values of 2.97 (ratio CVg CVe^{-1}) and 94.17% (h_r^2) were observed for grain productivity (Table 3). These observations suggest that the quinoa from the combined selection shows better condition because the situation is favorable once the relationship CVg CVe^{-1} achieves values close to or superior to 1.00 (Faluba et al., 2010).

Table 3. Genetic parameters estimated from the individual and combined analysis of the data obtained for productivity, growth cycle and plant height in experiments performed in Marechal Cândido Rondon, Paraná State, in the crop years 2010/11 and 2011/12.

Variable	Condition	σ^2g	h_r^2	CVg	Ratio CVg CVe^{-1}
Productivity	Envir. 1	192,580.30	95.58	40.78	3.29
	Envir. 2	15,208.24	70.65	48.90	1.10
	Comb.	126,977.70	94.17	46.45	2.97
Cycle	Envir. 1	9.05	76.74	2.24	1.28
	Envir. 2	3.55	92.04	1.87	2.41
	Comb.	7.10	80.01	2.19	1.40
Plants Height	Envir. 1	69.75	79.98	15.68	1.41
	Envir. 2	23.35	47.95	12.28	0.68
	Comb.	52.29	71.68	15.05	1.14

σ^2g: genotypic variance; h_r^2: heritability coefficient in narrow sense (%); CVg: coefficient of genetic variation (%); CVe: coefficient of experimental variation (%).

The observed results differ from the ones obtained by Vasconselos et al. (2012), who, in an experiment performed with soybean genotypes, aimed at choosing the best genotypes for selection, found greater values of the coefficient of genotypic determination (H^2) for each environment (varying from 49 to 66% in the tests performed for germination in each environment individually) than the value obtained by the combined variance analysis (43%), indicating a better possibility of gain if the selection is performed for each environment, separately.

The estimation of the heritability coefficient in the narrow sense (h_r^2) is a very useful parameter for breeders because it allows one to predict the possibility of success with the selection, as it reflects the proportion of phenotypic variation that can be inherited; in other words, the heritability coefficient measures the reliability of the phenotypic value as an indicator of genotypic value (Vasconcelos et al., 2012).

The obtained data for coefficient of heritability (h_r^2) within the growth cycle character were of 76.74 and 92.04%, and the ratio CVg CVe^{-1} reached results of 1.28 and 2.41 for environment 1 and environment 2, respectively. The estimation of the coefficient of heritability (h_r^2) for the combined analysis (80.01%), as well as the value obtained for the ratio CVg CVe^{-1} (1.4), showed a result closer to the one obtained for environment 1 (Table 3).

As cited by Spehar and Santos (2005), the plant height variable is directly related to the obtained grain yield. In contrast with this affirmation and the observed results, the pattern of values for the coefficient of heritability is similar to the one observed for the productivity variable; thus, the h_r^2 value resulting m the combined analysis is closer to the highest h_r^2 value that belongs to the first environment (Table 3).

When it is desirable to maximize the gains for a series of environments, it is appropriate to acquire estimates using the components of genetic variance from the combined analysis. The estimation of genetic variance was greater than the estimation of environmental variance, which is reflected in the higher estimation of heritability (Table 3).

Conclusion

The N24 and N08 genotypes were the only ones to surpass the productivity of the 46 other genotypes, presenting productivity values of 1446,23 and 1428,93 kg ha^{-1}, with a growth cycle of 117 and 111 days, respectively.

For the productivity and plant height variables, the selection of quinoa genotypes in which both environments are considered together, based on the results obtained for the heritability per place, may produce greater gains when compared to the selection in individual environments.

The determined heritability values demonstrate the possibility of genetic gain with the combined selection that involved both environments.

Acknowledgements

We wish to acknowledge the Araucaria Foundation and the State University of West Paraná (Unioeste) for the fellowship granted and for the conditions for developing this research.

References

Adolf, V. I., Shabala, S., Andersen, M. N., Razzaghi, F., & Jacobsen, S. (2012). Varietal differences of quinoa's tolerance to saline conditions. *Plant Soil, 357*(1), 117-129.

Cargnelutti Filho, A., Storck, L., & Ribeiro, N. D. (2009). Medidas da precisão experimental em ensaios com genótipos de feijão e de soja. *Pesquisa Agropecuária Brasileira, 44*(10), 1225-1231.

Cruz, C. D. (2013). Genes: um pacote de software para análise em estatística experimental e genética quantitativa. *Acta Scientiarum. Agronomy, 35*(3), 271-276.

Cruz, C. D. (2006). *Programa Genes: aplicativo computacional em genética e estatística*. Viçosa, MG: UFV.

Cruz, C. D., Regazzi, A. J., & Carneiro, P. C. (2012). *Modelos biométricos aplicados ao melhoramento genético* (4a ed.). Viçosa, MG: UFV.

Faluba, J. S., Miranda, G. V., Lima, R. O., Souza, L. V., Debem, E. A., & Oliveira, A. M. C. (2010). Potencial genético da população de milho UFV7 para o melhoramento em Minas Gerais. *Ciência Rural, 40*(6), 1250-1256.

Gonzalez, J. A., Konishi, Y., Bruno, M., Valoy, M., & Prado, F. E. (2012). Interrelationships among seed yield, total protein and amino acid composition of ten quinoa (*Chenopodium quinoa*) cultivars from two different agroecological regions. *Journal of the Science of Food and Agriculture, 92*(6), 1222-1229.

Pimentel-Gomes, F. (2009). *Curso de estatística experimental* (15a ed.). Piracicaba, SP: Fealq.

Ruiz, K. B., Biondi, S., Oses, R., Acuña-Rodríguez, I. S., Antognoni, F., Martinez-Mosqueira, E. A., ... Molina-Montenegro, M. A. (2014). Quinoa biodiversity and sustainability for food security under climate change. A review. *Agronomy for Sustainable Development, 34*(2), 349-359.

Silva, R. G., Chaves, M. C. L., Arnhold, E., & Cruz, C. D. (2009). Repetibilidade e correlações fenotípicas de caracteres do fruto de bacuri no Estado do Maranhão. *Acta Scientiarum. Agronomy, 31*(4), 587-591.

Silva, A. Z., Fuentes, F., Zamora, P., Jacobsen, S., & Schwember, A. R. (2014). Breeding quinoa (*Chenopodium quinoa* Willd.): potential and perspectives. *Molecular Breeding, 34*(1), 13-30.

Spehar, C. R. (2007). *Quinoa: alternativa para diversificação agrícola e alimentar*. Brasília, DF: Embrapa Cerrados.

Spehar, C. R., Rocha, J. E. S., & Santos, R. L. B. (2011). Desempenho agronômico e recomendações para cultivo de quinoa (BRS Syetetuba) no cerrado. *Pesquisa Agropecuária Tropical, 41*(1), 145-147.

Spehar, C. R., & Santos, R. L. B. (2005). Agronomic performance of quinoa selected in the Brazilian savannah. *Pesquisa Agropecuária Brasileira, 40*(6), 609-612.

Vasconcelos, E. S., Egewarth, J. F., Oliveira, G. A., & Piano, J. T. (2013). Características agronômicas de genótipos de quinoa. *Scientia Agraria Paranaensis, 12*(suplemento), 371-376.

Vasconcelos, E. S., Reis, M. S., Sediyama, T., & Cruz, C. D. (2012). Estimativas de parâmetros genéticos da qualidade fisiológica de sementes de genótipos de soja produzidas em diferentes regiões de Minas Gerais. *Semina: Ciências Agrárias, 33*(1), 65-76.

Vasconcelos, F. S., Vasconcelos, E. S., Balan, M. G., & Silvério, L. (2012). Desenvolvimento e produtividade de quinoa semeada em diferentes datas no período safrinha. *Ciência Agronômica, 43*(3), 510-515.

9

Genetic parameter estimates and identification of superior white maize populations

Sara Regina Silvestrin Rovaris[1*], André Luís Bombonato Oliveira[1], Eduardo Sawazaki[1], Paulo Boller Gallo[2] and Maria Elisa Ayres Guidetti Zagatto Paterniani[1]

*[1]Instituto Agronômico em Campinas, Centro de Grãos e Fibras, Fazenda Santa Elisa, Av. Theodureto de Almeida Camargo, 1500, Cx. postal 28, 13012-970, Vila Nova, Campinas, São Paulo, Brazil. [2]Agência Paulista de Tecnologia dos Agronegócios, Pólo Regional de Desenvolvimento do Nordeste Paulista, Mococa, São Paulo, Brazil. *Author for correspondence. E-mail: sarasr.rovaris@gmail.com*

ABSTRACT. In Brazil, there is a shortage of white maize cultivars and genetic studies for special maize breeding programs. This study aimed to identify populations and promising hybrid white maize for main agronomic traits and grits processing and to estimate the genetic parameters of parents and heterosis. In the 2012/13 growing season, fifteen hybrids were obtained by complete diallel crosses, and six parental and commercial check varieties were evaluated for: female flowering (FF), ear height (EH), grain yield (GY), ear length (EL), volumetric mass (VM) and grits processing (GP) in two locations in São Paulo State, Campinas and Mococa, using a randomized block design. Analyses of variance were carried out, and diallel crosses were performed using the Gardner and Eberhart model. The populations P3 and P6 stood out because of the estimated effects of the parents and of heterosis; the studied characters are promising for obtaining new lines and forming composites. For GP, the treatments showed no differences, implying the need to introduce new sources of germplasm.

Keywords: *Zea mays* L., complete diallel, Gardner and Eberhart, grits production.

Introduction

Maize is a versatile crop with wide genetic variability and the ability to successfully develop in tropical, subtropical and temperate regions under different agro-climatic conditions. The cultivated area of maize and its production have an upward growing trend globally, especially with the introduction of hybrids due to high yield potential (Izhar & Chakraborty, 2013).

In Brazil, there is increasing production of specialty corns, such as popcorn, mini-corn, sweet corn and white corn for grits because it adds value to the product, generates higher profits for the farmer and represents specific market niches. In particular, a 60 kg bag of white maize for grits can reach twice the commercial value of a 60 kg bag of maize, which provides a considerable economic advantage to the producer.

The so-called "canjica", a dessert appreciated in many regions of Brazil, originated in São Paulo in 1710 due to the restriction of the use of salt, which was monopolized by some of the king's agents at that time. This restriction persuaded people to prepare this nutritious and sweet dish, which was adapted from indigenous food (Ferreira, 2002).

The availability of registered cultivars for the production of corn grits is very scarce, highlighting the need to develop new hybrids for this sector. In 2014, among 467 cultivars of grain corn available for commercialization, only five white maize cultivars were recommended for production (Bignotto et al., 2015).

Specialty corn breeding programs aim to develop more productive and adapted cultivars to supply the market demand. However, these programs will only be successful if they are effective in the selection of populations to be used for crossings that result in hybrids that possess traits of interest (Rodrigues, Pinho, Albuquerque, Faria Filho, & Goulart, 2009).

The use of diallel crosses is one of the methods available in plant breeding that allows the selection of parents to improve the performance of the progeny (Ramalho, Abreu, & Santos, 2001). Moreover, diallel crosses are used to analyze the genetic design, providing estimates of useful parameters that allow for the selection of the best genitors to obtain hybrids, and to understand the genetic effects of a trait (Gonçalves et al., 2014; Melani & Carena 2005; Cruz, Regazzi, & Carneiro, 2004). This method has been used to evaluate several agronomic traits in hybrids obtained from common maize populations (Tonette & Carena, 2014; Bernini, & Paterniani, 2012; Doná, Paterniani, Gallo, & Duarte, 2011) and in specialty corns, such as popcorn (Cabral et al., 2015; Solalinde et al., 2014) and sweet corn (Assunção et al., 2010; Rodrigues et al., 2009), allowing the identification of the best parents and their hybrid combinations for the evaluated traits.

Cruz et al. (2004) reported several types of diallel analysis, such as balanced, partial, circulating, incomplete and unbalanced, highlighting that the most applicable type was the balanced diallel proposed by Hayman (1954), Griffing (1956) and Gardner and Eberhart (1966). The last method provides detailed information on the potential of populations *per se* as well as the heterosis manifested in the hybrids. This method presents the advantage of evaluating the specific effects of varieties and varietal heterosis separate from the effects of their general combining ability (Gardner & Eberhart, 1966).

In diallel crosses that are used to evaluate populations in Hardy-Weinberg equilibrium, the effects v_i can be translated as the performance of a given population "i" compared to the average of other populations, and the h_i effect is the contribution, in terms of heterosis, from crosses where "i" variety is involved. According to Bernini and Paterniani (2012), this method is superior to the others when open pollinated varieties are crossed, since it provides useful genetic information to the breeder from the mean values of the population and not from the variances.

This study aimed to identify populations and promising hybrids of white maize for main agronomic traits and grits processing and to estimate genetic parameters of the parents, as well as the heterosis and its components.

Material and methods

In the growing season 2011/12, six populations of white maize were crossed under a full diallel design (Table 1), thereby obtaining 15 intervarietal hybrids. The parent populations and their hybrids were evaluated in the growing season 2012/2013 at the Central Experimental Center of the Agronomic Institute (IAC) in Campinas, São Paulo State (latitude, 22° 54'S; longitude, 47° 3'W; altitude, 600 m) and at the Regional Pole of Paulista Northeast (APTA) in Mococa, São Paulo State (latitude, 21° 28'S; longitude, 47° 01'W; altitude, 665 m). The experiment was installed under a randomized block design with four replicates for agronomic traits and two replicates for grits processing. The plot consisted of four rows 5 m long, spaced apart 0.85 m between rows and 0.2 m between plants, with two plants per hole, totaling 25 plants per row after thinning. To collect agronomic data, the two center rows were used, and, in the two outside rows, industrial characters, such as grits processing yield, were measured. In both trial locations, fertilization was performed using 350 kg ha^{-1} of 8-28-16 (NPK), and the topdressing fertilization was performed using 150 kg ha^{-1} of N supplied as urea.

Table 1. Description of the white maize genotypes used for obtaining the intervarietal hybrids for this study.

Population (origin)	Company/Institution	Cultivar original type	Grain type
P1 (F_2 IPR 127)	IAPAR	Single-cross hybrid	Flint
P2 (IAC Nelore)	IAC	Intervarietal hybrid	Flint
P3 (F_2 Murano)	Syngenta México	Three-way cross hybrid	Semi Flint
P4 (Al Bianco)	CATI-DSMM	Variety	Semi Flint
P5 (RS 21)	FAPEAGRO	Variety	Semi Flint
P6 (F_2 IPR 119)	IAPAR	Double-cross hybrid	Semi Flint

The traits evaluated were as follows: female flowering (FF), measured in days after sowing when 50% of the plants in the plot presented the stigma-style off the ear; ear height (EH), sampled at 5 competitive plants per plot, taken from ground level to the insertion of the main ear, and expressed in centimeters (cm); grain yield (GY), total grains resulting from the mechanical threshing of ears of useful lines in the plot (corrected to 14% moisture and ideal stand), expressed in kg ha^{-1}; ear length (EL), obtained in 5 husked ears randomly sampled within each parcel and expressed in centimeters; volumetric mass (VM), measured in a volumetric mass determiner indicating the density of whole grains expressed in kilograms per 100 liters.

The collection of grits processing (GP) data was carried out in grams and then converted into percentage obtained in percentage (%) from 10 kg samples from each plot for two replications. For grits processing (GP), each sample was processed in a gritting processor for about 2 minutes and 45 seconds depending on the humidity of the sample. After chiseling, the resulting material was separated into bran and grits by the shaking motion of the gritting processor, using a Tyler5 sieve available on the machine. The sample was weighed on an electronic scale. GP assessment was conducted at the Experimental Farm of Iguatemi of the State University of Maringa (UEM) in Maringa, Paraná State, Brazil.

Grits processing data were transformed to percentage according to the formula below:

$$GP = \frac{mm}{ma} \times 100,$$

where GP is grits processing; mm: degerminated maize mass obtained by the gritting processor, expressed in grams (g); ma: total sample mass, expressed in grams (g).

Individual and joint analyses of variance were carried out, assuming the model as fixed. Means were compared using the Scott-Knott test (1974) at 5% probability.

The complete diallel analysis was conducted according to the model proposed by Gardner and Eberhart (1966):

$$Y_{ij} = \mu + (v_i + v_j)/2 + \theta(\bar{h} + h_i + h_j + s_{ij}) + \bar{e}_{ij},$$

where: Y_{ij} : average value observed in a parent ($i = j$) or in a hybrid combination (i); μ: mean of varieties; v_i: the effect of the i variety; v_j: the effect of the j variety; $\theta=0$, when $i=j$, and $\theta=1$ when $i \neq j$; h: average heterosis effect; h_i: the effect of the variety i heterosis; h_j: the effect of the variety j heterosis; s_{ij}: the effect of specific heterosis; and $\bar{e}_{ij}$: average experimental error.

The biometric analyses were performed using the software package Genes 2006 (Cruz, 2006).

Results and discussion

The joint analyses of variance detected significant differences with the F test ($p < 0.01$) for treatments and genotypes for most of the evaluated traits, indicating differences in the performance of the parent, hybrid and commercial check varieties. However, for volumetric mass (VM) and grits processing (GP), no significant differences were observed (Table 2). The effects of T x L and G x L interactions were significant only for GY and VM, so their means are presented by location (Table 2). For all other traits, significance was not observed; therefore, the discussion of the results is based on the mean of the experiments in both locations. Significance in these interactions implies that genotypes with good results at a given location may not have the same satisfactory results in a different location (Cruz & Regazzi, 1997).

The coefficients of variation obtained in this study ranged from low to medium, indicating that the experimental design was satisfactory in controlling environmental variation and allowed reliable data to be obtained (Fritsche-Neto, Vieira, Scapim, Miranda, & Rezende, 2012; Gomes, 2000).

The means for FF, EH, EL, VM and GP of the genotypes at two locations are presented in Table 3, along with their groupings by the Scott-Knott test ($p < 0.05$).

For FF, the formation of two groups (b and c) was observed between the hybrids, with means ranging from 55 to 58 days. Eight hybrids stood out as earlier in female flowering (group b) with no statistical differences from the check variety 'Al Bianco'. For the EH trait, the means of the hybrids ranged from 135.79 cm (P1xP3) to 180.27 cm (P4xP5), marking hybrids P1xP3 (135.79 cm) and P2xP3 (137.55 cm) due to the lower means presented. For GY, means were grouped into four distinct groups only at Mococa (a, b, c, and d). The most promising hybrids showed GY percentages ranging from 130% (P3xP6) to 109.4% (P2xP3) compared to the check varieties. For VM, the formation of two distinct groups was observed only at Campinas. Eleven hybrids significantly differed from the check variety, with means that ranged from 83.01 to 81.47%.

Table 2. Mean squares (MS) in the joint analysis of variance for female flowering (FF), ear height (EH), grain yield (GY), ear lenght (EL), volumetric mass (VM) and grits processing (GP) of hybrids and parental populations of white maize in Campinas and Mococa, São Paulo State. Season 2012/13.

Source of variation	df	Mean Squares (MS)					
		FF (days)	EH (cm)	GY[a] (kg ha^{-1})	EL (cm)	VM (kg 100 L^{-1})	GP (%)
Blocks	6	0.88	1280.07	4130.91	0.34	1.19	4.8
Treatments (T)	21	13.25 **	1771.41 **	5220.17 **	4.85 **	2.72	6.22
Genotypes (G)	20	13.9 **	1859.83 **	5455.78 **	4.62 **	2.69	6.48
Groups (Gr)	1	0.09	2.95	508.00	9.39 **	3.26 **	1.12 **
Environments (E)	1	2212.36 **	23360.30 **	10788.48	10.86 **	571.6 **	155.69 *
T x E	21	1.95	172.31	1719.17 **	1.18	4.29 **	4.25
G x E	20	2.03	172.27	1742.38 **	1.23	4.51 **	4.46
Gr x E	1	0.24	173.19	1254.91	0.04	0.04	0.09
Error	126	1.88	169.95	693.34	1.21	1.40	5.20
Total	175						
General means		57	147.32	7474	17.38	79.7	79.14
Population means		58	147.80	7336	17.20	79.57	78.98
Hybrids means		57	149.60	7806	17.40	79.65	78.94
CV (%)		2.39	8.85	11.14	6.32	1.48	2.88

[a] MS´s multiplied by 10^{-3}; *, ** significant at 5% and 1% probability.

Table 3. Means for female flowering (FF), ear height (EH), grain yield (GY), ear lenght (EL), volumetric mass (VM) and grits processing (GP) of the hybrids, parental and check variety of the white maize populations in Campinas and Mococa, São Paulo State. Season 2012/13.

Hybrids	FF (days)	EH (cm)	GY (kg ha^{-1})		EL (cm)	VM (kg 100 L^{-1})		GP (%)
	Average		Campinas	Mococa	Average	Campinas	Mococa	Average
P1xP2	56 c	143.41 b	6893 a	7259 c	17.03 b	81.6 a	77.91 a	79.96 a
P1xP3	55 c	135.79 b	7538 a	9173 a	16.69 b	83.01 a	77.32 a	76.73 a
P1xP4	57 c	143.54 b	7732 a	8272 b	17.76 a	81.60 a	77.73 a	78.80 a
P1xP5	57 c	158.54 a	7758 a	7542 c	16.39 b	82.31 a	78.13 a	79.63 a
P1xP6	58 b	138.61 b	7597 a	9588 a	17.61 a	81.56 a	77.00 a	79.39 a
P2xP3	56 c	137.55 b	7319 a	8294 b	17.45 a	82.28 a	78.04 a	78.74 a
P2xP4	57 c	148.93 b	7250 a	7774 c	18.16 a	82.40 a	78.29 a	79.99 a
P2xP5	59 b	168.08 a	7516 a	6088 d	19.92 a	81.60 a	76.88 a	76.56 a
P2xP6	58 b	146.25 b	7366 a	8176 b	17.43 a	80.11 b	76.86 a	80.25 a
P3xP4	56 c	138.89 b	7674 a	9014 a	17.35 a	81.81 a	76.40 a	77.69 a
P3xP5	58 b	150.31 b	7176 a	7316 c	16.73 b	81.37 a	78.24 a	78.94 a
P3xP6	56 c	139.76 b	8437 a	9860 a	16.37 b	80.67 b	78.02 a	79.85 a
P4xP5	58 b	180.24 a	7375 a	7613 c	17.30 a	81.47 a	78.73 a	78.76 a
P4xP6	58 b	152.35 b	7919 a	7970 c	17.37 a	80.89 b	78.25 a	79.91 a
P5xP6	58 b	161.40 a	7519 a	7345 c	17.76 a	79.63 b	79.27 a	78.85 a
Populations								
P1	57 c	127.67 c	5900 b	7237 c	16.25 b	82.56 a	78.47 a	79.16 a
P2	57 c	148.51 b	6625 b	7156 c	17.51 a	81.24 a	75.94 a	79.28 a
P3	55 c	113.56 c	5810 b	7733 c	15.8 b	82.90 a	78.12 a	81.29 a
P4	58 b	150.64 b	7817 a	6545 d	18.38 a	82.12 a	77.86 a	77.4 a
P5	61 a	173.16 a	5506 b	5004 e	17.62 a	80.36 b	78.90 a	78.88 a
P6	58 b	137.10 b	6382 b	7497 c	18.04 a	79.2 b	78.96 a	81.05 a
Check Variety								
Al Bianco	57 c	146.73 b	7859 a	7580 c	18.44 a	82.2	78.45	79.66 a
General Mean	57	147.32	7018	7196	79.70	81.5	77.9	79.14
CV(%)	2.39	8.85	11.07	11.19	1.48	1.08	1.82	2.88

P1: F_2 IPR 127, P2: IAC Nelore, P3: F_2 Murano, P4: Al Bianco, P5: RS 21, P6: F_2 IPR 119. Means followed by the same letter do not differ by the Scott-Knott test ($p < 0.05$).

Although not statistically different from some hybrids and parents, the hybrid P3xP6 stood out for the best performance of the traits: FF, EH, GY, and VM. However, for grits processing (GP), there was no separation of genotypes in the groups, precluding the identification of the most promising parent or hybrid. GP means presented in this paper do not differ with those presented by Castro, Naves, Oliveira, and Froes (2009), who obtained GP values ranging from 70 to 80%, and with those presented by Mestres, Matencio, and Dramé (2003), who observed values ranging from 78-80% for GP.

In the joint diallel analysis (Table 4), significant differences in the effects of treatments (T) and populations (P) were observed, indicating that there is variability among treatments for FF, EH, GY, VM, and EL traits. Heterosis (h) was significant only for GY and EL, implying that there is heterosis manifestation in the hybrids for these two traits.

The significance of the mean squares of heterosis effects on average heterosis ($\bar{h}$) for FF, EH, and GY indicates that there is enough genetic diversity among the parents, resulting in a favorable situation for the application of a breeding program in order to obtain hybrids and to detect heterogeneity in the values of gene frequencies (Kvitschal et al., 2004).

The GY trait showed significance in the heterosis parental effects (h_i), which reveals the existence of variation in the contribution of each parent to the performance of the hybrid, indicating dispersion of allele frequencies in the population (Oliveira et al., 2004). Similar h_i results for GY were found by other authors and enabled the selection of the best parents and their respective hybrids (Tonette & Carena, 2014; Bernini & Paterniani, 2012; Doná et al., 2011; Silva & Miranda Filho, 2003; Scapim et al., 2002).

The specific heterosis effects ($\hat{s}_{ij}$) showed significance only for the EL trait, indicating that there are epistasis and dominance effects in trait expression. According to Vencovsky and Barriga (1992), this significance shows that there is specific complementation between pairs of parents in locos with allelic dominance effects, contributing to the better performance of certain hybrid combinations where one can supplement for any deficiencies found in another.

For the other traits, no significant differences were observed for $\hat{s}_{ij}$. This absence of significance suggests the lack of differences among the degrees of complementation with each other, in relation to the frequency of alleles at loci with some dominance (Kvitschal et al., 2004) and predominating additive effects.

In relation to the decomposition in the sum of squares of heterosis in the analysis of variance, it was observed that for GY, the effects of $\bar{h}$, $\hat{h}_i$, and s_{ij} corresponded to 73, 17 and 9% of the total of their effects, respectively. A greater contribution from average heterosis in relation to the other effects was noted and provided evidence of the significant superiority of hybrids compared with the average of the parents.

For plant architecture traits such as plant height and ear height (Table 5), it is interesting to select genotypes that present negative estimates and contribute to the reduction of these traits (Solalinde et al., 2014). Thus, the parental P3 stood out with the lowest estimate of parental effect ($\hat{p}_i$) for FF, and the parents P1, P3 and P6 had the lowest estimates for EH. Studies presented by Dhasarathan, Babu, and Iyanar (2015), Bernini and Paterniani (2012), Doná et al. (2011), Garbuglio and Araújo (2006) and Gorgulho and Miranda Filho (2001) verified the significance of parental effects on EH, concluding that for the control of these traits, there is a predominance of the additive effects and only partial dominance effects.

For the traits GY, EL and VM, it is interesting that the parents show high and positive estimates. The parental P4 (1,476.81 kg ha^{-1}) and P2 (285.21 kg ha^{-1}) in Campinas and P3 (870.73 kg ha^{-1}), P6 (635.04 kg ha^{-1}) and P1 (375.18 kg ha^{-1}) in Mococa, stood out with high estimates for GY, showing their potential *per se* (Table 5). Promising $\hat{p}_i$ estimates for GY were presented by Doná et al. (2011), who evaluated intervarietal hybrids of maize and saw yield amounts of 1,388 kg ha^{-1} in the P3 genotype (DAS 2B710). Bernini and Paterniani (2012) observed yields of 901 kg ha^{-1} in Pop. 15, which are superior values to those found in this work.

The EL estimates of the parental effects ($\hat{p}_i$) ranged from -1.46 to 1.11 cm, and the parents P4 (1.11 cm) and P6 (0.76 cm) stood out with the highest estimates. For VM, the parents with highest $\hat{p}_i$ estimates in both locations were P1 and P3 (Table 5).

Table 4. Mean squares (MS) in the diallel analysis for female flowering (FF), ear height (EH), grain yield (GY), ear lenght (EL), volumetric mass (VM) and grits processing (GP) of hybrids and parental populations of white maize in Campinas and Mococa, São Paulo State. Season 2012/13.

Source of variation	df	MS					
		FF (days)	EH (cm)	GY[a] (kg ha^{-1})	EL (cm)	VM (kg L^{-1})	GP[b] (%)
Treatments (T)	20	13.90 **	1859.88 **	5455.77 **	4.62 **	2.69 *	6.48
Populations (P)	5	49.46 **	6594.54 **	8211.36 **	11.7 **	5.79 **	8.24
Heterosis (h)	15	2.69	281.66	4537.24 **	2.26 *	1.66	5.89
Mean heterosis ($\bar{h}$)	1	9.13 *	2088.53 **	49776.48 **	0.27	0.17	7.14
Parental heterosis (h_i)	5	1.08	95.41	2324.36 **	1.70	1.44	6.41
Specific heterosis (s_{ii})	9	2.87	184.37	740.04	2.78 *	1.95	5.46
Environments (E)	1	2258.70 **	23143.8 **	11887.92 **	10.2 **	543.74 **	150.10 **
T x E	20	2.27	172.26	1742.39 **	1.23	4.51 **	4.46
P x E	5	1.87	159.08	4957.35 **	1.41	13.29 **	2.54
h x E	15	2.40	176.66	670.73	1.18	1.58	5.1
$\bar{h}$ x E	1	4.23	296.72	1.63	1.15	0.91	14.81
h_i x E	5	1.99	146.58	969.15	0.76	0.72	6.26
s_{ii} x E	9	2.42	180.02	579.28	1.41	2.14	3.38
Combined error	120	1.94	177.03	704.97	1.23	1.39	5.28
Mean		57	147.34	7461.98	17.30	79.67	79.12
CV (%)		2.43	9.03	11.25	6.40	1.48	2.90

[a] MS´s multiplied by 10^{-3} *, ** significant at 5% and 1% probability [b]: Residual degrees of freedom = 42.

The parental P2 was the only one to show positive $\hat{p}_i$ estimate results for GY in both locations, Campinas (285.21 kg ha^{-1}) and Mococa (293.88 kg ha^{-1}), showing its potential for recurrent selection programs and inbred lines extraction. Conrado et al. (2014) evaluated 28 hybrids of white maize and observed results that agree with the use of P2 (IAC Nellore) for breeding programs.

The parents P3 and P6, with its promising $\hat{p}_i$ estimates for FF, EH, GY, EL, and VM, can be subjected to different breeding methods, such as intrapopulation recurrent selection, inbred lines obtainment and reciprocal recurrent selection, in maize breeding programs.

Estimates of parental heterosis (h_i) showed low values for the traits FF and EL in all parental populations (Table 6). Garbuglio and Araújo (2006), evaluating intervarietal hybrids, found estimates with higher values for FF, ranging from -0.7 to 0.6 days. According to Scapim et al. (2006), when there is a low level of heterosis, the parental selection used to make the hybrid should be carried out based on the parental means.

For EH, values ranged from -4.28 cm to 2.71 cm, and P2 parental (-4.28 cm) stood out as having a more negative estimate, indicating that their participation in crosses is promising for reducing ear height. Assunção et al. (2010), evaluating sweet corn cultivars, found $\hat{h}_i$ estimates of AE ranging from -1.889 to 1.556 days, which are lower values than those found in this study.

The parents P3, P6 and P1 stood out with promising $\hat{h}_i$ effects for GY, highlighting the P3 population estimate of 360.46 kg ha^{-1}, characterizing it as the most heterotic within the parental group (Table 6). These values demonstrated that these parents are promising in crosses for hybrids by raising the heterosis for GY. Silva, Amaral Jr, Gonçalves, Freitas Jr, and Ribeiro (2011) evaluated ten popcorn lines and obtained GY estimates that ranged from -492.14 to 525.37 kg ha^{-1}. Doná et al. (2011) observed high $\hat{h}_i$ estimates for GY while evaluating maize F_2 population hybrids in Campinas (1,079 kg ha^{-1}) and Mococa (429 kg ha^{-1}). For VM and GP traits, $\hat{h}_i$ estimates were not presented because there was no significance in the diallel analysis.

Table 6. Estimates of the effects of parental heterosis ($\hat{h}_i$), average heterosis ($\bar{h}$) and estimates standard deviations (sd) for female flowering (FF), ear height (EH), grain yield (GY) and ear lenght (EL) of white maize parental in Campinas and Mococa, São Paulo State. Season 2012/13.

Populations	FF (days)	EH (cm)	GY (kg ha^{-1})	EL (cm)
P1	-0.04	0.05	177.97	-1.00
P2	0.20	-4.28	-659.87	0.23
P3	0.20	2.71	360.46	-1.5
P4	0.20	-0.14	-222.93	1.11
P5	-0.41	1.97	71.35	0.35
P6	-0.16	-0.03	273.02	0.76
sd ($\hat{h}_i$)	0.45	4.29	270.98	0.35
sd ($\hat{h}_i$ - $\hat{h}j$)	0.70	6.65	419.81	0.55
Mean heterosis ($\bar{h}$)	-0.51	7.80	1204.91	0.08
sd ($\bar{h}$)	0.11	10.32	202.78	0.26

As emphasized by Jesus et al. (2008), one of the plant breeding objectives is to find hybrid combinations with differing parental populations and good genetic potential. Thus, if the objective of the breeding program is to obtain hybrids with high grain yields associated with smaller size and higher precocity, P6 parents have $\hat{h}_i$ effects that are more favorable for all studied traits. This fact was demonstrated by Cruz et al. (2004), who reported that crosses involving parental populations with higher $\hat{h}_i$ effects would result in hybrids that are more heterotic.

Table 5. Estimates of the effects of parental populations ($\hat{p}_i$) and estimates standard deviations (sd) for female flowering (FF), ear height (EH), grain yield (GY), ear lenght (EL) and volumetric mass (VM) of white maize parental in Campinas and Mococa, São Paulo State. Season 2012/13.

Populations	FF (days)	EH (cm)	GY (kg ha^{-1})		EL (cm)	VM (kg 100 L^{-1})	
	Average		Campinas	Mococa	Avg.	Campinas	Mococa
P1	-0.8	-14.1	-440.4	375.18	-1.02	1.14	0.42
P2	-0.3	6.74	285.21	293.88	0.23	-0.2	-2.09
P3	-2.8	-28.2	-529.61	870.73	-1.46	1.52	0.08
P4	-0.3	8.86	1476.81	-317.23	1.11	0.74	-0.18
P5	3.25	31.39	-834.14	-1857.6	0.35	-1	0.85
P6	0.75	-4.67	42.13	635.04	0.76	-2.2	0.91
sd ($\hat{p}_i$)	0.63	6.07	366	399	0.5	0.4	0.64
sd ($\hat{p}_i - \hat{p}_j$)	0.98	9.4	567	618	0.78	0.63	0.99

According to the diallel analysis (Table 4), significance in specific heterosis effects ($\hat{s}_{ij}$) was observed only for the EL trait. The $\hat{s}_{ij}$ estimates for EL ranged from -0.55 to 0.93 cm (Table 6). The hybrid combination that showed promising estimates for this trait was P2xP5 (0.93 cm) and was evidence of good genic complementation of parental populations in locos with a dominance effect.

According to the Gardner and Eberhart (1966) model, the best hybrids will be those that have a wider range of beneficial effects for the desired trait. The hybrid P3xP6 stood out for FF, EL, GY and VM traits and by collecting the $\hat{p}_i$ and $\hat{h}_i$ effects of both parental populations in the expression of these traits.

Conclusion

Significant effects in heterosis average, parental heterosis and specific heterosis is evidence of the heterotic potential of the parent populations in study.

Parents P3 (F_2 Murano) and P6 (F_2 IPR 119) stood out for $\hat{p}_i$ effects for female flowering, ear length, grain yield and volumetric mass traits, and parents P2 (IAC Nellore) stood out for grain yield traits in both locations, showing its potential to be used *per se* and for obtaining new maize lines.

Parents P3 (F_2 Murano), P6 (F_2 IPR 119) and P1 (F_2 IPR 127) appeared to be promising in hybrid combinations with high $\hat{h}_i$ effects for grain yield. These populations may be used for the synthesis of composite hybrids in breeding programs.

There was a predominance of additive effects in the expression of female flowering, ear height, grain yield and volumetric mass traits, while for the ear length trait, the dominance effects predominated.

For the grits processing trait, the genotypes showed no performance differences, implying the need to introduce new sources of germplasm that contribute to the diversity of this trait.

References

Assunção, A., Brasil, E. M., Oliveira, J. P., Reis, A. J. S., Pereira, A. F., Bueno, L. G., & Ramos, M. R. (2010). Heterosis performance in ndustrial and yield components of sweet corn. *Crop Breeding and Applied Biotechnology*, *10*(3), 183-190.

Bernini, C. B., & Paterniani, M. E. A. G. Z. (2012). Estimativas de parâmetros de heterose em híbridos de populações F_2 de milho. *Pesquisa Agropecuária Tropical*, *42*(1), 56-62.

Bignotto, L. S., Scapim, C. A., Pinto, R. J. B., Camacho, L. R. S., Kuki, M. C., & Amaral Jr, A. T. (2015). Evaluation of combining ability in white corn for special use as corn grits. *Crop Breeding and Applied Biotechnology*, *15*(4), 258-264.

Cabral, P. D. S., Amaral Jr, A. T., Viana, A. P., Vieira, H. D., Freitas, I. L. J., Vittorazzi, C., & Vivas, M. (2015). Combining ability between tropical and temperate popcorn lines for seed quality and agronomic traits. *Australian Journal of Crop Science*, *9*(4), 256-263.

Castro, M. V. L., Naves, M. M. V., Oliveira, J. P., & Froes, L. O. (2009). Rendimento industrial e composição química de milho de alta qualidade proteica em relação a híbridos comerciais. *Pesquisa Agropecuária Tropical*, *39*(3), 233-242.

Conrado, T., Scapim, C. A., Bignotto, L. S., Pinto, R. J. B., Freitas, I. L. J., Amaral Jr., & Pinheiro, A. C. (2014). Diallel analysis of corn for special use as corn grits: determining the main genetic effects for corn gritting ability. *Genetics and Molecular Research*, *13*(3), 6548-6556.

Cruz, C. D. (2006). *Programa Genes: Biometria. Análise dialélica completo*. Viçosa, MG: UFV.

Cruz, C. D., & Regazzi, A. J. (1997). *Modelos biométricos aplicados ao melhoramento genético* (2a. ed.). Viçosa, MG: UFV.

Cruz, C. D., Regazzi, A. J., & Carneiro, P. C. S. (2004). *Modelos biométricos aplicados ao melhoramento genético*, (3a. ed.). Viçosa, MG: UFV.

Dhasarathan, M., Babu, C., & Iyanar, K. (2015). Combining ability and gene action studies for yield and quality traits in baby corn (*Zea mays* L.) *Sabrao Journal of Breeding and Genetics*, *47*(1), 60-69.

Doná, S., Paterniani, M. E. A. G. Z., Gallo, P. B., & Duarte, A. P. (2011). Heterose e seus componentes em híbridos de populações F_2 de milho. *Bragantia*, *70*(4), 767-774.

Ferreira, A. C. (2002). *A epopeia bandeirante: letrados, instituições, invenção histórica (1870-1940)*. São Paulo, SP: Unesp.

Fritsche-Neto, F., Vieira, R. A., Scapim, C. A., Miranda, G. V., & Rezende, L. M. (2012). Updating the ranking of the coefficients of variation from maize experiments. *Acta Scientiarum. Agronomy*, *34*(1) 99-101.

Gardner, C. O., & Eberhart, S. A. (1966). Analysis and interpretation of the variety cross diall and related populations. *Biometrics*, *22*(3), 439-452.

Garbuglio, D. D., & Araújo, P. M. (2006). Avaliação de híbridos intervarietais de milho por meio de cruzamento dialélico parcial considerado quatro ambientes. *Semina: Ciências Agrárias*, *27*(3), 375-392.

Gomes, F. P. (2000). *Curso de estatística experimental* (14a. ed.). Piracicaba, SP: Livraria Nobel.

Gonçalves, L. S. A., Freitas Jr., S. P., Amaral Jr., A. T., Scapim, C. A., Rodrigues, R., Marinho, C. D., & Pagliosa, E. S. (2014). Estimating combining ability in popcorn lines using multivariate analysis. *Chilean Journal of Agricultural Research*, *74*(1) 10-15.

Gorgulho, E. P., & Miranda Filho, J. B. (2001). Estudo da capacidade combinatória de variedades de milho no esquema de cruzamento dialélico parcial. *Bragantia, 60*(1), 1-8.

Griffing, J. B. (1956). Concept of general and specific combining ability in relation to diallel systems. *Australian Journal of Biological Science, 9*(4), 463-493.

Hayman, B. I. (1954). The Teory and Analisys of Diallel Crosses. *Genetics, 39*(6), 789-809.

Izhar, T., & Chakraborty, M. (2013). Combining ability and heterosis for grain yield and its components in maize inbreds over environments (*Zea mays* L.). *African Journal of Agricultural Research, 8*(25), 3276-3280.

Jesus, W. C., Brasil, E. M., Oliveira, J. P., Pinto, G. R. C., Chaves, L. J., & Ramos, M. R. (2008). Heterose para teor de proteína no grão em cruzamentos entre populações de milho derivadas de híbridos comerciais. *Pesquisa Agropecuária Tropical, 38*(1), 32-38.

Kvitschal, M. V., Scapim, C. A., Tonet, A., Pinto, R. J. B., Retuci, V. S., Amaral Jr, A. T., & Braccini, A. L. (2004). Análise dialélica de populações de milho na região noroeste do Paraná, na "safrinha". *Revista Ceres, 51*(293), 19-32.

Mestres, C., Matencio, F., & Dramé, D. (2003). Small-scale production and storage quality of dry-milled degermed maize products for tropical countries. *International Journal of Food Science and Technology, 38*(2), 201-207.

Melani, M. D., & Carena, M. J. (2005). Alternative maize heterotic pattern for the northern corn belt. *Crop Science, 45*(6), 2186-2194.

Oliveira, J. P., Chaves, L. J., Duarte, J. B., Brasil, E. M., Ferreira Jr., L. F., & Ribeiro, K. O. (2004). Teor de proteína no grão em populações de milho de alta qualidade protéica e seus cruzamentos. *Pesquisa Agropecuária Tropical, 34*(1), 45-51.

Ramalho, M. A. P., Abreu, A. F. B., & Santos, J. B. (2001). Melhoramento de espécies autógamas. In: L. L. Nass, A. C. C. Valois, I. S. Melo, & M. C. Valadares, (Ed.). *Recursos Genéticos e melhoramento*. Plantas (201-230). Rondonópolis, MT: Fundação Mato Grosso.

Rodrigues, F., Pinho, R. G. V., Albuquerque, C. J. B., Faria Filho, E. M., & Goulart, J. C. (2009). Capacidade de combinação entre linhagens de milho visando à produção de milho verde. *Bragantia, 68*(1), 75-84.

Scapim, C. A., Braccini, A. L., Pinto, R. J. B., Amaral Júnior, A. T., Rodovalho, M. A., Silva, R. M., & Moterle, L. M. (2006). Componentes genéticos de médias e depressão por endogamia em populações de milho-pipoca. *Ciência Rural, 36*(1), p. 36-41.

Scapim, C. A., Pacheco, C. A. P., Tonet, A., Braccini, A. L., & Pinto, R. J. B. (2002). Análise dialélica e heterose de populações de milho-pipoca. *Bragantia, 61*,(3), 219-230.

Scott, A. J., & Knott, M. (1974). A cluster analysis method for grouping means in the analysis of variance. *Biometrics, 30*(3) 507-512.

Silva, V. Q. R., Amaral Jr, A. T., Gonçalves, L. S. A., Freitas Jr, S. P., & Ribeiro, R. M. (2011). Heterotic parameterizations of crosses between tropical and temperate lines of popcorn. *Acta Scientiarum. Agronomy, 33*(2), 243-249.

Silva, R. M., & Miranda Filho, J. B. (2003). Heterosis expression in crosses between maize populations: ear yield. *Scientia Agricola, 60*(3), 519-524.

Solalinde, J. M. Q., Scapim, C. A., Vieira, R. A., Amaral Jr, A. T., Vivas, M., Pinto, R. J. B., Mora, F., & Viana, A. P. (2014). Performance of popcorn maize populations in South American Avatí Pichingá using diallel analysis. *Australin Journal of Crop Science, 8*(12), 1632-1638.

Tonette, P. L., & Carena, M. J. (2014). Diallel analysis among 16 maize populations adapted ti the northert U.S. corn belt for grain yield and grain quality traits. *Euphytica, 200*(1), 29-44.

Vencovsky, R., & Barriga, P. (1992). *Genética biométrica no fitomelhoramento* (p. 469). Ribeirão Preto, SP: Sociedade Brasileira de Genética.

10

Karyotypic characterization of *Capsicum* sp. accessions

Willame Rodrigues do Nascimento Sousa[1], Angela Celis de Almeida Lopes[2], Reginaldo de Carvalho[2] and Regina Lúcia Ferreira Gomes[2] and Ana Paula Peron[1*]

[1]Programa de Pós-graduação em Genética e Melhoramento, Centro de Ciências Agrárias, Universidade Federal do Piauí, 64049-550, Bairro Iningá, Teresina, Piauí, Brazil. [2]Departamento de Biologia, Universidade Federal Rural de Pernambuco, Recife, Pernambuco, Brazil.
**Author for correspondence. E-mail: anpapegenpes@yahoo.com.br*

ABSTRACT. This study aimed to determine the chromosome number and the karyotype of *Capsicum annuum*, *Capsicum chinense*, *Capsicum frutencens* and *Capsicum baccatum* accessions in the active *Capsicum* sp. genebank at the Federal University of Piauí (BGC-UFPI). These species have great economic importance throughout the world, and their cytogenetic characterization can inform taxonomy and lead to improvement in the genus. Karyotypes were obtained from the rootlet meristems of the studied accessions using the squash method and Giemsa staining. The chromosome number 2n=2x=24 was verified for each of the four species. Chromosomal polymorphisms were observed for the *C. frutencens* accession BGC 37, which presented 12 pairs of metacentric chromosomes instead of 11 pairs of metacentric chromosomes and 1 submetacentric chromosome pair present in other accessions. In the accessions BGC 01 and BGC 37, secondary constrictions were observed in the 1 and 12 as well as in the 6 and 11 homologs, respectively. The karyotypes of the studied species were asymmetrical among themselves. The results obtained in this study confirm the high genetic diversity previously described in the literature for this genus.

Keywords: pepper, cytogenetics, genebank, morphometry.

Introduction

The *Capsicum* sp. peppers of the Solanaceae family, which are found in tropical and temperate regions worldwide, are valued as spices or vegetables by many different cultures. The genus has significant economic importance for the national and international condiment, seasoning and canning markets, and it is cultivated at scales ranging from family production to industrial systems (FONSECA et al., 2008). These peppers are also employed in phytotherapeutic medicine, primarily in South and Latin American countries. The genus contains high concentrations of vitamins A and C, β-carotene and capsaicins, which have been shown to provide antioxidant, antimicrobial, anti-inflammatory and hypocholesterolemic action among other benefits (COSTA et al., 2008; ZENI; BOSIO, 2011).

Four domesticated species of *Capsicum* are widely grown and consumed in Brazil: *C. annuum* L. (bell pepper, sweet pepper), *C. chinense* Jaqc. (Yellow Lantern Chili), *C. frutescens* L. (Malagueta Pepper) and *C. baccatum* L. (Bishop's Crown) (LANNES et al., 2007). These species are highly adapted to tropical climate conditions and contain high biological diversity, which is presented by the variety of fruit sizes, forms and coloration (HAVERROTH; NEGREIROS, 2011; ZENI; BOSIO, 2011).

Brazil is considered a natural habitat of these peppers, and the main pepper-producing states are Minas Gerais, São Paulo, Rio de Janeiro, Ceará and Bahia (ZENI; BOSIO, 2011). However, knowledge about the genetic diversity of these plants in Brazil is still rudimentary despite a significant research commitment to biochemical, molecular and cytogenetic studies of the species over the past two decades (HAVERROTH; NEGREIROS, 2011).

Studies of pepper chromosome number and morphology generate important data for *Capsicum* taxonomy (SOUZA et al., 2011), contribute to the understanding of the genetic variations involved in the evolution of the genus (MOSCONE et al., 2007), aid in the delineation of the cultivated, semi-cultivated and wild species (CARVALHO; BIANCHETTI, 2008; PEREIRA et al., 2006) and further plant diversity conservation by supplying information to aid genetic improvement programs for this genus (SOUZA et al., 2011). Rohani et al. (2010) mentioned that cytogenetic studies of different *Capsicum* genebank accessions comprise an important data source for breeders, which allows for better gene pool administration and a more efficient selection of genetic resources.

Because of the economic relevance and genetic variability described in the literature for *C. annuum*, *C. chinense*, *C. frutescens* and *C. baccatum* along with the relative lack of data for these species, this study aimed to characterize and analyze the karyotypes of accessions from the collection of the active *Capsicum* sp. genebank of the Federal University of Piauí (BGC-UFPI). In addition, the data on the agricultural production of peppers in the state of Píauí are imprecise and irregular as small producers dominate production in the region. Commercialization of this product involves small homemade or artisanal operations for sauces, preserves, jellies and powdered pepper, as well as small companies that directly sell their products to street markets and small supermarkets. However, Píauí possesses environmental conditions, such as temperature and soil type, that are quite favorable for the large-scale production of these four *Capsicum* species. Cytogenetic studies of these species may aid in the genetic enhancement and expansion of production of this genus in Piauí through better characterizing their variability.

Material and methods

Material

The accessions of the studied species were obtained from the active *Capsicum* sp. genebank at the Federal University of Piauí, which is located in the city of Teresina. Four accessions of the species *C. annum* (BGC 34, BGC 36, BGC 39 and BGC 59), two accessions of the species *C. chinense* (BGC 07 and BGC 49), two accessions of the species *C. frutencens* (BGC 01 and BGC 37) and four accessions of the species *C. baccatum* (BGC 21, BGC 26, BGC 27 and BGC 54) were cytogenetically characterized.

The geographical distances between the species were also taken into consideration. The accessions BGC 39, BGC 36, BGC 01, BGC 37, BGC 21 and BGC 26 were collected in Teresina, Piauí State; the accessions BGC 34, BGC 54 and BGC 07 were acquired in São Raimundo Nonato, Piauí State, Pedro II, Piauí State and Piripiri, respectively; the accession BGC 49 was collected in the city of São Paulo (SP); and the accessions BGC 59 and BGC 27 were collected in São Luís do Maranhão, Maranhão State and São Francisco, Maranhão State, respectively.

Seeds of these accessions were germinated in the Biochemical Genetics and DNA Sequencing laboratory (LGBS) of the Biology Department of the Pernambuco Rural Federal University (UFRPE) in January 2011. The seeds were placed in a growth chamber under room temperature and controlled lighting (12-hour days) in Petri dishes covered with filter paper and moistened daily with distilled water for five days.

Cytogenetic evaluation

The cytogenetic analyses were conducted in the same laboratory in the first semester (January-July) of 2011. The rootlets (approximately 0.5 cm) of the different accessions were collected and pre-treated with 8-hydroxyquinoline (2 mM) for four hours at room temperature. The rootlets were washed in distilled water, fixed in a solution of Carnoy-absolute ethanol and glacial acetic acid (3:1) for 24 hours and preserved at -20°C. For the preparation of the slides, the fixed rootlets were washed in distilled water and hydrolyzed in 1 N HCl at room temperature for 10 minutes.

After hydrolysis, the rootlets were washed again in distilled water and slides for each accession were prepared. For the preparation of each sheet, a rootlet tip was removed with a scalpel, macerated in a drop of 45% glacial acetic acid and covered with a coverslip. The coverslips were removed with nitrogen, and the slides were air-dried and stained with 2% Giemsa stain for 10 to 15 minutes and mounted with Entellan (Merck®).

Karyotypic analysis

The slides were analyzed at 100X magnification using a *Leica DM 2500* microscope equipped with a *DC 345 FX* digital microcamera coupled to a microcomputer for direct capture of the images of interest. The images were obtained using the Image-Pro Plus program (version 5.1).

Five cells in metaphase from each accession, each corresponding to one repetition, were used for the chromosome measurements, which were conducted with the aid of the MicroMeasure 3.3 Program (REEVES; TEAR, 2000). The following average values were obtained: a) length of the long arm (LA); b) length of the short arm (SL); c) total length of the chromosome (TCL = LA + SA); d) total length of the haploid batch (TLHB = $\sum$ TCL); and e) relative length (r = (TCL/TLHB) x 100).

The identification of the homologs was conducted using the selected parameters described above. The chromosomes were classified as metacentric, submetacentric, acrocentric or telocentric based on the ratio between the arms (r) and centromere index (ic), which was proposed by Guerra (1986). The accessions were also analyzed using the Asymmetry Index (AI) according to Romero-Zarco (1986).

Data analysis

The accessions were compared for two parameters: TLHB and TCL. Analysis was conducted as a randomized design with five repetitions using the following model:

Variance analysis was conducted using Tukey's test for the comparison of averages (1949) at a 5% level of significance. All analyses were calculated using the Genes software (CRUZ, 2006).

Results and discussion

Giemsa staining of the *C. annum*, *C. chinense*, *C. frutencens* and *C. baccatum* accessions allowed for accurate descriptions of the number, morphology and size of the chromosomes, type of interphase nucleus and chromosomal condensation pattern. The majority of the chromosomes did not display differentiated areas along their length aside from the centromeric region and visualization of secondary constrictions in the accessions of one species.

For each of the 12 accessions, a chromosome number of 2n = 2x = 24 was confirmed (Figure 1); this ploidy level is widely described in the literature for a number of *Capsicum* species (GUERRA, 2001; MOSCONE et al., 2007; PICKERSGILL, 1997; POZZOBON, WITTMANN, 2006; SOUZA et al., 2011; TEODORO-PARDO et al., 2007;) and is also common in the Solanaceae family (PICKERSGILL, 2007). However, in some wild *Capsicum*, such as *C. buforum*, *C. capylopodiume* and *C. cornutum*, a ploidy level of 2n = 2x = 26 has been reported (POZZOBON; WITTMANN, 2006). Pozzobon and Wittmann (2006) suggest that two different evolutionary lines emerged in the diversification of this genus, which is shown by a clear separation between the wild species (base number x = 13) and domesticated species (x = 12). These authors also hypothesized that the x = 13 lines are ancestral to the x = 12 plants.

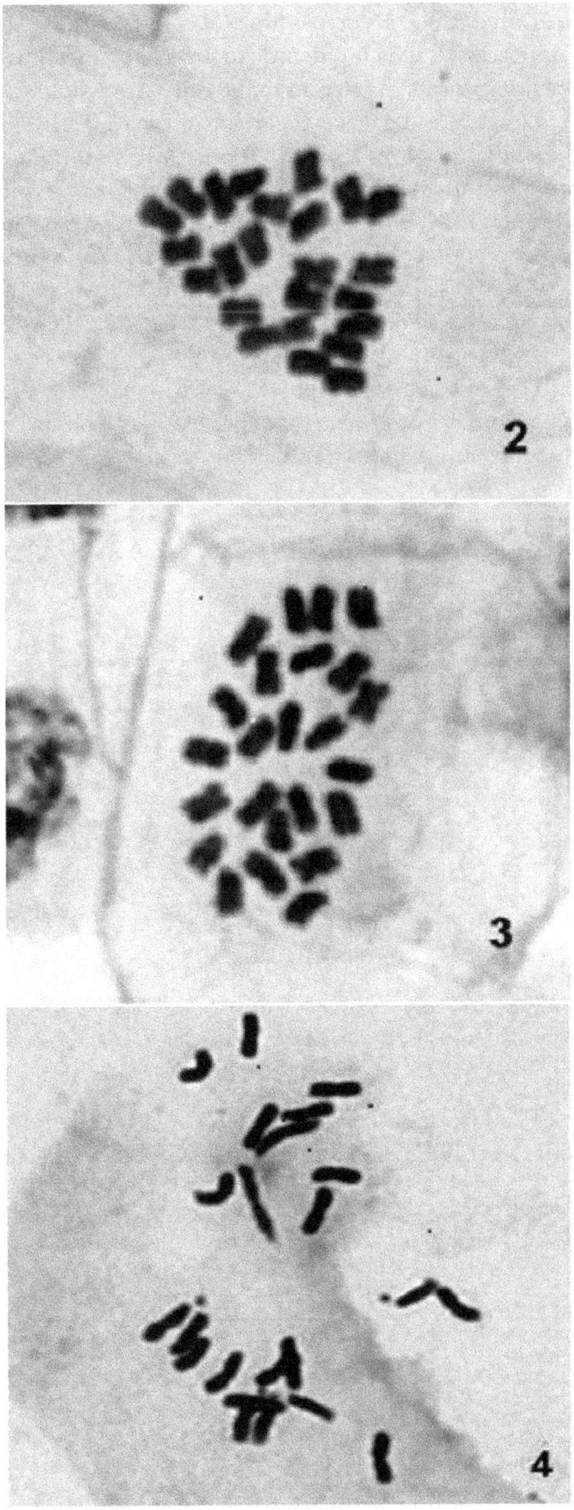

Figure 1. (1) Metaphasic chromosomes of *C. annuum* (BGC 39) (2) Metaphasic chromosome of *C. chinense* (BGC 49), (3) Metaphasic chromosomes of *C. frutescens* (BGC 01), and (4) Metaphasic chromosomes of *C. baccatum* (BGC 21). Bar = 10 mm.

Of the studied accessions, 11 presented the karyotypic formula 11M + 1SM (Figure 2), in which chromosome 12 was classified as submetacentric. The *C. frutescens* accession BGC 37 presented the karyotypic formula 12M (Table 1, Figure 3), which demonstrates chromosomal polymorphisms in relation to the other accessions.

However, the karyotypic formulas reported in the literature conflict with those observed in the present study. Guerra (2001) reported the karyotypic formula 11M + 1A for a number of Venezuelan accessions of *C. chinense* using Giemsa stain. Souza et al. (2011) also observed the formula 11M + 1A through conventional cytogenetics in *C. chinense* accessions from different states of Brazil. Sousa et al. (2011) and Moscone et al. (1996) analyzed the evolutionary patterns across species of *Capsicum* by chromosome banding and observed the karyotypic formulas of 11M + 1SM + 1A and 11M + 1A for *C. frutescens*, respectively.

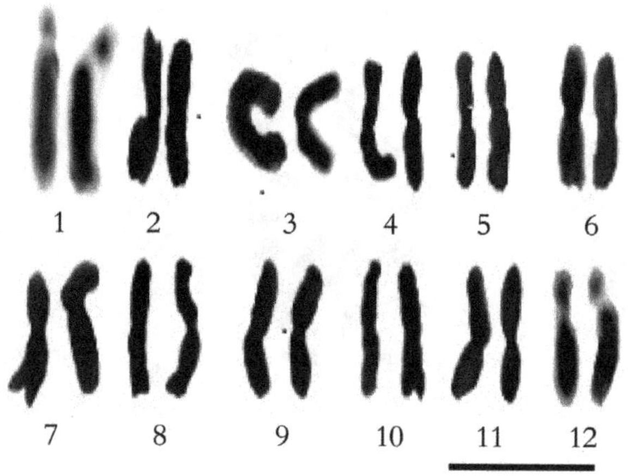

Figure 2. Diploid karyogram representative of the *Capsicum frutescens* (BGC 01) karyotype with karyotypic formula 12M. (bar = 10 μm).

Teodoro-Pardo et al. (2007) explained how different karyotypic formulas in a single species can occur due to genetic variations among populations, which are generated from the genomic response to different environments. The same authors discuss how the appearance of polymorphisms at the chromosomal level in individuals of the same population can alter the karyotypic pattern of those specimens and originate distinct chromosomal races.

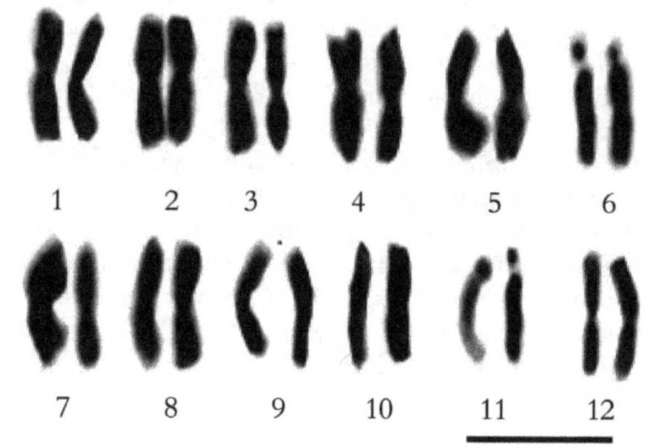

Figure 3. Diploid karyogram representative of the *Capsicum frutescens* (BGC 37) karyotype with the karyotypic formula 12M. (bar = 10 μm).

According to Moscone et al. (2007), differences in the morphology, size and number of chromosomes are common in populations of the same species or in interspecific taxa, and these differences are grouped into cytotypes or chromosomal races. These authors affirmed that such differences are frequent in the genus *Capsicum*, in which cytotypes differ mainly in karyotypic formula and chromosome size. For the *C. frutescens* accessions BGC 01 and BGC 37, secondary constrictions were observed in the homologous pairs 1 and 12 and 6 and 11, respectively (Figures 2 and 3).

Moscone et al. (1996) reported the presence of secondary constrictions in every *Capsicum* species, which ranged from 1 to 4 per karyotype.

Table 1. Average values, in micrometers, of CSI, CI, r, AI, and KF of four accessions of the species *C. annum* (BGC 34, BGC 36, BGC 39, BGC 59), two accessions of the species *C. chinense* (BGC 07, BGC 49), two accessions of the species *C. frutencens* (BGC 01, BGC 37) and four accessions of the species *C. baccatum* (BGC 21, BGC 26, BGC 27, BGC 54).

Species /Accession	2n	CSI	CI	r	AI (%)	KF
C. annuum - BGC 34	24	6.92 – 3.64	45.34	1.22	45.61	11M+1SM
C. annuum - BGC 39	24	6.49 – 3.59	45.64	1.20	45.81	11M+1SM
C. annuum - BGC 36	24	7.05 – 3.69	45.53	1.21	45.77	11M+1SM
C. annuum - BGC 59	24	7.42 – 4.42	44.63	1.25	44.68	11M+1SM
C. chinense - BGC 07	24	6.60 – 4.56	45.80	1.19	45.97	11M+1SM
C. chinense - BGC 49	24	5.82 – 3.29	45.88	1.07	46.05	11M+1SM
C. frutescens - BGC 01	24	7.17 – 4.21	45.19	1.12	45.46	11M+1SM
C. frutescens - BGC 37	24	5.73 – 3.60	46.55	1.06	46.54	12M
C. baccatum - BGC 21	24	7.00 – 4.45	47.45	1.18	47.64	11M+1SM
C. baccatum - BGC 54	24	7.48 – 4.31	44.90	1.22	45.08	11M+1SM
C. baccatum - BGC 26	24	6.92 – 3.49	47.60	1.22	45.51	11M+1SM
C. baccatum - BGC 27	24	6.56 – 3.75	45.70	1.19	46.02	11M+1SM

CSI: chromosome size interval, CI: centromere index, r: chromosome arm ratio, AI: asymmetry index and KF: karyotypic formula.

The average size of the chromosomes observed in the present work varied from 3.29 (BAC 49) to 7.48 μm (BGC 54) (Table 1). This result diverges from the data found in the literature: Souza et al. (2011) reported an average chromosome size of 2.59 to 4.12 μm in four Brazilian *C. chinense* accessions, and Teodoro-Pardo et al. (2007) reported values of 1.6 to 8.4 μm, 2.4 to 2.9 μm and 2.1 to 5.2 μm for *C. annum* karyotypes from three Mexican states.

A high frequency of metacentric chromosomes was observed in the complements of the 12 accessions, which is indicated by the average values of the centromere indices (CI) and ratios obtained between the arms of the chromosomes (r) (Table 1). The results indicate symmetrical karyotypes in the studied accessions, especially in the *C. frutescens* accession BGC 37, which presented only metacentric chromosomes. According to Stebbins (1958), plants with higher karyotypic symmetry than others of the same genus are evolutionarily ancestral to those that possess lower symmetry.

Wadt et al. (2004) reported that although most of *Capsicum* species are 2n = 24 and present high similarity in chromosome morphology, the genus possesses high intra- and interspecific karyotypic variability. This observation is supported by the results of the asymmetry index (AI%) among the karyotypes of the species studied in this work (Table 1), which varied from 44.68 (BGC 59) to 47.64% (BGC 21). These values are under 50% and demonstrate that the karyotypes are asymmetrical amongst themselves according to Romero-Zarco (1986).

According to Moscone et al. (2007), a higher asymmetry index among karyotypes of species in the same genus is correlated to higher genetic variability among them. In the species studied here, the various fruit shapes and sizes can be cited as examples of variability: the fruit of *C. annuum* is a small berry with a smooth texture (MOSCONE et al., 2007); the fruits of *C. chinense* are elongated, which measure approximately 1.0 to 1.5 cm in diameter and possess a smooth texture (NUEZ-VIÑALS et al., 2003); the fruit of *C. frutencens* is a smooth fusiform berry (BENTO et al., 2007); and the fruits of *C. baccatum* are smooth and round but variable in size and shape and average 0.6 cm in diameter and 10 cm in length (CARVALHO et al., 2009).

According to Carvalho and Bianchetti (2008), the flowers of *Capsicum* can also be used to observe the genetic variability of the genus because there is significant diversity in the number of flowers per node, flower and peduncle position, corolla and anther color, presence or absence of spots on the petal lobes and cup edge.

The studied accessions were grouped using Tukey's test at an error rate of 5% (Table 2), and the results showed that significant differences do not exist among the accessions BGC 07, BGC 21, BGC 54, BGC 36, BGC 34 and BGC 27 in terms of the TCL and TLHB parameters.

Table 2. Average TCL and TLHB values, expressed in micrometers (μm), for the accessions of the species *C. annum* (BGC 34, BGC 39, BGC 36, BGC 59), *C. chinense* (BGC 07, BGC 49), *C. frutescens* (BGC 01, BGC 37) and *C. baccatum* (BGC 21, BGC 54, BGC 26, BGC 27).

Accession	Species	TCL	Group	TLHB	Group
BGC 01	*C. frutescens*	6.02	a	72.22	a
BGC 59	*C. annuum*	5.82	ab	69.89	ab
BGC 07	*C. chinense*	5.78	abc	69.35	abc
BGC 21	*C. baccatum*	5.69	abc	68.26	abc
BGC 54	*C. baccatum*	5.63	abc	67.51	abc
BGC 36	*C. annuum*	5.42	abc	65.01	abc
BGC 34	*C. annuum*	5.34	abc	64.11	abc
BGC 26	*C. baccatum*	5.28	c	63.38	c
BGC 27	*C. baccatum*	5.24	abc	62.84	abc
BGC 39	*C. annuum*	4.89	bc	58.65	bc
BGC 49	*C. chinense*	4.65	ab	55.86	ab
BGC 37	*C. frutescens*	4.41	ab	52.96	ab

TCL: Total chromosome length; TLHB: Total Length of Haploid Batch. Accessions followed by the same letter were grouped by Tukey test at 5%.

The accession BGC 01 differs significantly from the accessions BGC 39 and BGC 26, which were all collected in Piauí, the accession BGC 26, which was from Piauí, and differs significantly from the accessions BGC 37 and BGC 49, which were collected in Piauí and São Paulo, respectively, in relation to these parameters (Table 2). Davide et al. (2007) claim that the presence of karyotypic asymmetry that is associated with significant differences of TLHB and TCL among individuals of the same or closely related species may be the result of chromosomal alterations, such as Robertsonian translocations, inversions, unequal translocations, deletions and duplications. The same authors explain that these alterations can occur because of the environmental conditions, including climate, soil, temperature and moisture, that the individuals are subjected. According to Livingstone et al. (1999) chromosomal alterations, such as translocations, duplications and deletions, have been observed in a large number of plants in the species *C. annuum* and *C. chinense*.

The results obtained in this work demonstrate variation among the studied species, which is in agreement with the scientific literature; however, the species presented symmetrical karyotypes and identical chromosomal numbers. Thus, wider sampling and more detailed characterization of the chromosomes, through heterochromatin distribution and identification of sequences by *in situ* hybridization, will be necessary to determine if the differences among the species characterize intra- and/or interspecific differentiation. Such approaches will be able discriminate among species with the same karyotypic formula.

Conclusion

The results of the present study, when combined with future cytogenetic studies using the BGC - UFPI accessions, will be of significant value to genetic enhancement programs for *Capsicum* in the state of Piauí. This research will aid in the development of higher quality and more productive varieties that are more resistant to pests and diseases, which will encourage farmers to produce more of these vegetables. Increased production will generate economic benefits by employing more labor during planting and meet the market demands of the state.

References

BENTO, C. S.; SUDRÉ, C. P.; RODRIGUES, R.; RIVA, E. M.; PEREIRA, M. G. Descritores qualitativos e multicategóricos na estimativa da variabilidade fenotípica entre acessos de pimentas. **Scientia Agraria**, v. 8, n. 2, p. 664-679, 2007.

CARVALHO, S. I. C.; BIANCHETTI, L. B.; REISFSCHENEIDER, J. B. Registro e proteção de cultivares pelo setor público: a experiência do programa de melhoramento de *Capsicum* da Embrapa. **Horticultura Brasileira**, v. 27, n. 2, p. 135-138, 2009.

CARVALHO, S. I. C.; BIANCHETTI, L. B. **Botânica e recursos genéticos**. Pimentas *Capsicum*. Brasília: Embrapa Hortaliças, 2008.

CRUZ, C. D. **Programa GENES**: análise multivariada e simulação. Viçosa: UFV, 2006.

COSTA, L. V.; LOPES, T. G.; LOPES, R.; ALVES S. R. M.; GAMA, A. S. Efeito das polinizações no vingamento de frutos de *Capsicum chinense* Jacq. **Horticultura Brasileira**, v. 26, n. 2, p. 1685-1689, 2008.

DAVIDE, L. C.; TECHIO, V. H.; NENUES, J. D.; PEREIRA, A. V. Variação cromossômica numérica em *Pennisetum*. **Ciência e Agrotecnologia**, v. 31, n. 2, p. 398-405, 2007.

GUERRA, N. Estudios cromossómicos de cuatro seleciones de *Capsicum chinense* Jacq. **Revista Científica UDO Agrícola**, v. 1, n. 1, p. 34-41, 2001.

FONSECA, R. M.; LOPES, R.; BARROS, W. S.; LOPES, M. T. G.; FERREIRA, F. M. Morphologic characterization and genetic diversity of *Capsicum chinese* Jacq. Accessions along the upper Rio Negro – Amazonas. **Crop Breending and Applied Biotechnology**, v. 81, n. 3, p. 187-197, 2008.

HAVERROTH, M.; NEGREIROS, P. R. M. Calendário agrícola, agrobiodiversidade e distribuição espacial de roçados Kulina (Madeja), Alto Rio Envira, Acre, Brasil. **Sitientibus série Ciências Biológicas**, v. 11, n. 2, p. 299-308, 2011.

LANNES, S. D.; FINGER, F. L.; SCHUELTER, A. R.; CASALI, V. W. D. Growth and quality of Brazilian accessions of *Capsicum chinense* fruits. **Scientia Horticulturae**, v. 112, n. 3, p. 266-270, 2007.

LIVINGSTONE, K. D.; LACKNEY, V. K., BLAUTH, J. R.; VANWIJK, R. I. K.; JAHN, M. K. Genome mapping in *Capsicum* and the evolution og genome structure in the Solanaceae. **Genetics**, v. 152, n. 3, p. 1183-1202, 1999.

MOSCONE, E. A.; SCALDAFERRO, M. A.; GABRIELE, M.; CECCHINNI, N. M.; EHRENDORFER, F. The evolution of chili peppers (Capsicum – Solanaceae): a cytogenetic perspective. **Acta Horticulturae**, v. 745, n. 2, p. 137-169, 2007.

MOSCONE, E. A.; LAMBROU, M.; EHRENDORFER, F. Fluorescent chromosome banding in the cultivated species of *Capsicum* (Solanaceae). **Plant Systematics and Evolution**, v. 202, n. 1-2, p. 37-63, 1996.

NUEZ-VIÑALS, F. N.; ORTEGA, R. G.; COSTA, J. C. **El cultivo de pimientos, chiles y ajies**. Madri: Mundi-Prensa, 2003.

PEREIRA, L. P.; LUZ, L. P.; TEDESCO, S. B.; SILVA, A. C. F. Número de cromossomos em populações de *Achyrocline satureioides* Lam. **Ciência Rural**, v. 36, n. 2, p. 678-681, 2006.

PICKERSGILL, B. Domestication of plants in the Americas: insights medelian and molecular genetics. **Annals of Botany**, v. 100, n. 5, p. 925-940, 2007.

PICKERSGILL, B. Genetic resources and breeding of *Capsicum* spp. **Euphytica**, v. 96, n. 1, p. 129-133, 1997.

POZZOBON, M. T.; WITTMANN, M. T. A meiotic study of the wild and semi-domesticated Brazilian species of genus *Capsicum* L. (Solanaceae). **Cytologia**, v. 71, n. 3, p. 275-287, 2006.

REEVES, A.; TEAR, J. **MicroMeasurefor windows version 3.3**. Free program distributed by the authors over the internet. 2000. Available from: <http://www.colostate.edu/depts/biology/micromeasure>. Access on: Set 10, 2013.

ROHANI, M.; MOHAMMADI, A.; KHOSRO-SHAHLI, M.; AHMADI, H.; DARANDEH, N. Karyotype analysis of several ecotypes of *Capsicum annuum* L. in Iran. **Notulae Botanicae Horticultu and Agrobotanica Cluj**, v. 38, n. 1, p. 177-180, 2010.

SOUZA, S. A. M.; MARTINS, K. C.; PEREIRA, T. N. S. Polimorfismo cromossômico em *Capsicum chinense* Jacq. **Ciência Rural**, v. 41, n. 10, p. 1777-1783, 2011.

ROMERO-ZARCO, C. A. A new method for estimating karyotype. **Taxon**, v. 35, n. 3, p. 526-530, 1986.

STEBBINS, G. L. Longevity, habitat and release of genetic variability in the higher plants. **Cold Spring Harbor Symposia Quantitative Biology**, v. 23, n. 1, p. 365-378, 1958.

TEODORO-PARDO, C. V.; GARCÍA-VELÁZQUEZ, A.; CORONA-TORRES, T. Polimorfismo cromossômico en *Capsicum annuum* L.(Solanaceae) em recolectas de Puebla, Morelos y Quartenário, México. **Agrociencia**, v. 41, n. 10, p. 873-881, 2007.

TUKEY, J. W. One degree of freedon for non additivy. **Biometrics**, v. 5, n. 3, p. 232-242, 1949.

ZENI, A. L.; BOSIO, F. O uso de plantas medicinais em uma comunidade rural de Mata Atlântica – Nova Rússia, SC. **Neotropical Biology and Conservation**, v. 6, n. 1, p. 55-63, 2011.

WADT, L. H. O.; EHRINGHAUS, C.; KAGEYAMA, P. Y. Genetic diversity of Pimenta Longa' genotypes (*Piper* spp., Piperaceae) of Embrapa Acre germplasm collection. **Genetics and Molecular Biology**, v. 27, n. 1, p. 74-82, 2004.

11

Genetic responses of traits relationship to components of nitrogen and phosphorus use efficiency in maize

Júlio César DoVale[1*], Ciro Maia[2], Roberto Fritsche-Neto[1], Glauco Vieira Miranda[1] and Paulo Cezar Cavatte[3]

*[1]Departamento de Fitotecnia, Universidade Federal de Viçosa, Av. Peter Henry Rolfs, s/n, 36570-000, Viçosa, Minas Gerais, Brazil. [2]Departamento de Biologia Geral, Universidade Federal de Viçosa, Viçosa, Minas Gerais, Brazil. [3]Departamento de Biologia Vegetal, Universidade Federal de Viçosa, Viçosa, Minas Gerais, Brazil. *Author for correspondence. E-mail: julio.vale@ufv.br*

ABSTRACT. This study had three objectives: i) to observe maize inbred lines response to nitrogen (N) and phosphorus (P) acquisition as well as their utilization efficiencies and traits of root morphology in contrasting levels of these nutrients, ii) to study the relationship between root morphology and nutrient use efficiencies for both nutrients, and iii) to identify contrasting parents with components N and P use efficiency for an inheritance study. We evaluated 15 inbred lines in two experiments conducted in contrasting conditions of N and P. We evaluated the shoot, root traits and nutritional efficiencies and observed the genetic variability for most traits. Selection can be practiced for most of these traits in specific environments. Under conditions of nutritional stress, average and heritability estimates were of lesser magnitude. In this study, the shoot and root morphology traits were shown to be associated with the acquisition efficiency of both N and P in all of the environments that were evaluated.

Keywords: *Zea mays*, acquisition and utilization efficiencies, root morphology, abiotic stress, REML/BLUP.

Introduction

Considering the current scenario of global climate change and that the major advances in agriculture have been designed for favorable environments during the "Green Revolution", the performance of crops under adverse conditions in marginal environments remains a subject of constant discussion. However, the use of fertilizers in agriculture, together with an improvement in crop systems, provides an adequate food supply for human and animal consumption.

The production of nitrogen (N) fertilizers by the Haber-Bosch process was one of the most important inventions of the 20th century enabling food supplies to world production (SMIL, 1999). However, with the high cost of fertilizers, declining natural resources, and the expectation that the world population will reach 10 billion people by the mid-21st century, an increase of approximately 43% of the world's current population (FEDOROFF et al., 2010), the future challenge will be to accommodate the population's necessities through the development of a high agriculture productivity while preserving environmental quality. Thus, the effective utilization of genetic plant resources will be needed to answer the challenge.

Research of plant species or genotypes that are able to absorb and assimilate nutrients efficiently is extremely important (SOUZA et al., 2008). Genetic variability exists in the capture of soil nutrients proven by genotypes that are increasingly responsive to high amounts of fertilizers, especially N and phosphorus (P), which are limiting factors to agricultural production (TILMAN et al., 2002). This responsive research has been accompanied by decreased nutritional efficiency, contributing to increased pollution in many regions (CUI et al., 2009).

Phosphorus is a limiting nutrient in many regions of the world, especially in the highly weathered soils of the tropics, while N has not yet been invested in by developing country producers because of its high price (HIREL et al., 2007). Thus, it is necessary to adopt different genotypes when subjected to minimum nutrients to maintain satisfactory production.

Many definitions and methodologies exist that describe the calculation of the nutritional efficiency of a crop. However, researchers have recently adopted a more plausible definition specifically; is the ability of a given genotype to produce biomass or grains under certain nutritional supplementation (PARENTONI; SOUZA JUNIOR, 2008; WHANG et al., 2010). Nutrient use efficiency is composed of the acquisition efficiency (AE) and the internal utilization efficiency (UtE). Acquisition efficiency is defined as the ability of a genotype to acquire specific nutrients from the soil; the UtE refers to the ability of this genotype to produce biomass or grain using the absorbed nutrient. Thus, greater N use efficiency (NUE) and P (PUE) can be achieved by increasing the AE and UtE (CHEN et al., 2009; PRESTERL et al., 2002).

Several biochemical and physiological traits are involved in increasing UtE in plants. For example, in P acquisition the most important traits are morphology and root architecture, because P availability is unevenly distributed throughout the soil profile (WHANG et al., 2010). Root architecture, defined as the space configuration of the root system (LYNCH; BROWN, 2001), determines the extent explored through the soil profile. As a result of seeking to adapt to adverse conditions, the root architecture genetics may suffer changes. For example, low moisture stress can increase root system depth and lateral root formation (CHUN et al., 2005). Under conditions of low P availability, the changes occur in the upper root layers, providing increased nutrient acquisition (HAMMER et al., 2009).

Genotypic variation in the extension of lateral roots and thrust can lead to the selection of genotypes that are more efficient in nutrients acquisition. The identification of quantitative trait loci (QTL) in the initial stage of maize to root traits, including extension of lateral and axial root under water stress (RUTA et al., 2010), response to low P and N availability (GALLAIS; HIREL, 2004; ZHU et al., 2005) revealed the importance of studying root morphology at an early stage to predict the performance of genotypes in later stages. Thus, early evaluation or indirect selection using correlated traits are of great interest in plant improvement under abiotic stress because they accelerate the process of selection by immediately discarding less promising genotypes, thus concentrating on potentially accessible resources.

The objectives of this study were as follows: i) Identify differential behavior of maize inbred lines for nutrient acquisition efficiency, nutrient utilization and traits of root morphology in contrasting N and P environments, ii) Study the relationship between the traits of root morphology and the acquisition efficiency and utilization efficiency of N and P in contrasting levels and, iii) Identify contrasting parents with components N and P use efficiency for an inheritance study.

Material and methods

Plant material and experimental design

We used 15 maize inbred lines from the *Programa Milho®*, Federal University of Viçosa - UFV, Minas Gerais State, Brazil. These inbred lines were evaluated in two experiments with contrasting conditions of N (Experiment 1) and P (Experiment 2). The experiments were performed in a randomized block design in a factorial structure in triplicate. The plot was one plant per pot.

Growth conditions

The experiments were conducted in the greenhouse of the Crop Science Department, UFV, Minas Gerais State, Brazil (20°45'14"S, 42°52'53"W) in October 2009. The day to night average temperature was 33 to 25°C. The seeds were sterilized for 1 minute in a 0.5% solution of sodium hypochlorite (NaOCl) and then washed with deionized water. The seeds were pre-germinated in a box with individual cells for germination, three seeds per inbred line. Soon after emergence, seeds were transplanted. One seedling from each inbred line was transplanted in cylindrical PVC pots 4 dm3 in size (10 cm diameter x 50 cm height). The substrate for Experiment 1 was prepared by sand and vermiculite mixing at a ratio of 1:1, previously describe by Walk et al. (2006). For experiment 2, a mixed of 50% sand, 37.5% vermiculite and 12.5% soil with very low P was used. The soil was used to absorb the phosphorus and to prevent it from being readily available to the plant.

From the seventh day after seedling emergence, the nutrient solution was replenished every 2 days. In Experiment 1, the solution used contained (in mmol L^{-1}): 2.0 $Ca(NO_3)_2$, 0.75 K_2SO_4, 0.65 $MgSO_4$, 0.1 KCl, 0.25 KH_2PO_4, 1.0×10^{-3} H_3BO_3, 1.0 $MnSO_4 \times 10^{-3}$, 1.0×10^{-4} $CuSO_4$, 1.0×10^{-3} $ZnSO_4$, 5.0×10^{-6}

$(NH_4)_6Mo_7O_{24}$, 0.1 Fe-EDTA and one of two N levels, low N (LN) or high N (HN), describe previously by Chun et al. (2005). The LN solution contained 0.2 mmol L^{-1} Ca $(NO_3)_2$, (i.e., 10 times less and Ca was offset by $CaCl_2$). For Experiment 2, P was added in the form of triple superphosphate. For the low P condition (LP), 20 mg dm^{-3} was added; 104 mg dm^{-3} of P was added for the high P (HP). The other nutrients were supplied by the nutrient solution without phosphorus addition containing 1.0M $Ca(NO_3)_2x4H_2O$, 1.0 M $MgSO_4x7H_2O$, 0.5M K_2SO_4, 0.32 mM $CuSO_4x5H_2O$, 46.0 mM H_3BO_3, 0.073 mM $(NH_4)_6Mo_7O_{24}$, 9.1 mM $MnCl_2x4H_2O$, 0.76 mM $ZnSO_4x7H_2O$ and 38.0 mM Fe-EDTA.

The seedlings were harvested at the vegetative stage when 4 leaves were completely expanded (V4), 25 days after sowing. The seedling shoots were separated from the root system, packaged in paper bags and dried in a circulating forced air oven at 60°C for 72 hours.

Measurements of root and shoot traits and nutritional efficiencies

The N and P shoot dry weights (SDW_N and SDW_P, respectively) were determined after drying for 72 hours at 60°C.

The root system was evaluated by image analysis using the software WinRhizo Pro 2009a (Basic, Reg, Pro & Arabidopsis for Root Measurement) coupled to an Epson Perfection V700/V750 scanner equipped with additional light (TPU), with a resolution of 400 dpi, as described by Bouma et al. (2000). The lateral and axial root lengths were extracted from the all class of diameter (d), obtained by WinRhizo as lateral root length (LLAT) with diameter less than or equal to 0.5mm and axial root length (LAXI) with diameter greater than 0.5mm, according Trachsel et al. (2009). We also evaluated for the two components of nitrogen and phosphorus use efficiency as proposed by Hirel et al. (2007): i) acquisition efficiency (AE) – NAE or PAE = N or P in the plant/N or P applied, ii) utilization efficiency (UtE) – NUtE or PUtE = SDW/N or P in the plant.

Statistical analyses

Because some data were not normally distributed, data transformation was performed. For the variables NAE, NUtE, PAE and PUtE that represented ratio values expressed between zero and one, transformations were performed using the follow equation $\sqrt{x+0.5}$. For the variables LLAT and LAXI that demonstrated positive asymmetry, the transformation equation that was used $log(x+1)$.

From these data, statistical analyses were performed using the equations of mixed models, as described by Bernardo (2002):

$$y = Xr + Zg + Wi + e$$

where:

y is the vector of phenotypic data; r is the vector of repetition effects within the environments (fixed); g is the vector of genotypic effects (random), where $g \sim N(0,G)$ being $G = A\sigma_g^2$; i is the vector of the effects of genotype interaction x environment (random), and e the error vector, being that $e \sim N(0,R)$ and $R = I\sigma_e^2$. The letters X, Z and W are incidence matrices that relate the effects of r, g and i to vector y, respectively.

In the focus of mixed models, G refers to the genetic covariance matrix between the inbred lines, being denoted by $A\sigma_g^2$. In this study, the coefficient of coancestry (A) was disregarded and, consequently, the matrix G was assumed as $A\sigma_g^2$, i.e., A=I. Thus, σ_g^2 equals the genetic variance between the inbred lines and the corresponding vector $\hat{g}$ corresponds to the predictions of the genotypic values of inbred lines.

To obtain the above solutions, the components of genetic and non-genetic variance were assumed as unknown and were estimated by restricted maximum verisimilitude (DF-REML). The predicted responses to the selection were conducted by making the average breeding values of selected inbred lines for each environment. Based on the predicted genotypic values of the inbred lines, the genetic correlations among the traits studied were estimated using the Pearson coefficient. The selection gains were obtained by the average genotypic values of selected inbred lines. All tests were performed using the computer statistical package Statistical Analysis System (SAS) version 9.1 (SAS, 2003).

Results and discussion

The genotypes differed significantly by analysis of deviance for the majority of the traits evaluated, with the exception PUtE (Table 1). Differential genotypic behaviors were observed when subjected to conditions of low and high N and P, evidenced by the significant effect of genotype-by-environment interactions; the exception was for root traits in the high N environment. Therefore, evaluated in this way, the environments were contrasted sufficiently to identify the variability between genotypes.

Table 1. Values of verisimilitude (LRT) of deviance analysis, estimates of genotypic variance ($A\sigma_g^2$), the variance of genotype-by-environment interaction (σ_{ge}^2), residual variance (σ_e^2), broad sense heritability (h_g^2), overall average and coefficient of variation (CV%) of shoot dry matter of N (SDWN) and P (SDWP), N acquisition efficiency (NAE) and P (PAE), N utilization efficiency (NUtE) and P (PUtE), lateral root length in the environment of N (LLAT N) and P (LLAT P), axial root length in the environment of N (LAXI N) and P (LAXI P) from 15 maize inbred lines, Viçosa, Minas Gerais State, Brazil.

Effect	SDW_N	SDW_P	NAE	PAE	NUtE	PUtE	L_{LAT} N	L_{LAT} P	L_{AXI} N	L_{AXI} P
Genotype[1]	45.6^{***}	35.27^{***}	18.13^{***}	5.63^{**}	3.02^{*}	0.93^{NS}	23.97^{***}	30.51^{***}	22.05^{***}	35.79^{***}
G x E[1]	8.23^{**}	39.91^{***}	5.5^{**}	10.35^{***}	2.79^{*}	11.35^{***}	0^{ns}	50.28^{***}	0.01^{ns}	20.78^{***}
σ_g^2	0.053	0.035	$4x10^{-4}$	$1x10^{-5}$	$3.4x10^{-3}$	0.023	0.015	$6x10^{-3}$	0.019	$5.2x10^{-3}$
σ_{ge}^2	0.019	0.053	$4x10^{-4}$	$1x10^{-5}$	0.016	2.556	$1x10^{-4}$	0.013	$4x10^{-4}$	$4.8x10^{-3}$
σ_e^2	0.028	0.018	$8x10^{-4}$	$1x10^{-5}$	0.052	3.814	0.023	$3x10^{-3}$	0.026	$3.2x10^{-3}$
h_g^2	0.79	0.55	0.58	0.07	0.17	0.01	0.82	0.44	0.80	0.64
Mean	0.83	1.17	0.81	0.72	3.51	23.27	3.74	3.43	3.26	2.85
CV%	20.27	11.35	3.39	0.34	6.50	8.39	3.79	1.63	4.98	2.00

[1] Values of LRT; Significant at ***p = 0.01, **p = 0.05, * p = 0.10 and ns non-significant at test χ^2.

The traits of shoot dry matter and root morphology for both P and N environments presented medium to high magnitudes of heritability. The broad sense heritability of medium to high magnitude for the majority of traits indicated that direct or indirect selection enables selection with satisfactory gains. However, PAE, NUtE and, PUtE showed low magnitude coefficients of heritability. Thus, for these traits that suffer high environmental influence - because they depend on the expression of many enzymes (as in the case of N use) and specific element nutrition carries, multiple progenies or replicates should be used to soften the residue effect of the heritability problem.

The significant differences observed between inbred lines indicated genetic variability allowing the selection of possible genetic gains with hybrid combinations. The significant difference observed between stress and no stress, for both N and P, indicated that they were adequate in simulating contrasting environments. As different behaviors were observed inbred lines according to the environments, indicated by the significant genotype-by-environment interaction, selection for specific environments was practicable for most traits. The exceptions to this selection ability were LLATN and LAXIN. Trachsel et al. (2009), by mapping QTLs for lateral and axial root growth of tropical maize, identified that genotype selection, based in these same traits, may be only performed with nutrient availability. However, despite the genetic variability for each N availability condition, the results indicated that the genotype performances for LLAT and LAXI were similar in those two conditions because the interaction G x E is not significant. Thus, selection could be based on performance of only one nutrient level.

For almost all traits evaluated, the highest general average values were observed in the experiment with N compared to P; the only exception was for PUtE. This result demonstrated that more N than P is required to produce biomass in maize seedlings.

Considering the relationship between the traits studied, the presence of a significant and positive relationship between LLAT and LAXI and both low and high N and P was observed (Table 2). Moreover, both low and high N and P also had a significant positive correlation between AE with LLAT and LAXI. UtE did not significantly correlate with any radical morphological trait for both levels of N and P; however, UEt correlated negatively with AE in the LP condition. The SDW presented significant correlated with high magnitude with radical morphological traits and AE in high and low N and P, with the exception of the trait LLAT in HP.

Correlations among these same traits were observed previously by Trachsel et al. (2009) in tropical maize. According to that study, this significant correlation between the types of roots occurred by hierarchical dependence among them, i.e., the extent of axial length increases the points of emergence of lateral roots. Hodge et al. (2009) reported that extension of the root system is genetically controlled, but the growth rate of axial and lateral roots is modulated by environmental conditions. This was observed by Zhang et al. (2007) in Arabidopsis, which when subjected to high N availability inhibited the development of lateral root tissue while under low availability the opposite effect was observed.

Hund et al. (2009) also worked with maize inbred lines and demonstrated that under high nutrient availability, the plants invest in the largest axial root preferentially over the lateral roots. However, this relationship was not observed in our study. Under both high and low availability of N and P, only the AE were correlated significantly and positively with the types of roots.

Table 2. Genotypic correlations between shoot dry matter (SDW), acquisition efficiencies (AE), utilization (UtE), lateral (LLAT) and axial root length (LAXI), in high (HN) and low (LN) N (lower diagonal) and high (HP) and low (LP) P (higher diagonal) from 15 maize inbred lines, Viçosa, Minas Gerais State, Brazil.

	AE		UtE		L_{LAT}		L_{AXI}		
	LP	HP	LP	HP	LP	HP	LP	HP	
	0.68**	0.87**	0.24 ^{ns}	0.16 ^{ns}	0.63**	0.33 ^{NS}	0.72**	0.58*	SDW
			-0.52*	-0.35 ^{ns}	0.58*	0.56*	0.60*	0.72**	AE
AE	0.92**	0.97**			-0.10 ^{ns}	-0.46 ^{ns}	0.00 ^{ns}	-0.33 ^{ns}	UtE
UtE	0.46 ^{ns}	0.42 ^{ns}	0.13 ^{ns}	0.25 ^{ns}			0.72**	0.82**	L_{LAT}
L_{LAT}	0.80**	0.81**	0.80**	0.87**	0.17 ^{ns}	0.21 ^{ns}			
L_{AXI}	0.79**	0.84**	0.81**	0.88**	0.15 ^{ns}	0.22 ^{ns}	0.90**	0.90**	
	LN	HN	LN	HN	LN	HN	LN	HN	
	SDW		AE		UtE		L_{LAT}		

Significant at **p = 0.01, *p = 0.05, ^{ns} non-significant at test t.

The significant correlation between LLAT and NAE in LN conditions indicates that the lateral roots have an important role in N acquisition, with a trait that can be used for early indirect selection. This was also observed by Lazof et al. (1992) in maize, where nitrate acquisition by lateral roots was 1.5 times greater than in the axial roots. The elongation of lateral roots requires less biomass per root unit than axial roots. According to Forde and Lorenzo (2001) smaller diameter roots explore the soil volume more intensively. In contrast, the correlation between the types of roots and the efficiency of utilization were not significant in both conditions of nutrient availability, both for N and P. Thus, the radical traits showed no important role in P and N utilization and in this way are not suitable for indirect selection of more efficient genotypes. Therefore, the roots only have significance for the acquisition of N and P and are not related to the incorporation of these nutrients in the plant mass, which is instead performed by internal mechanisms of the plant, such as enzyme activity and specific carriers.

According to Liu et al. (2004), axial roots determine root architecture, defined as the space configuration of the root system, and they allow a better geometric development. Plants with fewer angles in the root system in relation to the soil surface have better handling of mainly P in surface soil and, consequently, are more tolerant to low nutrient supplementation (RUBIO et al., 2003). It was also observed in present study for HP condition that there was a significant and negative relation between AE and UtE, demonstrating that plants that absorb a readily are not necessarily tolerant or efficient because they cannot utilize all the absorbed nutrients.

The genotypic values of inbred lines for AE and UtE, in both experiments (N and P) are shown in Figure 1, with QI and QII in the right and left upper quadrants and QIII and QIV in the lower left and lower right quadrants, respectively. QI is characterized by the high genotype value of the components of use efficiency of N and P, with the inverse observed in QIII. It was observed that some inbred lines had very contrasting performance in the studied environments. For example, the inbred line L8 presented positive genotypic values for the components of use efficiency in the majority of environments evaluated, except PUtE in HP, while the L3 line presented negative genotypic values for almost all environments, except for HN.

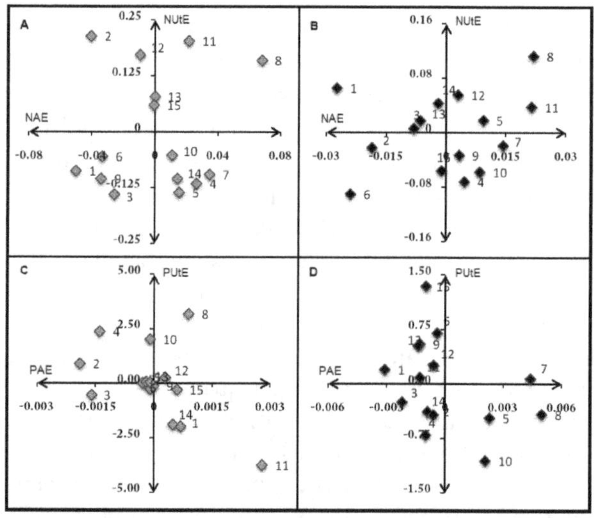

Figure 1. Genotypic values between N utilization efficiency (NUtE) and N acquisition efficiency (NAE) in LN (A) and HN (B), P utilization efficiency (PUtE) and P acquisition efficiency (PAE) in LP (C) and HP (D) from 15 maize inbred lines, Viçosa, Minas Gerais State, Brazil.

Gain estimates with selection were obtained by genotypic values of the inbred line considered to be superior. For this, we simulated the 20% selection of the superior inbred line for each of the traits evaluated in N and P environments (Table 3). It was observed that it is possible to gain with selection for all the traits studied; the predicted gains with direct selection were superior to indirect gains for all traits in all environments considered. However, for some of these traits, situations existed where the indirect gains were equal to or similar to the direct gains. This can be observed for the AE in all environments for the SDW in HN and LP when selected indirectly by SDW in LN. Additionally, NUtE in low and high N presented satisfactory gains with indirect selection by AE at low P.

The trait that provided the greatest gains with indirect selection was SDW in LN (12.05%). However, the trait that generated positive gains with selection for all traits was the SDW in HP. Moreover, this presented indirect gains at satisfactory levels (10.49%).

Considering the experimental accuracy, all the traits presented good accuracy in the estimates of genotypic values. This indicated that few repetitions were necessary for the assessment of genotypes. However the same was not observed for the SDW, because a higher number of observations were required than what were used in this study. Considering the coefficient of coancestry among genotypes, an attempt was made to minimize the effects of uncontrolled factors in the experiment.

In most environments, the same inbred lines were higher for AE and UtE, observed for both N and P. This indicated that these inbred lines can be used in the formation of more efficient N and P use in synthetic or hybrid combinations. This strong relationship between the genotypes performance indicated that the mechanisms that control AE and UtE for N and P may be pleiotropic or epistatic. This may occur due to the dependence of metabolic pathways on ATP in the acquisition and internal transport of N in the plant, which requires considerable energy availability. Thus, plants more efficient in the use of N must also be the same for P.

To achieve increases in the UE of a particular nutrient, it is necessary to increase its components. Thus, to increase the nutrient use efficiency, the inbred lines L8 and L11 presented themselves as the most promising, with similar performance in LN and HN. For PUE in the LP, the inbred line L8 distinguished itself again. The L7 line also presented higher genotypic values for components of PUE in HP conditions; the genetic potential of these inbred lines for efficient N and P use is clear. Moreover, based on the dispersion of genotypes in the graphs, it reiterates the lack of relationship between AE and UtE in all N and P environments.

Considering the results of selection gains, it was possible to obtain significant gains through direct selection; however, using indirect selection, it was possible to distinguish SDW that provided satisfactory gains in two environments. In HP, SDW provided gains with all traits, albeit reduced. Additionally, in LN, SDW provided the largest selection gains with selection, but also generated negative gains, especially for the trait of UtE in LP and HP. According to Whang et al. (2010), when the availability of a particular nutrient is limiting, AE is more important than UtE, although the effect of the latter trait has been neglected in some studies. According to Corrales et al. (2007), higher PUE in maize occurs because of major increases in PAE in relation to PUtE in LP. Similar results were found by Parentoni and Souza Junior (2008) under LP and HP. Thus, the SDW in LN conditions should be the trait used in improvement programs aimed at selecting plants efficient in the use of both N and P.

For that this to occur, the genes that are being screened for controlling the trait should be the same in both environments and/or be strongly linked. Moreover, genetic variation and, heritability in the environment for indirect selection, needs to be greater than in the environment of direct selection. In breeding for nutritional use efficiency is common and was observed in this study. Thus, under stress, there is a decrease in phenotypic variety and heritability as well, making it harder to select because the different genotypes tend to have similar results. Although there are significant genotype-by-environment interactions, genetic gains for nutrient use efficiency under low availability can be obtained by selecting the most efficient in ideal farming situations.

Table 3. Estimates of direct gains (diagonal) and indirect with selection (%) for the traits, shoot dry matter (SDW), acquisition efficiency (AE) and utilization efficiency (UtE) in high and low N and P for 15 maize inbred lines, Viçosa, Minas Gerais State, Brazil.

		LN			HN			LP			HP			
Trait		AE	UtE	SDW	AE	UtE	SDW	AE	UtE	SDW	AE	UtE	SDW	Mean
	AE	5.30	-0.48	27.34	1.68	0.20	24.65	-0.03	8.06	7.55	0.39	-1.52	32.57	8.81
LN	UtE	-1.14	5.61	6.51	0.25	0.66	13.07	0.06	-3.72	-3.04	-0.13	-0.07	-12.17	0.49
	SDW	5.11	2.54	36.78	2.36	1.24	47.10	0.16	-0.71	12.98	0.38	-0.37	36.97	12.05
	AE	5.11	2.54	36.78	2.36	1.24	47.10	0.16	-0.71	12.98	0.38	-0.37	36.97	12.05
HN	UtE	0.37	2.34	9.83	-0.09	2.20	7.14	0.09	2.12	12.69	0.06	0.04	6.40	3.60
	SDW	4.33	2.17	32.84	2.16	1.59	47.29	0.16	-0.78	12.06	0.28	-1.15	20.78	10.14
	AE	1.65	2.62	19.27	0.66	2.04	20.71	0.20	-3.62	12.41	0.03	-0.20	-0.15	4.64
LP	UtE	4.33	-0.07	23.12	1.44	-0.17	19.98	-0.03	10.86	10.21	0.28	-3.13	8.98	6.32
	SDW	2.86	2.67	23.11	1.38	1.04	28.87	0.05	7.84	15.92	0.30	-1.75	19.92	8.52
	AE	4.86	-0.66	26.60	1.87	1.05	35.40	0.02	4.66	10.07	0.54	-1.17	49.25	11.04
HP	UtE	-2.77	-0.94	-16.13	-0.88	-1.71	-19.19	0.02	-0.97	2.12	-0.13	3.71	10.24	-2.22
	SDW	4.20	1.19	25.78	1.44	0.34	23.45	0.06	4.25	13.02	0.39	1.43	50.31	10.49

Conclusion

Maize inbred lines exhibit differential behavior on the components of nitrogen and phosphorus use efficiency. The lengths of axial and lateral roots showed a strong association with nitrogen and phosphorus acquisition and utilization efficiency, independent of the availability of these nutrients. Inbred lines 8, 11, 3 and 12 showed contrasting performance to components N and P use efficiency and thus can be used to study inheritance.

References

BERNARDO, R. **Breeding for quantitative traits in plants**. Woodbury: Stemma Press, 2002.

BOUMA, T. J.; NIELSEN, K. L.; KOUTSTAAL, B. Sample preparation and scanning protocol for computerised analysis of root length and diameter. **Plant and Soil**, v. 218, n. 1-2, p. 185-196, 2000.

CHEN, J. Y.; XU, L.; CAI, Y. L.; XU, J. Identification of QTLs for phosphorus utilization efficiency in maize (*Zea mays* L.) across P levels. **Euphytica**, v. 167, n. 2, p. 245-252, 2009.

CHUN, L.; MI, G. H.; LI, J. S.; CHEN, F. J.; ZHANG, F. S. Genetic analysis of maize root characteristics in response to low nitrogen stress. **Plant and Soil**, v. 276, n. 2, p. 369-382, 2005.

CORRALES, I.; AMENÓS, M.; POSCHENRIEDER, C.; BARCELÓ, J. Phosphorus efficiency and root exudates in two contrasting tropical maize varieties. **Journal of Plant Nutrition**, v. 30, n. 6, p. 887-900, 2007.

CUI, Z. L.; ZHANG, F. S.; MI, G. H.; CHEN, F. J.; LI, F.; CHEN, X. P.; LI, J. L.; SHI, L. F. Interaction between genotypic difference and nitrogen management strategy in determining nitrogen use efficiency of summer maize. **Plant and Soil**, v. 317, n. 1-2, p. 267-276, 2009.

FEDOROFF, N. V.; BATTISTI, D. S.; BEACHY, R. N.; COOPER, P. J. M.; FISCHHOFF, D. A.; HODGES, C. N.; KNAUF, V. C.; LOBELL, D.; MAZUR. B. J.; MOLDEN, D.; REYNOLDS, M. P.; RONALD, P. C.; ROSEGRANT, M. W.; SANCHEZ, P. A.; VONSHAK, A.; ZHU, J. K. Radically rethinking agriculture for the 21st century. **Science**, v. 327, n. 5967, p. 833-834, 2010.

FORDE, B. G.; LORENZO, H. The nutritional control of root development. **Plant and Soil**, v. 232, n. 1-2, p. 51-68, 2001.

GALLAIS, A.; HIREL, B. An approach to the genetics of nitrogen use efficiency in maize. **Journal of Experimental Botany**, v. 55, n. 396, p. 295-306, 2004.

HAMMER, G. L.; DONG, Z. S.; MCLEAN, G.; DOHERTY, A.; MESSINA, C.; SCHUSLER, J.; ZINSELMEIER, C.; PASZKIEWICZ, S.; COOPER, M. Can changes in canopy and/or root system architecture explain historical maize yield trends in the US Corn Belt? **Crop Science**, v. 49, n. 1, p. 299-312, 2009.

HIREL, B.; LE GOUIS, J.; NEY, B.; GALLAIS, A. The challenge of improving nitrogen use efficiency in crop plants: towards a more central role for genetic variability and quantitative genetics within integrated approaches. **Journal of Experimental Botany**, v. 58, n. 9, p. 2369-2387, 2007.

HODGE, A.; BERTA, G.; DOUSSAN, C.; MERCHAN, F.; CRESPI, M. Plant root growth, architecture and function. **Plant and Soil**, v. 321, n. 1-2, p. 153-187, 2009.

HUND, A.; TRACHSEL, S.; STAMP, P. Growth of axile and lateral roots of maize: I development of a phenotying platform. **Plant and Soil**, v. 325, n.1-2, p. 335-349, 2009.

LAZOF, D. B.; RUFTY, T. W.; REDINBAUGH, M. G. Localization of nitrate absorption and translocation within morphological regions of the corn root. **Plant Physiology**, v. 100, n. 3, p. 1251-1258, 1992.

LIU, Y.; MI, G. H.; CHEN, F. J.; ZHANG, J. H.; ZHANG, F. S. Rhizosphere effect and root growth of two maize (*Zea mays* L.) genotypes with contrasting P efficiency at low P availability. **Plant Science**, v. 167, n. 2, p. 217-223, 2004.

LYNCH, J. P.; BROWN, K. M. Topsoil foraging - an architectural adaptation of plants to low phosphorus availability. **Plant and Soil**, v. 237, n. 2, p. 225-237, 2001.

PARENTONI, S. N.; SOUZA JUNIOR, C. L. Phosphorus acquisition and internal utilization efficiency in tropical maize genotypes. **Pesquisa Agropecuária Brasileira**, v. 43, n. 7, p. 893-901, 2008.

PRESTERL, T.; GROH, S.; LANDBECK, M.; SEITZ, G.; SCHMIDT, W.; GEIGER, H. H. Nitrogen uptake and utilization efficiency of European maize hybrids developed under conditions of low and high nitrogen input. **Plant Breeding**, v. 121, n. 6, p. 480-486, 2002.

RUBIO, G.; LIAO, H.; YAN, X. L.; LYNCH, J. P. Topsoil foraging and its role in plant competitiveness for phosphorus in common bean. **Crop Science**, v. 43, n. 2, p. 598-607, 2003.

RUTA, N.; LIEDGENS, M.; FRACHEBOUD, Y.; STAMP, P.; HUND, A. QTLs for the elongation of axile and lateral roots of maize in response to low water potential. **Theoretical and Applied Genetics**, v. 120, n. 3, p. 621-631, 2010.

SAS Institute. **SAS/STAT software versão 9.1**. Cary: SAS 2003. (CD-ROM).

SMIL, V. Detonator of the population explosion. **Nature**, v. 400, n. 6743, p. 415, 1999.

SOUZA, L. V.; MIRANDA, G. V.; GALVÃO, J. C. C.; ECKERT, F. R.; MANTOVANI, E. E.; LIMA, R. O.; GUIMARÃES, L. J. M. Genetic control of grain yield and nitrogen use efficiency in tropical maize. **Pesquisa Agropecuária Brasileira**, v. 43, n. 11, p. 1517-1523, 2008.

TILMAN, D.; CASSMAN, K. G.; MATSON, P. A.; NAYLOR, R.; POLASKY, S. Agricultural sustainability and intensive production practices. **Nature**, v. 418, n. 6898, p. 671-677, 2002.

TRACHSEL, S.; MESSMER, R.; STAMP, P.; HUND, A. Mapping of QTLs for lateral and axile root growth of tropical maize. **Theoretical and Applied Genetics**, v. 119, n. 8, p. 1413-1424, 2009.

WHANG, X.; SHEN, J.; LIAO, H. Acquisition or utilization, which is more critical for enhancing phosphorus efficiency in modern crops? **Plant Science**, v. 179, n. 4, p. 302-306, 2010.

WALK, T. C.; JARAMILLO, R.; LYNCH, J. P. Architectural tradeoffs between adventitious and basal roots for phosphorus acquisition. **Plant and Soil**, v. 279, n. 1-2, p. 347-366, 2006.

ZHANG, L. X.; LI, S. X.; ZHANG, H.; LIANG, Z. S. Nitrogen rates and water stress effects on production, lipid peroxidation and antioxidative enzyme activities in two maize (*Zea mays* L.) genotypes. **Journal of Agronomy and Crop Science**, v. 193, n. 6, p. 387-397, 2007.

ZHU, J. M.; KAEPPLER, S. M.; LYNCH, J. P. Mapping of QTLs for lateral root branching and length in maize (*Zea mays* L.) under differential phosphorus supply. **Theoretical and Applied Genetics**, v. 111, n. 4, p. 688-695, 2005.

12

Genetic variability and progeny selection of peach palm via mixed models (REML / BLUP)

Haroldo Silva Rodrigues[1*], Cosme Damião Cruz[1], Jeferson Luis Vasconcelos de Macêdo[2], Marcos Deon Vilela de Resende[3], Ricardo Lopes[2] and Aluízio Borém[1]

*[1]Departamento de Fitotecnia, Universidade Federal de Viçosa, Av. Peter Henry Rolfs, s/n, 36570-900, Campus Universitário, Viçosa, Minas Gerais, Brazil. [2]Embrapa Amazônia Ocidental, Manaus, Amazonas, Brazil. [3]Embrapa Florestas, Colombo, Paraná, Brazil. *Author for correspondence. E-mail: roldoagro@hotmail.com*

ABSTRACT. Heart of palm harvesting and exploitation has reduced the natural reserves of açaí and juçara; hence, the peach palm is an excellent option as source of heart of palm that can reduce the pressure on these two species. The peach palm has high tillering and no post-harvest oxidation. An experiment in a randomized block design with 72 progenies of peach palm, 36 replications and a single tree plot was performed for five years at the experimental field of Embrapa in the Western Amazon. Genetic parameters were estimated using two mixed models, one based on progeny means and the second on joint information from crops. A repeatability analysis, genotype selection, and progeny selection were performed; the Mulamba and Mock selection index was determined; and Tocher's clustering method was applied. The repeatability analysis revealed the existence of genetic variability, low repeatability, which ranged from 0.3307 to 0.6698 for five years. The analysis performed through the progeny means showed a heritability that varied from 51% to 23% for in all variables; the clustering revealed two groups. The progeny selection should be the main selection strategy, and crosses of superior progenies of different groups should be emphasized.

Keywords: heart of palm, perennial breeding, amazon.

Introduction

Brazil is the world's largest producer and consumer of heart of palm, the main way to obtain such good is through exploitation of juçara (*Euterpe edulis* Mart.) and açaí (*Euterpe oleracea* Mart.) (Sousa, Soares, Cordeiro, & Silva, 2011). Because the natural reserves of these two species in southern Brazil are running out, the exploitation of Amazonian acai and juçara has increased; these plants may also be depleted by predatory exploitation without good management practices.

The peach palm (*Bactris gasipaes* Kunth.) is proving to be a promising species and an alternative for heart of palm production. It shows desirable features for this sector. It is a native species, has high rusticity, is heat resistant, and has high tillering capacity, which reduces the need for land mechanization and replanting after several harvests. The heart of palm extracted from peach palm has a sweetish taste, is slightly yellowish in color, has good palatability, a doughy texture, low fibrousness, and does not show post cut darkening by oxidative enzymes, as noted in heart of palm extracted from juçara and açaí, which contributes to its commercialization as a minimally processed product (Kapp, Pinheiro, Silva Raupp, & Chaimsohn, 2003).

Breeding programs to increase heart of palm productivity are conducted through successive cycles of selection and require evaluations over several years in field conditions, and a precise estimation of the variance components is imperative to maximize the accuracy of progeny selection. Then, it becomes necessary to use estimation methods that may fit certain experimental conditions and unbalanced data (Farias Neto & Resende, 2001; Farias Neto et al., 2007; Farias Neto, Clement, & Resende, 2013).

The restricted maximum likelihood/best linear unbiased prediction procedure (REML / BLUP) has been successfully used in peach palm breeding (Farias Neto & Resende, 2001). According to Resende (2000), the prediction of additive genetic values and genotypic values through mixed model values is essential. Comparing this method with the least squares estimation method for variance components and the multi-effects indexes method for genotypes genetic values estimation, this method leads to similar results in cases of small imbalances. However, in conditions of highly unbalanced data, it provides more accurate results.

This study aimed to estimate genetic parameters in individual and joint analysis, conduct heritability and repeatability studies to guide the progeny selection, and estimate genotype genetic values of half-sibs of peach palm through BLUP and the application of Mulamba and Mock selection index, as well as study the genetic diversity by grouping progenies through Tocher's clustering method (Neto et al., 2013; Barbosa et al., 2014).

Material and methods

Experiment location

The experiment was conducted at the experimental station of the Agroforestry Research Center of the Western Amazon - Embrapa/CPAA, located at the AM-010 highway, km 29, in the coordinates of 2° 51 'S latitude and 59° 52' W longitude. The data were collected in 2009, 2010, 2011, 2012, and 2013. The area's soil is classified as clayey alic yellow latosol; it has a clay content of approximately 80% on the surface horizon, is acidic and shows low cation exchange capacity, high aluminum saturation and low phosphorus content. To characterize the area, two soil compound samples (20 single samples in each half of the area) were collected for physical (particle size) and chemical analysis (fertility; Table 1).

In October 2006, the area was tilled to incorporate plant residues and limed (1.5 tons of dolomitic limestone ha^{-1}). In January 2007, the seedlings were transplanted *in situ*; fertilization at planting and five topdressing fertilizations at 2, 6, 12, 24, and 36 months were performed according to Embrapa's technical recommendations.

Genetic resource

The progenies used in the study originated from the city of Benjamin Constant, Amazonas, Brazil, selected for two generations, and from Yurimaguas, a city in the Peruvian Amazon; all came from the RECA Project (Economical Condensed Intercropped Reforestation) in the municipality of Extrema, RO - Brazil.

Number of progenies

The number of progenies was selected by number of seedlings acquired at the end of the production period, considering vigor, nutritional aspects, the absence of pests and diseases and especially the total absence of thorns on the stipe. Thus, from 100 progenies, 72 progenies were selected and transplanted *in situ*.

Table 1. Physical and chemical characteristics, base saturation, micronutrients, total cation exchange capacity (T), effective cation exchange capacity (t), base saturation index (V) and aluminum saturation index (m) in the experimental area soil for peach palm progeny selection.

Sample	Depth (cm)	Particle size (g kg^{-1})					pH (H_2O)	Base saturation ($cmol_c$ dm^{-3})				Micronutrients (mg dm^{-3})			
		Sand		Silt	Clay	Silt/Clay		Ca	Mg	Al	H+Al	Fe	Zn	Mn	Cu
		Coarse	Fine												
A1	20	110.15	31.45	211.90	646.50	0.33	4.22	0.60	0.33	0.84	7.11	264	1.93	2.98	0.63
A2	20	111.45	30.43	230.62	627.50	0.37	4.32	0.56	0.33	0.82	7.75	191	1.78	2.58	0.56
		P	K	Na	SB	T	t		V	m		Carbon		Organic matter	
		(mg dm^{-3})						(%)				(g kg^{-1})			
A1	20	10	72	5	1.14	8.25	1.98		13.77	42.63		17.41		29.95	
A2	20	8	70	5	1.09	8.84	1.91		12.34	43.03		16.64		28.61	

Number of replications, experimental sketch and analyzed traits

The peach palms plants were arranged in a 2.0 m x 1.0 m space (density of 5,000 plants ha^{-1}). The experiment was conducted in a randomized block design with single tree plots. The treatments consisted of 72 progenies of peach palms in 36 replications (a total of 2,592 plants), which occupied an area of 0.8 hectares. Six traits directly related to the productivity and yield of heart of palm were measured: heart of palm basal weight (HPBW), heart of palm apical weight (HPAW), and heart of palm weight (HPW), which were measured in grams (g) using a high precision scale; billets number (BN), which is the number of billets produced in the tussock with a standard size of 9 cm; heart of palm diameter (HPD), measured in millimeters (mm); and heart of palm length (HPL), expressed in centimeters (cm) and measured with an electronic paquimeter. The measurements were taken after harvest and cleaning of the product in the field

Genotype and progeny selection

To estimate the genotype genetic parameters, we used the SELEGEN-REML/BLUP model 15. The estimations of the variance components and genetic values predictions via a suitable linear mixed model were equivalent to:

$y = Xf + Za + Wc + e$, where:

y, f, a, c, e are the data vectors, fixed effects (general average), additive effects (random), block effects (random events) and random errors, respectively, and X, Z and W are the incidence matrices associating f, a and c to the data vector y, respectively.

The f value was estimated through generalized least squares prediction and a and c through BLUP, the values were obtained by solving the linear equations system, which is called mixed model equation (MME):

$$\begin{bmatrix} \hat{f} \\ \hat{a} \\ \hat{c} \end{bmatrix} = \begin{bmatrix} X'X & X'Z & X'W \\ Z'X & Z'Z + A^{-1}\lambda_1 & Z'W \\ W'X & W'Z & W'W + I\lambda_2 \end{bmatrix}^{-1} \begin{bmatrix} X'y \\ Z'y \\ W'y \end{bmatrix}$$

where,

$$\lambda_1 = \frac{1-h^2-c^2}{h^2} = \frac{\sigma_e^2}{\sigma_a^2}; \quad \lambda_2 = \frac{1-h^2-c^2}{c^2} = \frac{\sigma_e^2}{\sigma_c^2};$$

A and I = additive kinship and identity matrix, respectively.

To obtaining the random effects variance, REML estimators applying the EM algorithm (Expectation-Maximization) are:

$$\hat{\sigma}_e^2 = [y'y - \hat{f}'X'y - \hat{a}'Z'y - \hat{C}'W'y]/[N - r(x)]$$

$$\hat{\sigma}_a^2 = [\hat{a}'A^{-1}\hat{a} + \sigma_e^2 trA^{-1}C^{22}]/q$$

$$\hat{\sigma}_c^2 = [\hat{c}'c + \hat{\sigma}_e^2 trC^{33}]/S$$

where:
tr = matrix trace;
$r(x)$ = X matrix's rank;
N- $r(x)$= Error degrees of freedom;
q = individuals number;
s = block number;
N = total data number.

C^{22} e C^{33}= comes from: $C = \begin{bmatrix} C^{11} & C^{12} & C^{13} \\ C^{21} & C^{22} & C^{23} \\ C^{31} & C^{32} & C^{33} \end{bmatrix}$, which is the coefficient matrix's generalized inverse of the MME.

The joint analysis, along with the repeatability analysis and progeny BLUP, were obtained through model 8 (Resende, 2007).

$y = Xm + Za + Wp + Qi + Ts + e$, where:

y, m, a, p, i, s, e are the data vectors, vector of the effects of the measuring-repetition combinations (fix) added to the general mean, vector of the individual additive genetic effects (random), plot effects vector (random), effects of the genotype measuring interaction vector (random), vector of the permanent effects (random), vector of residues (random), respectively, and X, Z, W, Q, and T are the incidence matrices associating m, a, p, i and s to the data vector y, respectively.

Simultaneous trait selection

The Mulamba and Mock selection index was used (Resende, Silva, & Azevedo, 2014).

Genetic diversity study

To study the genetic diversity among genotypes, we adopted a multivariate methodology. Through this methodology, a dissimilarity matrix was generated by a modified Mahalanobis generalized distance applied to the genotypic values estimated by:

$D^2_{ii'} = \delta R \delta$ where:

$D^2_{ii'}$ = Mahalanobis distance between i and i' genotypes;

R = genotypic variance and covariance matrix;

$\delta = [d_1 d_2 \ldots d_v]$, where $d_j = Y_{ij} - Y_{i'j}$;

Y_{ij}= genotypic mean of the ith progeny in the jth trait.

The cluster of progeny was performed using Tocher's method, as described by Cruz, Ferreira, and Pessoni (2011), where the estimated intragroup distance must be smaller than the intergroup distance. Because a large group was established, the same method was applied to split it into subgroups.

All analysis were performed in SELEGEN-REML / BLUP (Resende, 2007).

Results and discussion

The genetic parameters estimation for individual analysis are shown in Table 2; low values for additive genetic variance (σ^2a) can be observed. The residual variance (σ^2) was greater than σ^2a for all traits, which may mask the phenotypic selection. This fact is reflected in the narrow sense heritability based on genotype selection (h^2a); it is observed that the highest heritability was estimated in the years 2010 and 2013, in which we also observed the highest concentration of significant additive genetic variance. The heritability based in progeny selection (h^2pa) seems more suitable to selection; this heritability ranged from 51 to 23% for the traits with statistically significant additive genetic variance. Padilha, Oliveira, and Mota (2003) found CV% values of 49.25, 46.20, 56.45 8.15, 22.50 and 18.14 for HPBW, HPAW, HPW, BN, HPD, and HPL, respectively. These authors analyzed 14-year-old plants; the high coefficients of variation found in this study may be due to the use of young plants that have not reached production stability yet.

A joint analysis showed low h^2a for all studied traits, which ranged from 1% to 3% for HPBW, HPAW, BN, HPD, and HPL (Table 2). The low heritability values found can be attributed to the fact that the tested progeny were at the third selection cycle. The heritability for progeny selection is much higher than the heritability based on the genotypes either in the individual or joint analysis; thus, progeny selection is the best way to obtain meaningful gains. Aiming at the rapid availability of superior genetic materials for farmers, genotypes from the best progenies can be selected and cloned. The difference observed between means was due to the destruction required for the data collection: the trees were cut to gain accesses to the heart of palm.

The analyzed correlations between years (r) showed a low magnitude (Table 2), which indicates a complex genotype-environment interaction (Vencovsky & Barriga, 1992) and means that the genotype ranking changes over time, which is not a positive situation if the genotypes show unstable behavior over time.

Table 2. Individual analysis, joint analysis and genetic parameters estimative for the traits heart of palm basal weight (HPBW), heart of palm apical weight (HPAW), heart of palm weight (HPW), billets number (BN), heart of palm diameter (HPD) and heart of palm length (HPL) of half-sibs of peach palm (Bactris gasipaes Kunth.) progenies grown in EMBRAPA Western Amazon in Manaus, Amazonas.

	2009							2010					
	HPBW	HPAW	HPW	BN	HPD	HPL		HPBW	HPAW	HPW	BN	HPD	HPL
σ^2a	76.46^{ns}	2.63^{ns}	426.26★	0.03^{ns}	0.03★	2.48^{ns}		7354.46★	118.09★	2254.05^{nss}	2.38★	0.01^{ns}	193.16★
σ^2b	833.71	53.61	454.45	0.05	0.01	4.42		9789.11	131.54	1074.77	1.49	0.03	120.68
B													
σ^2	2850.68	880.06	6842.85	1.22	0.16	99.14		96387.40	1841.60	43315.08	21.04	0.40	1704.03
σ^2p	3760.84	936.29	7723.56	1.31	0.19	106.05		113530.97	2091.22	46643.89	24.91	0.43	2017.86
F													
h^2a	0.02±0.01	0.002±0.06	0.05±0.02	0.02±0.01	0.13±0.04	0.02±0.01		0.06±0.03	0.05±0.03	0.05±0.02	0.10±0.04	0.01±0.01	0.10±0.04
h^2pa	0.16	0.02	0.30	0.15	0.51	0.15		0.30	0.25	0.23	0.38	0.08	0.38
CVe%	28.51	63.71	54.53	36.99	16.28	36.99		79.59	80.19	77.81	69.95	26.89	69.95
$\bar{X}$	189.14	46.61	155.21	3.02	2.59	27.17		401.07	54.79	272.64	6.83	2.36	61.47
	2011							2012					
	HPBW	HPAW	HPW	BN	HPD	HPL		HPBW	HPAW	HPW	BN	HPD	HPL
σ^2a	87.38^{ns}	14.25^{ns}	230.05^{ns}	0.02^{ns}	0.01^{ns}	1.72^{ns}		18.94^{ns}	30.39^{ns}	251.25^{ns}	0.01^{ns}	0.01^{ns}	0.87^{ns}
σ^2b	1823.79	27.23	727.15	0.03	0.00	2.66		251.89	54.07	1500.18	0.03	0.02	2.03
B													
σ^2	13624.22	1599.08	34431.77	2.19	0.21	177.58		6702.92	1404.04	43433.63	3.20	1.14	259.15
σ^2p	15535.40	1640.56	35388.97	2.25	0.23	181.97		6973.75	1488.51	45185.05	3.24	1.16	262.05
F													
h^2a	0.01±0.01	0.01±0.01	0.01±0.01	0.01±0.01	0.06±0.04	0.01±0.01		0.00±0.01	0.02±0.02	0.01±0.01	0.00±0.01	0.01±0.01	0.00±0.01
h^2pa	0.02	0.03	0.02	0.03	0.17	0.03		0.01	0.07	0.02	0.01	0.03	0.01
CVe%	62.49	73.67	60.52	57.64	16.29	57.64		57.22	68.64	61.48	57.99	38.56	57.99
$\bar{X}$	187.23	54.46	307.36	2.58	2.87169	23.20		143.23	55.03	339.71	3.09	2.77	27.79
	2013							Joint analysis					
	HPBW	HPAW	HPW	BN	HPD	HPL		HPBW	HPAW	HPW	BN	HPD	HPL
σ^2a	154.33★	9.51^{ns}	536.71^{ns}	0.20★	0.01^{ns}	16.46★		540.72★	38.01★	822.89★	0.19★	0.01★	15.48★
σ^2b	406.43	111.28	1227.05	0.59	0.01	48.19		1372.157	32.21	390.69	0.11	0.01	9.03
B													
σ^2	1504.53	1564.62	30898.42	2.14	0.36	173.38		26057.087	1378.22	27766.39	6.74	0.34	546.11
σ^2p	2065.28	1685.41	32662.19	2.94	0.39	238.03		30111.129	1514.44	31636.27	7.41	0.47	600.43
F													
h^2a	0.07±0.03	0.01±0.01	0.02±0.01	0.07±0.03	0.03±0.02	0.07±0.03		0.01±0.01	0.03±0.01	0.02±0.01	0.03±0.01	0.03±0.01	0.03±0.01
h^2pa	0.35	0.03	0.09	0.33	0.17	0.33		0.31	0.36	0.38	0.39	0.36	0.39
CVe%	61.41	77.00	69.64	59.63	22.99	59.63		82.33	71.02	62.91	72.08	21.97	72.08
$\bar{X}$	65.55	51.49	254.04	2.54	2.65	22.86		196.07	52.28	264.87	3.60	2.64	32.42
-	-	-	-	-	-	-	r	0.02	0.09	0.12	0.09	0.28	0.09

σ^2a: additive genetic variance, σ^2b: block variance, σ^2: residual variance, σ^2p: individual phenotypic variance, h^2a: narrow sense heritability based in genotypes and its standard errors represented by ±, h^2pa: heritability based in progeny means, CV(%): experimental coefficient of variation, $\bar{X}$: average, ★/ns: statistically significant through Deviance test [$\chi^2(1, 5\%)$= 3.,84] and not statistically significant, respectively, r: correlation between years.

Repeatability for five measurements (five years) presented determination coefficients that ranged from 0.3307 to 0.6698, which are associated with accuracies that ranged from 0.2416 to 0.3182 (Table 3). Using five measures, the h^2a based on the yearly averages were relatively low for all the analyzed traits. Padilha et al. (2003) concluded that repeatability coefficients for all heart of palm production traits are negligible and that several measurements are needed to express a reliable progeny value. Moreover, Bergo, Negreiros, Miqueloni, and Lunz (2013) indicated that seven measurement cycles was sufficient to predict the real value of the progenies and reach an accuracy of 85% by the principal components method. The use of seven measures for all the analyzed traits showed determination coefficient values that ranged from 0.4086 to 0.7396 and accuracies from 0.2616 to 0.3426, while the use of 10 measures showed values that ranged from 0,4987 to 0.8023 for the coefficient of determination and 0.2814 to 0.3802 for accuracy.

The ranking of the best selected progenies (20%) is shown in Table 4. These progenies can compose a seed production orchard and meet the market demand for improved seeds.

The selection based on joint analysis (Table 4) shows that the three best progenies for HPBW, BN and HPL did not change: progenies 157, 158 and 49. These progenies also ranked well in HPAW an HPW. The traits HPBW, BN, and HPL showed constant statistically significant σ^2a over the years. Progeny 157 has high additive genetic value (a) and is well ranked in most of the studied traits; it is expected that its offspring inherit half of the a. Selecting 20% of the analyzed progenies (15 progenies) to assemble a seed production orchard is expected to achieve a gain of 18.43 grams for HPBW, 5.55 grams HPAW, HPW to 25.99 grams, 0.712 billets for BN, and 0.09 cm to 6.38 cm for HPD and HPL. The use of a high number of traits makes selection difficult to perform; using selection indexes that maximize the predicted gain may be a good option in many cases.

The adoption of a selection strategy without overlapping generations in seed production can be positive due to the heterosis, possibly obtained through hybridization. Urpí and Solis (1980) described peach palm as an allogamous species and proposed the existence of a genetic self-incompatibility system. Mossanek, Wendling, Koehler, and Zuffellato-Ribas (2014) mentioned that peach palm pollination happens through crosses, which also suggests allogamy.

Table 3. Repeatability study for the traits heart of palm basal weight (HPBW), heart of palm apical weight (HPAW), heart of palm weight (HPW), billets number (BN), heart of palm diameter (HPD) and heart of palm length (HPL) of half-sibs of peach palm (Bactris gasipaes Kunth.) progenies grown in EMBRAPA Western Amazon in Manaus, Amaazonas, analyzed during 5 years.

HC	$h^2_a ya$	R^2	PA	REHU	HC	$h^2_a ya$	R^2	PA	REHU
		HPBW					HPAW		
1	0.0180	0.1346	0.1340	1.00	1	0.0251	0.0899	0.1584	1.00
2	0.0317	0.2373	0.1779	1.33	2	0.0461	0.1650	0.2146	1.35
3	0.0424	0.3182	0.2060	1.54	3	0.0638	0.2287	0.2526	1.59
4	0.0512	0.3836	0.2262	1.69	4	0.0791	0.2833	0.2812	1.77
5	0.0584	0.4375	0.2416	1.80	5	0.0923	0.3307	0.3038	1.92
6	0.0644	0.4828	0.2538	1.89	6	0.1039	0.3723	0.3223	2.03
7	0.0695	0.5213	0.2637	1.97	7	0.1141	0.4089	0.3378	2.13
8	0.0740	0.5545	0.2720	2.03	8	0.1232	0.4416	0.3510	2.22
9	0.0778	0.5834	0.2789	2.08	9	0.1314	0.4708	0.3624	2.29
10	0.0812	0.6087	0.2849	2.13	10	0.1387	0.4971	0.3724	2.35
		HPW					BN		
1	0.0260	0.1223	0.1613	1.00	1	0.0258	0.0905	0.1606	1.00
2	0.0464	0.2180	0.2153	1.33	2	0.0473	0.1659	0.2174	1.35
3	0.0627	0.2948	0.2504	1.55	3	0.0655	0.2298	0.2559	1.59
4	0.0761	0.3579	0.2759	1.71	4	0.0811	0.2846	0.2848	1.77
5	0.0873	0.4107	0.2955	1.83	5	0.0946	0.3322	0.3076	1.92
6	0.0968	0.4554	0.3112	1.93	6	0.1065	0.3738	0.3263	2.03
7	0.1050	0.4938	0.3240	2.01	7	0.1170	0.4105	0.3420	2.13
8	0.1121	0.5272	0.3348	2.08	8	0.1263	0.4431	0.3553	2.21
9	0.1183	0.5564	0.3440	2.13	9	0.1346	0.4724	0.3669	2.28
10	0.1238	0.5822	0.3519	2.18	10	0.1421	0.4987	0.3769	2.35
		HPD					HPL		
1	0.0285	0.2886	0.1688	1.00	1	0.0360	0.0801	0.1615	1.00
2	0.0442	0.4480	0.2103	1.25	2	0.0462	0.1660	0.2257	1.37
3	0.0542	0.5490	0.2328	1.38	3	0.0756	0.2301	0.2693	1.61
4	0.0611	0.6187	0.2471	1.46	4	0.0711	0.2798	0.2771	1.80
5	0.0661	0.6698	0.2571	1.52	5	0.1016	0.3402	0.3182	1.96
6	0.0700	0.7088	0.2645	1.57	6	0.1103	0.3837	0.3315	2.10
7	0.0730	0.7396	0.2702	1.60	7	0.1190	0.4086	0.3426	2.19
8	0.0755	0.7645	0.2747	1.63	8	0.1282	0.4518	0.3622	2.25
9	0.0775	0.7850	0.2784	1.65	9	0.1339	0.4697	0.3691	2.27
10	0.0792	0.8023	0.2814	1.67	10	0.1419	0.4972	0.3802	2.33

HC: harvest numbers, $h^2_a ya$: individual narrow sense heritability based in year's average, R^2: determination coefficient, PA: phenotypic accuracy and REHU: relative efficiency of one harvest necessary use to achieve certain selective accuracy values.

Table 4. Selection of the fifteen best half-sib progenies of peach palm (Bactris gasipaes Kunth.) for the traits heart of palm basal weight (HPBW), heart of palm apical weight (HPTW), heart of palm weight (HPW), billets number (BN), heart of palm diameter (HPD) and heart of palm length (HPL), grown in EMBRAPA Western Amazon in Manaus, Amazonas.

HPBW				HPAW			
Genitor	*a*	Gain	New average	Genitor	*a*	Gain	New average
157	46.65	46.65	242.72	113	9.69	9.69	61.97
49	31.50	39.07	235.14	40	8.45	9.07	61.35
158	29.83	35.99	232.06	6	8.23	8.79	61.07
137	25.74	33.43	229.50	131	8.21	8.65	60.92
48	24.87	31.72	227.79	48	7.15	8.35	60.62
163	22.42	30.17	226.24	85	6.91	8.11	60.38
47	13.39	27.77	223.84	114	6.16	7.83	60.10
131	13.35	25.97	222.04	162	4.82	7.45	59.73
162	11.80	24.39	220.47	157	4.31	7.10	59.38
44	10.81	23.04	219.11	12	4.03	6.80	59.07
85	10.45	21.89	217.96	18	3.45	6.49	58.77
51	10.00	20.90	216.97	137	3.01	6.20	58.48
113	8.80	19.97	216.04	75	2.94	5.95	58.23
12	8.78	19.17	215.24	68	2.92	5.73	58.01
41	8.06	18.43	214.50	163	2.91	5.55	57.82
HPW				BN			
Genitor	*a*	Gain	New average	Genitor	*a*	Gain	New average
158	43.78	43.78	308.66	157	0.78	0.78	4.38
88	42.92	43.35	308.23	158	0.65	0.72	4.32
137	41.00	42.57	307.44	49	0.61	0.68	4.29
157	40.63	42.08	306.96	48	0.48	0.63	4.23
48	36.26	40.92	305.79	137	0.47	0.60	4.20
131	29.43	39.00	303.88	41	0.41	0.57	4.17
47	28.26	37.47	302.34	163	0.40	0.54	4.15
46	24.53	35.85	300.73	47	0.38	0.52	4.13
154	19.91	34.08	298.95	162	0.32	0.50	4.10
152	18.84	32.56	297.43	131	0.29	0.48	4.08
49	17.89	31.22	296.10	88	0.29	0.46	4.07
40	12.43	29.66	294.53	85	0.28	0.45	4.05
36	12.05	28.30	293.18	164	0.27	0.43	4.04
41	11.36	27.09	291.97	46	0.26	0.42	4.02
163	10.53	25.99	290.86	51	0.19	0.41	4.01
HPD				HPL			
Genitor	*a*	Gain	New average	Genitor	*a*	Gain	New average
156	0.36	0.36	3.00	157	7.04	7.04	39.46
131	0.13	0.24	2.89	158	5.89	6.46	38.89
17	0.10	0.20	2.84	49	5.52	6.15	38.57
81	0.10	0.17	2.82	48	4.29	5.68	38.11
18	0.09	0.16	2.80	137	4.27	5.40	37.82
152	0.09	0.14	2.79	41	3.67	5.11	37.54
88	0.08	0.13	2.78	163	3.63	4.90	37.32
114	0.07	0.13	2.77	47	3.40	4.71	37.14
9	0.06	0.12	2.76	162	2.90	4.51	36.93
123	0.06	0.11	2.76	131	2.64	4.33	36.75
137	0.06	0.11	2.75	88	2.61	4.17	36.59
158	0.06	0.11	2.75	85	2.50	4.03	36.45
75	0.06	0.10	2.75	164	2.45	3.91	36.33
154	0.05	0.10	2.74	46	2.36	3.80	36.22
155	0.05	0.09	2.74	51	1.73	3.66	36.08

a:additive genetic value.

Farias Neto and Resende (2001) selected 31 and 53 genotypes for a short and a long peach palm breeding program, respectively. This type of selection is reflected in the effective population size (Ne). Table 5 shows the selection of genotypes based on different Ne values. To achieve a selective ceiling where Ne varies between 19 (Farias Neto & Rezende, 2001), 30 and 60 (Pires, Resende, Silva, & Resende Jr., 2011), a number of genotypes and progenies can be selected. It appears that the genotype of progeny 157 located in block 10 showed phenotypic values much higher than other genotypes for at least half of the studied traits.

As cited above, Farias Neto and Resende (2001) presented a selection of 31 genotypes from 9 progenies that resulted in an Ne of 19 for a short-term breeding program; in this study, the selection that results in an Ne of 19 is the use of 48 to 170 genotypes from 11 to 19 progenies. Table 5 shows that the selection of 226 to 303 genotypes leads to 22 to 32 progenies, maintaining the Ne near 30; in the work of Farias Neto and Resende (2001), 53 genotypes were selected and, consequently, 15 progenies reached the same Ne in all traits. Focusing on a selective ceiling to reach an Ne of 60, 552 to 600 genotypes should be selected, thus selecting 35 to 49 progenies. The use of a large number of progenies retains the genetic diversity within the breeding population for a longer amount of time; otherwise, results will be obtained in the long term.

To select superior genotypes that simultaneously possess high performance in the desirable traits is usually a difficult task for a breeder. Thus, using selection index theory can be a good effective alternative (Cruz Regazzi, & Carneiro, 2014). The analysis performed on the Mulamba and Mock selection index (Resende et al., 2014) classified the 72 analyzed progenies (Table 6). The best rated progeny was progeny 157, with a gain of 563.6% in relation to all other studied progenies; the last in the rank was progeny 151. Using a selection rate of 20%, it resulted in selecting 15 progenies, which led to an expected gain of 166%.

Table 7 shows the result of Tocher's clustering based on Mahalanobis distance, where the formation of two distinct groups can be observed. Group one is the largest, containing 71 progenies; it was divided into 32 subgroups, also using Tocher's method. The formed group can point to future crosses in breeding programs; an intra-population recurrent selection program could also be started, evaluating general and specific combining ability.

Overall, based on the analyses performed in this study, it is easy to conclude that progeny 157 had a superior performance; it is the best ranked in progeny selection (Table 4), individual BLUP (Table 5), and the selection index (Table 6) and is isolated in a group on the cluster analysis (group 2). Negreiros, Bergo, Miqueloni, and Lunz (2013) found 26 groups when they analyzed 100 progenies of peach palm; they also reported that they found a strong genetic divergence using Tocher's method.

Table 5. Peach palm (Bactris gasipaes Kunth.) genotype selections grown in EMBRAPA Western Amazon in Manaus, Amazonas for the traits heart of palm basal weight (HPBW), heart of palm apical weight (HPAW), heart of palm weight (HPW), billets number (BN), heart of palm diameter (HPD) and heart of palm length (HPL) based on the effective population size.

NGen	Prog	Block	*f*	*a*	*u+a*	Gain	New avarege	Ne	*d*	*g*	NP
HPBW											
1	157	10	967.3	53.5	250	53.46	250	1	20	73.6	1
157	158	18	202.4	15.3	211.4	21.1	217.1	19.0	0.3	15.6	19
303	48	34	44.5	11.3	207.4	17.3	213.3	30.0	-0.7	10.6	32
588	11	20	341.0	6.0	202.1	12.9	209.0	60.0	4.6	10.6	48
HPAW											
1	6	29	273.5	11.7	64.0	11.7	64.0	1.0	5.1	16.8	1
48	162	18	88.8	5.6	57.9	6.9	59.1	19.0	2.1	7.7	14
226	131	17	40.0	3.8	56.1	5.0	57.3	30.0	-0.2	3.6	22
573	18	35	70.0	2.1	54.4	3.7	56.0	60.1	0.3	2.4	47
HPW											
1	158	2	535.4	40.5	305.3	40.5	305.3	1.0	12.4	52.8	1
76	18	4	579.8	24.5	289.4	30.2	295.1	19.1	12.9	37.5	11
278	51	2	491.8	17.5	282.4	23.1	287.9	30.1	11.7	29.2	25
600	49	17	259.7	10.2	275.1	17.9	282.8	60.0	0.9	11.1	48
BN											
1	157	10	14.4	1.0	4.6	1.0	4.6	1.0	0.4	1.4	1
117	48	27	4.7	0.3	4.0	0.4	4.0	19.1	0.1	0.4	19
244	40	31	11.3	0.3	3.9	0.4	4.0	30.2	0.2	0.4	30
552	74	31	8.5	0.2	3.8	0.3	3.9	60.0	0.1	0.3	49
HPD											
1	156	32	38.5	1.0	3.6	1.0	3.6	1.0	0.5	1.5	1
170	18	21	2.9	0.1	2.7	0.1	2.8	19.1	0.0	0.1	17
266	88	24	3.0	0.0	2.7	0.1	2.7	30.1	0.0	0.1	23
569	155	24	2.8	0.0	2.7	0.1	2.7	60.0	0.0	0.0	35
HPL											
1	157	10	129.4	8.9	41.4	8.9	41.4	1.0	3.6	12.5	1
117	48	27	42.0	3.1	35.6	3.9	36.3	19.1	0.7	3.8	19
244	40	31	101.3	2.5	34.9	3.3	35.7	30.2	1.4	3.9	30
552	74	31	76.5	1.5	33.9	2.5	35.0	60.0	0.8	2.3	48

NGen: number of genotypes, Prog: progeny, f: Phenotypic value, a: additive effect, u+a: additive genetic value, Ne: effective population size, d: dominance effect, g: effect genotypic total NP: number of progenies.

Table 6. Mulamba and Mock selection index for 72 half-sib progenies of peach palm (Bactris gasipaes Kunth.) grown in EMBRAPA Western Amazon in Manaus, Amazonas, based on heart of palm basal weight (HPBW), heart of palm apical weight (HPAW), heart of palm weight (HPW), billets number (BN), heart of palm diameter (HPD) and heart of palm length (HPL).

Prog	Rank	Gain	Gain (%)	Prog	Rank	Gain	Gain (%)	Prog	Rank	Gain	Gain(%)
157	5.5	5.50	563.6	36	27.3	18.11	101.51	118	45.7	28.24	29.3
131	6.7	6.08	500.0	155	28.7	18.52	97.09	123	46.0	28.59	27.7
137	6.7	6.28	481.4	154	30.3	18.96	92.54	68	47.3	28.96	26.0
158	7.8	6.67	447.5	139	32.3	19.43	87.81	74	47.8	29.32	24.5
48	11.5	7.63	378.2	78	32.5	19.89	83.55	73	48.0	29.68	23.0
49	14.5	8.78	315.8	17	33.3	20.33	79.51	35	48.5	30.02	21.6
41	15.5	9.74	274.8	9	34.5	20.79	75.56	81	48.5	30.36	20.2
162	15.5	10.46	249.0	63	35.3	21.24	71.81	165	49.2	30.70	18.9
163	15.7	11.04	230.7	102	36.0	21.69	68.27	145	51.3	31.06	17.5
85	16.3	11.57	215.6	136	38.2	22.18	64.59	43	51.8	31.42	16.2
44	16.8	12.05	203.0	126	38.3	22.64	61.23	125	52.0	31.77	14.9
47	17.0	12.46	193.0	6	38.5	23.08	58.15	71	52.7	32.11	13.7
46	18.5	12.92	182.4	11	39.2	23.51	55.23	107	55.3	32.49	12.3
88	18.7	13.33	173.8	54	40.2	23.95	52.39	84	55.3	32.86	11.1
113	19.2	13.72	166.0	144	40.5	24.38	49.74	39	55.5	33.22	9.9
114	19.3	14.07	159.4	31	41.0	24.79	47.23	156	55.8	33.58	8.7
40	21.2	14.49	151.9	160	41.3	25.20	44.87	7	56.0	33.92	7.6
164	23.7	15.00	143.3	83	42.0	25.60	42.60	69	56.7	34.27	6.5
18	23.8	15.46	136.0	153	43.0	26.00	40.38	70	56.8	34.60	5.5
12	24.0	15.89	129.7	25	43.5	26.40	38.27	14	58.7	34.96	4.4
51	26.5	16.40	122.6	37	43.7	26.78	36.29	65	59.3	35.31	3.4
24	26.8	16.87	116.3	16	43.8	27.15	34.43	149	61.0	35.68	2.3
75	27.2	17.32	110.8	60	44.2	27.51	32.66	108	63.5	36.07	1.2
152	27.2	17.73	105.9	55	44.8	27.88	30.94	151	67.2	36.50	0.0

Prog: progeny, Rank: average progeny rank.

Table 7. Clustering of 72 half-sib progenies of peach palm (Bactris gasipaes Kunth.) through Tocher's method based on the Mahalanobis distance for the traits heart of palm basal weight (HPBW), heart of palm apical weight (HPAW), heart of palm weight (HPW), billets number (BN), heart of palm diameter (HPD) and heart of palm length (HPL).

Group	Progenies													
1	6	9	35	63	164	43	71	55	114	81	74	83	44	156
	40	73	153	25	165	151	131	46	113	68	17	78	139	7
	16	136	160	41	118	47	155	51	158	39	88	11	137	145
	149	154	108	125	24	54	85	75	102	107	163	49	123	162
	144	48	60	37	18	31	84	36	12	65	69	126	14	70
	152													
2	157													

Conclusion

The studied population has low genetic variability. However, the progeny selection can result in significant gains in breeding programs.

A large number of measurements of the production traits of the heart of palm from peach palm are necessary to obtain reliable repeatability values.

The progeny selection by BLUP and the selection index confirmed the superiority of some progenies.

Crossing between genotypes of superior progenies located in different genetic groups can result in heterosis or high specific combined ability.

Acknowledgements

The authors thank the Embrapa (Brazilian Agricultural Research Corporation) Western Amazon - CPAA for permission to conduct the experiment, the bioinformatics laboratory of the Federal University of Viçosa for providing the necessary structure for this work and FAPEMIG (Research Support Foundation of Minas Gerais) for the scholarship support.

References

Barbosa, M. H. P., Ferreira, A., Peixoto, L. A., Resende, M. D. V., Nascimento, M., & Silva, F. F. (2014). Selection of sugar cane families by using BLUP and multi-diverse analyses for planting in the Brazilian savannah. *Genetics and Molecular Research, 13*(1), 1619-1626.

Bergo, C. L., Negreiros, J. D. S., Miqueloni, D. P., & Lunz, A. M. P. (2013). Estimativas de repetibilidade de caracteres de produção em pupunheiras para palmito da raça putumayo. *Revista Brasileira de Fruticultura, 35*(3), 829-836.

Cruz, C. D., Ferreira, F. M., & Pessoni, L. A. (2011). *Biometria aplicada ao estudo da diversidade genética.* Visconde do Rio Branco, MG: Suprema.

Cruz, C. D., Regazzi, A. J., & Carneiro, P. C. S. (2014). *Modelos biométricos aplicados ao melhoramento de plantas* (Vol. 1., 4th ed.). Viçosa, MG: UFV.

Farias Neto, J. T. D., & Resende, M. D. V. (2001). Aplicação da metodologia de modelos mistos (REML/BLUP) na estimação de componentes de variância e predição de valores genéticos em pupunheira (Bactris gasipaes). *Revista Brasileira de Fruticultura*, *23*(2), 320-324.

Farias Neto, J. T., Resende, M. D. V., Oliveira, M. D. S. P., Santos, N. S. A., Canuto, E. L., Nogueira, O. L., & Muller, A. A. (2007). Genetic evaluation of progenyes of acaizeiro (Euterpe oleracea) and estimates of genetic parameters. *Cerne*, *13*(4), 376-383.

Farias Neto, J. T., Clement, C. R., & Resende, M. D. V. D. (2013). Estimates of genetic parameters and selection gain for fruit production in open-pollinated progenies of peach palm in the State of Pará, Brazil. *Bragantia*, *72*(2), 122-126.

Kapp, E. A., Pinheiro, J. L., Silva Raupp, D., & Chaimsohn, F. P. (2003). Tempo de preservação de tolete de palmito pupunha (Bactris gasipaes) minimamente processado e armazenado sob refrigeração. *Publicatio UEPG: Ciências Exatas e da Terra, Agrárias e Engenharias*, *9*(3), 51-57.

Urpí, J. M., & Solís, E. M. (1980). Polinización en Bactrisgasipaes HBK (palmae). *Revista Biologia. Tropical*, *28*(1), 153-174.

Mossanek, E. A. O., Wendling, I., Koehler, H. S., & Zuffellato-Ribas, K. C. (2014). Indução de perfilhos em mudas de pupunheira. *Brazilian Journal of Forestry Research*. *37*(77), 39-48.

Negreiros, J. R. S., Bergo, C. L., Miqueloni, D. P., & Lunz, A. M. P. (2013). Divergência genética entre progênies de pupunheira quanto a caracteres de palmito. *Pesquisa Agropecuária Brasileira*. *48*(5), 496-503.

Neto, H. Z., Daros, E., Bespalhok Filho, J. C., Scapim, C. Gonçalves Vidigal, M. C., & Vidigal Filho, P. S. (2013). Selection of families and parents of sugarcane (Saccharum spp.) through mixed models by joint analysis of two harvests. *Euphytica*, *193*(3), 391-408.

Padilha, N. C. C., Oliveira, M. D. S. P. D., & Mota, M. G. D. C. (2003). Estimativa da repetibilidade em caracteres morfológicos e de produção de palmito em pupunheira (Bactris gasipaes Kunth). *Revista Árvore*, *27*(3), 435-442.

Pires, I. E., Resende, M. D. V., Silva, R., & Resende Jr, M. F. R. (2011). *Genética florestal*. Viçosa, MG: Arka.

Resende, M. D. V. (2000). *Análise estatística de modelos mistos via REML/BLUP na exper imentação em melhoramento de plantas perenes* (Documentos, 47). Colombo, PR: Embrapa Florestas.

Resende, M. D. V. (2007). *SELEGEN-REML/BLUP: sistema estatístico e seleção genética computadorizada via modelos lineares mistos*. Colombo, PR: Embrapa Florestas.

Resende, M. D. V., Silva, F. F., & Azevedo, C. F. (2014). *Estatística matemática, biométrica e computacional*. Viçosa, MG: UFV.

Sousa, E. P. D., Soares, N. S., Cordeiro, S. A., & Silva, M. L. D. (2011). Competitividade da produção de palmito de pupunha no Espírito Santo e em São Paulo. *Revista de Economia e Sociologia Rural*, *49*(1), 157-179.

Vencovsky, R., & Barriga, P. (1992). *Genética biométricano fitomelhoramento*. Ribeirão Preto, SP: Sociedade Brasileira de Genética.

13

Adaptability and stability of strawberry cultivars using a mixed model

Andréa Ferreira Costa[1*], Nilton Rocha Leal[2], José Aires Ventura[1], Leandro Simões Azeredo Gonçalves[3], Antonio Teixeira do Amaral Júnior[2] and Hélcio Costa[1]

[1]*Instituto Capixaba de Pesquisa, Assistência Técnica e Extensão Rural, Estação Experimental Mendes da Fonseca, BR-262, km 94, Domingos Martins, Espírito Santo, Brazil.* [2]*Centro de Ciências e Tecnologias Agropecuárias, Laboratório de Melhoramento Genético Vegetal, Universidade Estadual do Norte Fluminense "Darcy Ribeiro", Campos dos Goytacazes, Rio de Janeiro, Brazil.* [3]*Centro de Ciências Agrárias, Departamento de Agronomia, Universidade Estadual de Londrina, Londrina, Paraná, Brazil. *Author for correspondence. E-mail: andreacosta_2000@yahoo.com.br*

ABSTRACT. Although strawberry crops have a strong socio-economic impact on the agricultural sector of Espírito Santo State, there are few studies on the performance of strawberry cultivars in different locations and years under a low tunnel management system (LT). Therefore, this study aimed to estimate the parameters of adaptability and stability of strawberry cultivars under LT-protected cultivation using the harmonic mean of the relative performance of genetically predicted values. Seven strawberry cultivars were assessed ('Dover', 'Camino Real', 'Ventana', 'Camarosa', 'Seascape', 'Diamante' and 'Aromas') in the agricultural years 2006/7, 2007/8 and 2008/9 in three locations in the mountainous region of Espírito Santo State. The experiment was arranged in a randomized block design, with 3 replications and 15 plants per plot. The analysis of deviance for the yield (ton. ha^{-1}) demonstrated that only the effects of genotype and the triple interaction genotype x location x year were significant, which indicates the presence of genetic variability among the cultivars and the inconsistency of the position among the genotypes for combinations of year and location. Considering the selection for yield, adaptability and stability, in LT-protected cultivation, cultivars Camarosa and Aromas are highlighted for expressing average values that were 22% higher than the overall mean of the cultivars (24.55 ton. $há^{-1}$) in the environments assessed.

Keywords: *Fragaria x ananassa*, REML/BLUP, MHPRVG method, genotype x environment interaction.

Introduction

Strawberry (*Fragaria x ananassa* Duch.) is an accidental hybrid of two native species, *F. chiloensis* and *F. virginiana*, and is one of the most appreciated fruits in the world due to its organoleptic properties and health benefits (BOMBARELY et al., 2010; KURAS; KORBIN, 2010). In 2010, the world strawberry production reached 4.17 million tons, with the United States of America as the largest producer, accounting for 30% of the global production. In Brazil, the annual production is approximately of 100 thousand tons and is distributed in temperate and subtropical regions where it is used both for fresh and industrial consumption (RADIN et al., 2011).

In Espírito Santo State, strawberry cultivation has increased over the years and has played an important socio-economic role, mainly due to the increased income of small farms and the establishment of workers in rural areas because much manual labor is required during the harvest season. Approximately 7200 tons are produced per year, involving more than 240 hectares and generating approximately 2500 direct jobs (INCAPER, 2012).

Espírito Santo producers have gradually been using protected cultivation to increase production. This management system has a number of advantages, particularly the protection of the crop against abiotic (winds, hail, rain, frost and low temperatures) and biotic (pests and diseases) factors (ANTUNES et al., 2007; CALVETE et al., 2008; WITTER et al., 2012). However, there are few studies on the performance of strawberry cultivars in different locations and years under this management system.

Therefore, a study on the genotype x environment interaction (GE) plays an important role in determining the adaptation and fitness of genotypes to the physical environment. GE interaction can be classified as simple, such as when it results only from genetic variability in a genotype, and complex, such as when there is no correlation between the measures of the same genotype in different environments. The latter indicates inconsistency of genotype superiority under environmental variations, which impedes the recommendation of cultivars with wide adaptability (CRUZ et al., 2004).

To minimize the effects of GE interactions and to achieve a higher performance predictability, the identification of the most stable genotypes and cultivars adapted to the specific conditions of each environment is necessary. Several methods have been described for examining stability and adaptability, as based on the analysis of variance, linear regression, nonlinear regression, multivariate analysis and nonparametric statistics (BASTOS et al., 2007; PENA et al., 2012).

Within the context of mixed models, the method of the harmonic mean of the relative performance of genetically predicted values proposed by Resende (2004) has shown good results for studies on stability and adaptability (BASTOS et al., 2007; CARBONELL et al., 2007; SILVA et al., 2011; SILVA et al., 2012). The advantages of this procedure include the following: i) allowing the simultaneous selection of yield, adaptability and stability; ii) considering the genotypic effects as random and, therefore, providing estimates of stability and adaptability for predicted and non-predicted genotypic values; iii) allowing the use of unbalanced data, non-orthogonal designs and heterogeneity of variance; iv) allowing the consideration of correlated errors within the locations in addition to stability and adaptability in the selection of individuals within progeny; v) providing genetic values already discounted (penalized) with regard to instability and allowing the application of any number of environments; and vi) generating results at the unit or scale of the evaluated trait, which can be directly interpreted as genetic values (RESENDE, 2002a).

Therefore, the present study aimed to estimate the adaptability and stability of strawberry cultivars under protected cultivation using the method of the harmonic mean of the relative performance of genetically predicted values.

Material and methods

The experiments were conducted during the agricultural years of 2006/7, 2007/8 and 2008/9 in three locations in the mountainous region of Espírito Santo State (municipalities Domingos Martins, Iúna and Muniz Freire). The elevations of the experimental sites in the municipalities were 700 m in Iúna, 950 m in Domingos Martins and 1100 m in Muniz Freire. The planting for the experiments was performed in late May, and the experiment was conducted until early January, with the exception of the experiment in 2007/8, which was conducted until March 2008. In all years, the harvests were performed twice a week, from the early harvest (usually in August) to the end of the cycle.

The experiment was arranged in a randomized block design, with 3 replications and 15 plants per plot. The study involved seven strawberry cultivars ('Dover', 'Camino Real', 'Ventana', 'Camarosa', 'Seascape', 'Diamante' and 'Aromas'). Plant spacing was 0.4 x 0.4 m, arranged in three rows on 30-cm high and 14-m long beds covered with black mulching. The environmental conditions for cultivation consisted of covering the plants with white plastic film, which was suspended over arcs of galvanized iron, at a height of approximately 1.0 m.

The values of productivity (ton. ha^{-1}) in the different environments were subjected to statistical analyses using the following linear model:

$$y_{ijkn} = \mu + g_i + b_{jkn} + a_k + l_n + ga_{ik} + gl_{kn} + + gal_{ikn} + \varepsilon$$

where:

Y_{ijkn} is the observation value relating to treatment i in repetition j in year k within location n; μ is the general average; g_i is the effect of genotype i; b_{jkn} is the effect of block j within year k within location n; a_k is the effect of the planting year k; l_n is the effect of the location n; ga_{ik} is the effect of the genotype x planting year interaction; gl_{kn} is the effect of the genotype x location interaction; gal_{ikn} is the effect of the genotype x year x location interaction; and ϵ is the random error or residue.

In matrix form, the corresponding model is as follows:

$$Y = Xb + Z_g + Q_{ga} + T_{gl} + U_{gla} + \varepsilon$$

where:

Y is the vector of the observations; b is the vector of the effects of the block-local-year combinations (fixed effects) plus the overall average; g is the vector of the genotypic effects, g (assumed as random); ga is the vector of random effects of the genotype x year interaction; gl is the vector of random effects of the genotype x location interaction; gla is the vector of random effects of the triple interaction genotype x location x year; and ϵ is the vector of errors (random). X, Z, Q, T and U refer to the incidence matrices for these effects, respectively. The effects of the year, local and block-local-years are grouped into effect b, as they are environmental effects (RESENDE, 2007).

The values of the harmonic mean of the genotypic values (MHVG) to evaluate stability, the relative performance of the genotypic values (PRVG) for adaptability and the harmonic mean of the relative performance of the genotypic values (MHPRVG) for stability, adaptability and productivity, simultaneously, for all genotypes, were obtained according to the expressions:

$$MHVG = \frac{1}{\sum_{i=1}^{l}\frac{1}{VG_j}}, \; PRVG = \frac{1}{l}\left(\frac{\sum VG_j}{M_j}\right) \text{ and}$$

$$MHPRVG = \frac{1}{\sum_{i=1}^{l}\frac{1}{PRVG_j}}$$

where:

l is the number of locations, VG is the genotypic value, and j is the genotype.

The significance of the effects of the model was estimated by an analysis of deviance, as recommended by Resende (2002a). The deviances were obtained through analyses with and without the effects of g, ga, gl and gla. The deviance without the mentioned effect was then subtracted from each full-model deviance and was compared to the value of chi-square, with one degree of freedom, at 1 and 5% probability. The Selegen software system (RESENDE, 2002b) was used for the analyses.

Results and discussion

The analysis of deviance in relation to productivity (ton. ha^{-1}) showed that the effects of genotype and the triple interaction genotype x location x year were statistically significant ($p \leq 0.05$ and $p \leq 0.01$, respectively). In contrast, no significance was observed for the genotype x location and genotype x year interactions (Table 1). These results demonstrate the presence of genetic variability among the strawberry cultivars tested and that the genotype x location and genotype x year interactions did not cause changes in the performance of the genotypes in the different locations and years. However, for the triple interaction genotype x location x year, an inconsistency in the position among the genotypes was verified for all the combinations of year and location. According to Allard and Bradshaw (1964), this type of interaction is caused by unpredictable environmental variations, such as precipitation, temperature, relative humidity and the occurrence of pests and diseases.

Table 1. Deviance, variance components and coefficients of determination related to the global joint analysis involving seven strawberry cultivars evaluated in nine environments.

Effect	Deviance	LRT ($\chi 2$)	Component of variance	Coefficient of determination
Genotype	825.79⁺	5.64*	Vg = 25.465	$h^2_g = 0.306$
Genotype x Location	820.17⁺	0.02^{ns}	Vgl = 0.206ns	$c^2_{gl} = 0.002^{ns}$
Genotype x Year	821.24⁺	1.09^{ns}	Vga = 7.186ns	$c^2_{ga} = 0.086^{ns}$
Genotype x Location x Year	858.34⁺	38.19**	Vgla = 28.014	$c^2_{gla} = 0.337$
Residue			Ve = 22.289	$c^2res = 0.269$
Complete Model	820.15			$C^2total = 1.00$

⁺Deviance of the adjusted model without the corresponding effects. * and **Significant by the chi-square test at 5% (3.84) and 1% (6.63), respectively.

The coefficients of determination (c^2_{gl} for the genotype x location interaction, c^2_{ga} for the genotype x year interaction, and c^2_{gla} for the genotype x location x year interaction) indicate how much each component contributes to the total phenotypic variance. Thus, the genotype x location, genotype x year, and genotype x location x year interactions contributed 0.24, 8.60 and 33.70%, respectively, and demonstrated the importance of the triple interaction in the total phenotypic variance. The heritability estimate of the broad-sense individual plots (h_g^2) for productivity was 30.60%. This result is explained by the fact that productivity is a polygenic character and is therefore greatly influenced by the environment.

The genotypic correlation of the cultivars across the locations and years was 0.4183, showing a moderate level of complex interaction, which indicates that the ranking of the cultivars across environments (locations and years) will not necessarily be the same, i.e., a cultivar rated with excellent productivity in environment 1 may not be excellent in environment 2 or in any other environment.

Table 2 shows the cultivars with the respective genotypic values (μ+g+ge) for each cultivation location in the different agricultural harvests. The Dover cultivar presented the lowest genotypic values for most environments evaluated, whereas cultivar Camarosa presented the best genotypic values for most environments and occupied the first or second position in eight of the nine environments assessed. Resende et al. (2010) evaluated the effect of three cultivation systems (high tunnel, low tunnel and field) on strawberry productivity in Guarapuava, Paraná State, and reported that the cultivars Camorosa and Oso Grande achieved the highest productivity rates (56.66 and 53.76 ton. ha^{-1}, respectively) and differed statistically from cultivars Sweet Charlie (45.70 ton. ha^{-1}) and Dover (36.98 ton. ha^{-1}) under low tunnel-protected cultivation.

According to the expected results (in ton. ha^{-1}) for stability (MHVG), adaptability (PRVG), and stability and adaptability (MHPRVG) for the set of locations, either harming or favoring cultivars, according to their performance, it was observed that cultivar Camorosa was superior, followed by cultivars Aromas, Diamante, Camino Real, Seascape, Ventana and Dover (Table 3). Therefore, the two best cultivars (Camarosa and Aromas) by the MHPRVG criterion were, on average, 22% superior to the overall average of the nine environments (24.55 kg ha^{-1}). The MHPRVG method has the advantage of providing results for the scale of trait measurement, which can be directly interpreted as predicted genetic values for productivity, stability and adaptability (BASTOS et al., 2007; CARBONELL et al., 2007; VERARDI et al., 2009).

Silva et al. (2011) compared the methods of adaptability and stability and verified that the REML/BLUP-MHPRVG analysis stood out from the others (AMMI and GGE biplot). Although the method does not provide information about the grouping of similar sites, the results are according to the genotypic values when considering the parameters of adaptability and stability.

Table 2. Estimates of the genotypic values (μ + g + ge) related to the yield trait of seven strawberry cultivars evaluated in nine environments.

Cultivar	Domingos Martins		Muniz Freire		Iúna	
	μ + g + ge	Gain	μ + g + ge	Gain	μ + g + ge	Gain
			Agricultural Year 2006/2007			
Dover	17.78	0.00	9.25	0.00	5.67	0.00
Camino Real	25.88	2.80	19.73	4.08	11.54	2.85
Ventana	26.12	3.35	12.54	1.37	6.19	1.23
Camarosa	30.05	5.76	23.74	6.27	16.49	4.36
Seascape	18.72	1.25	16.18	2.63	17.00	4.81
Diamante	26.30	4.19	19.66	3.61	15.71	3.93
Aromas	32.00	6.74	21.19	4.99	18.72	5.67
			Agricultural Year 2007/2008			
Dover	19.83	0.00	27.87	0.00	7.41	0.00
Camino Real	38.07	7.66	39.53	9.03	10.56	2.15
Ventana	21.53	2.19	33.28	5.67	12.19	3.39
Camarosa	42.72	9.72	55.49	17.79	18.33	5.25
Seascape	35.14	6.29	32.31	3.46	17.65	4.92
Diamante	41.20	8.96	45.19	12.64	8.64	1.00
Aromas	32.50	4.93	30.23	1.63	18.95	5.56
			Agricultural Year 2008/2009			
Dover	32.06	0.00	3.10	0.00	19.32	1.54
Camino Real	41.82	5.19	7.36	2.61	30.14	5.07
Ventana	46.99	7.62	6.00	0.95	18.97	0.00
Camarosa	51.20	9.72	12.42	5.25	37.74	9.51
Seascape	37.98	1.57	10.24	3.97	32.71	7.00
Diamante	39.57	2.58	6.92	1.71	32.61	6.12
Aromas	40.70	3.70	15.73	6.90	26.10	3.63

It is important to mention that the three cultivars with the highest values for MHVG, PRVG and MHPRVG have excellent performance for firmness and a good balance between the total soluble solids and acidity, such that they will be widely accepted by both consumers and industry. Within this context, such information on the stability and adaptability of strawberry cultivars under protected cultivation (low tunnel) are important because they allow recommendations for cultivation and the verification of the performance of these cultivars in different strawberry-producing regions in the state of Espírito Santo.

Table 3. Stability of genotypic values (MHVG), adaptability of genotypic values (PRVG), stability and adaptability of genotypic values (MHPRVG) for the yield trait in seven strawberry cultivars evaluated in nine environments.

Cultivar	Genotypic value		
	MHVG	PRVG	MHPRVG
Dover	14.65	16.23	16.03
Camino Real	23.40	24.92	24.92
Ventana	19.09	20.61	20.57
Camarosa	29.97	31.63	31.54
Seascape	22.73	24.23	24.23
Diamante	24.56	26.10	26.10
Aromas	26.61	29.13	28.50

Conclusion

Based on selection for yield, adaptability and stability, the Camarosa and Aromas cultivars may be indicated for a low tunnel management system.

Acknowledgements

The author thanks FAPES (Fundação de Amparo à Pesquisa do Espírito Santo) for funding the project and Agricola Llahuen and Bioagro for donating a portion of the seedlings.

References

ALLARD, R. W.; BRADSHAW, A. D. Implications of genotype environment interaction applied plant breeding. **Crop Science**, v. 4, n. 5, p. 503-508, 1964.

ANTUNES, O. T.; CALVETE, E. O.; ROCHA, H. C.; NIENOW, A. A.; CECCHETTI, D.; RIVA, E.; MARAN, R. E. Produção de cultivares de morangueiro polinizadas pela abelha jataí em ambiente protegido. **Horticultura Brasileira**, v. 25, n. 1, p. 94-99, 2007.

BASTOS, I. T.; BARBOSA, M. H. P.; RESENDE, M. D. V.; PETERNELLI, L. A.; SILVEIRA, L. C. I.; DONDA, L. R.; FORTUNATO, A. A.; COSTA, P. M. A.; FIGUEIREDO, I. C. R. Avaliação da interação genótipo x ambiente em cana-de-açucar via modelos mistos. **Pesquisa Agropecuária Tropical**, v. 37, n. 4, p. 195-203, 2007.

BOMBARELY, A.; MERCHANTE, C.; CSUKASI, F.; CRUZ-RUS, E.; CABALLERO, J. L.; ESCOBAR N. M.; PORTALES, R. B.; BOTELLA, M. A.; BLANCO, J. M.; SEVILLA, J. F. S.; VALPUESTA, V. Generation and analysis of ESTs from strawberry (*Fragaria x ananassa*) fruits and evaluation of their utility in genetic and molecular studies. **BMC Genomics**, v. 11, n. 1, p. 1-17, 2010.

CALVETE, E. O.; MARIANI, F.; WESP, C. L.; NIENOW, A. A.; CASTILHOS, T.; CECCHETTI, D. Fenologia, produção e teor de antocianinas de cultivares de morangueiro em ambiente protegido. **Revista Brasileira de Fruticultura**, v. 30, n. 2, p. 396-401, 2008.

CARBONELL, S. A. M.; CHIORATO, A. F.; RESENDE, M. D. V.; DIAS, L. A. S.; BERALDO, A. L. A.; PERINA, E. F. Estabilidade de cultivares e linhagens de feijoeiro em diferentes ambientes no estado de São Paulo. **Bragantia**, v. 66, n. 2, p. 193-201, 2007.

CRUZ, C. D.; REGAZZI, A. J.; CARNEIRO, P. C. S. **Modelos biométricos aplicados ao melhoramento genético**. 3. ed. Viçosa: UFV, 2004, v. 1, 2004.

INCAPER-Instituto Capixaba de Pesquisa, Assistência Técnica e Extensão Rural. Available from: <http://www.incaper.es.gov.br>. Access on: Aug. 2, 2012.

KURAS, A.; KORBIN, M. PCo-A analysis of strawberry germplasm used in European breeding programs, based on evaluation of DNA polymorphism of investigated plants. **Journal of Fruit and Ornamental Plant Research**, v. 18, n. 2, p. 7-16, 2010.

PENA, G. F.; AMARAL JÚNIOR, A. T.; GONÇALVES, L. S. A.; CANDIDO, L. S.; VITTORAZZI, C.; RIBEIRO, R. M.; FREITAS JÚNIOR, S. P. Stability and adaptability of popcorn genotypes in the state of Rio de Janeiro, Brazil. **Genetics and Molecular Research**, v. 11, n. 3, p. 3042-3050, 2012.

RADIN, B.; LISBOA, B. B.; WITTER, S.; BARNI, V.; REISSER JUNIOR, C.; MATZENAUER, R.; FERMINO, M. H. Desempenho de quatro cultivares de morangueiro em duas regiões ecoclimáticas do Rio Grande do Sul. **Horticultura Brasileira**, v. 29, n. 3, p. 287-291, 2011.

RESENDE, M. D. V. **Genética biométrica e estatística no melhoramento de plantas perenes**. Brasília: Embrapa Informação Tecnológica, 2002a.

RESENDE, M. D. V. **Software Selegen-REML/BLUP**. Colombo: Embrapa Florestas, 2002b.

RESENDE, M. D. V. **Métodos estatísticos ótimos na análise de experimentos de campo**. Colombo: Embrapa Florestas, 2004.

RESENDE, M. D. V. **Matemática e estatística na análise de experimentos e no melhoramento genético**. Colombo: Embrapa Florestas, 2007.

RESENDE, J. T. V.; MORALES, R. G. F.; FARIA, M. V.; RISSINI, A. L. L.; CAMARGO, C. K. Produtividade e teor de sólidos solúveis de frutos de cultivares de morangueiro em ambiente protegido. **Horticultura Brasileira**, v. 28, n. 2, p. 185-189, 2010.

SILVA, G. O.; CARVALHO, A. D. F.; VIEIRA, J. V.; BENIN, G. Verificação da adaptabilidade e estabilidade de populações de cenoura pelos métodos AMMI, GGE biplot e REML/BLUP. **Bragantia**, v. 70, n. 3, p. 494-501, 2011.

SILVA, G. O.; CARVALHO, A. D. F.; VIEIRA, J. V.; FRITSCHE-NETO, R. Adaptabilidade e estabilidade de populações de cenoura. **Horticultura Brasileira**, v. 30, n. 1, p. 80-83, 2012.

VERARDI, C. K.; RESENDE, M. D. V.; COSTA, R. B.; GONÇALVES, P. S. Adaptabilidade e estabilidade da produção de borracha e seleção em progênies de seringueira. **Pesquisa Agropecuária Brasileira**, v. 44, n. 1, p. 1277-1282, 2009.

WITTER, S.; RADIN, B.; LISBOA, B. B.; TEIXEIRA, J. S. G.; BLOCHTEIN, B.; IMPERATRIZ-FONSECA, V. C. Desempenho de cultivares de morango submetidas a diferentes tipos de polinização em cultivo protegido. **Pesquisa Agropecuária Brasileira**, v. 47, n. 1, p. 58-65, 2012.

14

Estimates of the genetic parameters, optimum sample size and conversion of quantitative data in multiple categories for soybean genotypes

Éder Matsuo[1*], Tuneo Sediyama[1], Cosme Damião Cruz[2], Rita de Cássia Teixeira Oliveira[3] and Luiz Renato Cadore[1]

*[1]Departamento de Fitotecnia, Universidade Federal de Viçosa, Campus Universitário, 36570-000, Viçosa, Minas Gerais, Brazil. [2]Departamento de Biologia Geral, Universidade Federal de Viçosa, Viçosa, Minas Gerais, Brazil. [3]Bacuri Pesquisa e Melhoramento, Viçosa, Minas Gerais, Brazil. *Author for correspondence. E-mail: matsuoeder@yahoo.com.br*

ABSTRACT. The objective of this study was to estimate the genetic parameters and optimal sample size for the lengths of the hypocotyl and epicotyls and to analyze the conversion of quantitative data in multiple categories for soybean genotypes. A total of 85 soybean genotypes were analyzed in four experiments in a completely randomized design under greenhouse conditions at the Federal University of Viçosa, Brazil. The magnitude of the genetic parameter estimates characterized the influence of the genetic components in the phenotypic expression of the length of the hypocotyls and epicotyls of different soybean genotypes. The optimal size of the sample varied between the genotypes and the conversion of the quantitative data in multiple categories provided estimates of correlation coefficients greater than 0.80 and coincidence greater than 81% between the converted data and the original data.

Keywords: *Glycine max*, soybean improvement, hypocotyl length, epicotyl length.

Introduction

Soybean (*Glycine max* [L.] Merrill) shows awide adaptation to tropical and subtropical climates (SEDIYAMA et al., 2005). During the 2010/2011 season in Brazil, it was estimated that a cultivated area of 24,165,000 hectares would produce 72,227,800 tons and an average yield of 2,989 kg ha^{-1} (CONAB, 2011). In 2005, because of the wide acceptance of new technologies by soybean producers in Brazil and the associated efforts of soybean breeding programs, it was estimated that the average productivity in some regions of the central-western region could exceed 3,000 kg ha^{-1} (SEDIYAMA et al., 2005). Six years later, Conab (2011) estimated that the average productivity in the central-western states (Mato Grosso, Mato Grosso do Sul, Goiás and Distrito Federal) for the 2010/2011 harvest would be 3,031 kg ha^{-1}.

The many uses of soybeans, such as in the oil and bran industries, agribusiness and human consumption, are incentives for breeders to develop superior cultivars (SEDIYAMA et al., 2005, 2009). In addition, there has been an intense development of new cultivars in Brazil, especially after 1997, when the Law of Plant Varieties Protection number 9,456 of April 25, 1997, regulated by the Act number 2,366 of November 5, 1997, was enacted (NETO et al., 2005).

For a particular cultivar to be granted protection, it must meet three basic requirements: it must be distinct, uniform and stable. A cultivar is considered a distinct plant when it is clearly distinct from any other cultivar in existence on the date that the application for protection is recognized (FERRAZ; CAMPOS et al., 2009). Currently, approximately 38 descriptors between obligatory and additional are used to differentiate soybean cultivars. However, these descriptors are insufficient to distinguish the cultivars, clearly showing the need to expand the list of descriptors (NOGUEIRA et al., 2008). According to Nogueira et al. (2008), the length of the hypocotyl and epicotyl remain useful characters to distinguish soybean genotypes.

The knowledge of genetic parameters, such as the genotypic determination coefficient, the component of genotypic variability and the relationship between the coefficients of genetic variation by environmental variation (which control certain characters), is of great importance for the breeder because it allows the determination of the best method of improvement for the culture (CRUZ et al., 2004). Several studies have been performed to estimate genetic parameters in soybean (COSTA et al., 2008; FARIAS NETO; VELLO, 2001; GOMES et al., 2004; MAURO et al., 1995; NOGUEIRA et al., 2008; REIS et al., 2002) and in other cultures (ANDRADE et al., 2010; ARNHOLD; MILANI, 2011; BORGES et al., 2010; DAHER et al., 2004; ROCHA et al., 2010).

The estimate of the optimal size of the sample, with regard to the number of individuals, is based on the simulation, which evaluates the variation of the average, variance and coefficient of variation in relation to the maximum size of the test, that is, the original population (CRUZ, 2006b). Furthermore, to determine the optimal sample size, one can take into account the minimum size capable of representing the average, variance and coefficient of variation of the original population for all or a certain set of characteristics that are the most important (CRUZ, 2006b). Cargnelutti Filho et al. (2009) reported that the sample size is directly proportional to the variability of the population data and studies related to the determination of the sample size have been conducted in various cultures, with different objectives (ADAMI et al., 2010; CARGNELUTTI FILHO et al., 2011; ESTEFANEL et al., 1984; MARTIN et al., 2005; PEREIRA et al., 2009; SILVA et al., 2011).

Based on the aforementioned considerations and the fact that there are few studies in the literature involving the estimation of genetic parameters, the determination of the optimum sample size and the conversion of quantitative data in multiple categories of the lengths of the hypocotyl and epicotyl in soybean genotypes, the need to study the genotypes of soybeans to elucidate these topics is evident. Therefore, this study aimed to estimate the genetic parameters and the optimal sample size for the lengths of the hypocotyl and epicotyls and to convert the quantitative data into multiple categories of soybean genotypes.

Material and methods

The experiments were conducted under greenhouse conditions at the Soybean Genetic Improvement Program of the Federal University of Viçosa in Viçosa – Minas Gerais State (20°45'14"S, 42°52'54"W, 649 m of altitude). A total of 85 soybean genotypes were evaluated in four experiments. For planting, a random sample of seeds consisting of different sizes was used and the depth was set at 3.0 cm. After the germination, the plants were managed according to the recommendations of the culture.

In experiment 1, 45 genotypes were evaluated (BCR109B01, BCR127B01, BCR121B01, BCR203P, BCR119B01, BCR166P, DM 309, BCR123B01, BCR119P, BCR111B01, BCR124B01, BCR165P, BCR115B01, BCR112B01, BCR114B01, P98C81, BCR107B01, BCR152P, DM 339, BCR213P, BCR158P, BCR196P, BCR146P, MG/BR-46 (Conquista), BCR164P, BCR144P, Emgopa 313 (Anhanguera), BCR209P, UFV 18 (Patos de Minas), BCR104B01, BCR113B01, BCR105B01, BCR110E, BCR103F, BCR110A, BCR110B, BCR110C, BCR110D, BCR105G, BCR113H, FT-Estrela, FT-12 (Nissei), UFV 16 (Capinópolis), BCR132390HM and BCR1346142HP). In experiment 2, four genotypes were evaluated (BRS Valiosa RR, Água-Marinha RR, UFVS 2010 and NK 7059 RR). In experiment 3, 22 genotypes were evaluated (BCR1-295, BCR2-296, BCR3-297, BCR4-298, BCR5-299, BCR6-300, BCR7-301, BCR8-302, BCR9-303, BCR10-304, BCR11-305, BCR12-306, BCR13-307, BCR14-308, BCR15-309, BCR16-310, BCR17-311, BCR18-312, BCR19-313, BCR20-314, Bossier and MG/BR-46 (Conquista)). Lastly, in experiment 4, 14 genotypes were evaluated (BRC10-01, BRC10-02, BRC10-03, BRC10-04, BRC10-05, BRC10-06, BRC10-07, BRC10-08, BRC10-09, BCR10-10, BCR10-11, BCR10-12, Bossier and BRS Valiosa RR). The genotypes used for experiments 1, 3 and 4 were grown in pots containing 3 dm^3 of soil with 1/3 organic matter;for each genotype, five plants were assessed with regard to the length of the hypocotyl at the V2 stage (FEHR; CAVINESS, 1977) and the length of the epicotyl at the V3 stage (FEHR;

CAVINESS, 1977). The plants for experiment 2 were planted in a plastic tray (39 x 26 x 5 cm) containing soil with 1/3 organic matter; for each genotype, 100 plants were evaluated for the lengths of the hypocotyl and epicotyl at the V3 stage (FEHR; CAVINESS, 1977). The measurements were made using digital calipers. A completely randomized design was adopted, with each replicate (experimental unit) consisting of one plant.

Initially, the Lilliefors or Kolmogorov-Smirnov's tests were performed to assess whether the data followed the normal distribution. Posteriorly, the data were analyzed for variability in the lengths of the hypocotyl and epicotyl in each experiment, adopting a fixed statistical model. Additionally, the following genetic parameters were estimated: the genotypic quadratic component ($\widehat{\Phi}_g$), coefficient of variability ($\widehat{CV}_{e\%}$), coefficient of genotypic variation ($\widehat{CV}_{e\%}$), ($\widehat{CV}_{g\%}/\widehat{CV}_{e\%}$) ratio and coefficient of genotypic determination (H^2). The equations are described below:

$$\widehat{\Phi}_g = \frac{QMG - QMR}{r}$$

where:

QMG = Average square of the genotype;

QMR = Average square of the residue;

r = Number of experimental replications.

$$\widehat{CV}_{e\%} = \frac{\left[100\sqrt{QMR}\right]}{\bar{X}}$$

where:

$\bar{X}$ = General average of the general trait.

$$\widehat{CV}_{g\%} = \frac{\left[100\sqrt{\Phi_g}\right]}{\bar{X}}$$

$$H^2 = \left[\frac{\widehat{\Phi}_g}{\frac{QMG}{r}}\right]100$$

Based on the data from Experiment 2 for each genotype, we determined the optimal sample size of the sample that represented the average and the variance of the original character, on the basis of the simulation, graphical analysis and confidence intervals for the average and the variance at a 95% level of probability.

In addition, the conversion of the quantitative data (lengths of hypocotyl and epicotyl) of four experiments into multiple categories (3 and 5 classes) was based on the number of predefined classes from the fair division of amplitude. Although it would be interesting to evaluate the plants directly in each class, because this work can be considered an initial effort in categorizing the lengths of the hypocotyl and epicotyl, in 3 and 5 classes, it was necessary to first measure the plants and then categorize them. The converted data were analyzed by Spearman's correlationand the coefficient of coincidence, considering a sample of 50%. According to Cruz (2006a), the correlation of occurrence of accessin the coefficient of coincidence is calculated in the set formed by the observations above or below for each pair of characters regarding a predetermined sample size, n_1, with $n_1<n$. The analyses were performed using the Genes Program: Experimental Statistics and Matrices (CRUZ, 2006a) and Genes Program: Multivariate Analysis and Simulation (CRUZ, 2006b).

Results and discussion

The normality tests showed that it was possible to analyze the data through a normal distribution. Significant differences were observed ($p < 0.01$) by the F test for the lengths of the hypocotyl and epicotyl in the four experiments (Table 1), indicating the existence of genetic variability among the genotypes, which enables success when selecting promising materials. Similar results for the hypocotyl length were obtained by Nóbrega and Vieira (1995) and Costa et al. (1999) and for the lengths of the hypocotyl and epicotyl by Nogueira et al. (2008). The estimates of the genotypic quadratic component ($\widehat{\Phi}_g$) of the length of the epicotyl were 7.1 times greater than the estimates for the hypocotyl in the average of four experiments (Table 1).

For the experimental coefficient of variation ($CV_{e\%}$), the estimates ranged from 8.0413 to 18.2839% for the hypocotyl length and 11.1836 to 13.7596% for the length of the epicotyl in the four experiments. The magnitudes of the $CV_{e\%}$ were found to be consistent with those reported by Bays et al. (2007) and Nogueira et al. (2008). The coefficient of the genetic variation ($CV_{g\%}$) ranged from 7.8205 to 25.1840% and 10.8847 to 36.4231% for the lengths of the hypocotyl and epicotyl, respectively, in the four experiments. The $CV_{g\%}/CV_{e\%}$ ratios were higher than the unit in experiments 1 and 3 (Table 1). In the experiments 1 and 3, there was the possibility of successfully identifying superior genotypes because the genotypic variation exceeded the environmental (VENCOVSKY, 1987) and because the ratio of the genetic variance coefficient ($CV_{g\%}$) and experimental coefficient of variance ($CV_{e\%}$) demonstrates how the experimental variance is explained by the genotypic variance.

The magnitude of the estimates in the determination of the genotypic coefficient (H^2) for both characters that were analyzed were above 82% and values above 98% were observed for the length of the hypocotyl in experiment 2 and for the lengths of the epicotyls in experiments 1 and 2 (Table 1). Similar results were obtained by Nogueira et al. (2008). These results indicate that the genotypes had a higher genetic variability for the length of the epicotyl and that there was a large genetic influence over the environmental effects on the lengths of the hypocotyl and epicotyl. This is because, according to Cruz (2005), the H^2 is a measure that is analogous to the heritability and expresses the phenotypic variance because of the genetic variability between the treatment averages. Thus, high estimates of H^2 indicate that most of the variation between the averages of the genotypes is genetic in nature (VENCOVSKY, 1987). In addition, the highest importance of the heritability in genetic studies of metric characters is its predictive role, which is expressed by the confidence of the phenotypic value as a guide of the genetic value or the degree of correspondence between the phenotypic value and the value of a population or a set of genotypes (FALCONER, 1987). Moreover, Andrade et al. (2010) suggested that, for the characters that have a high genetic component in their phenotypic expressions, obtaining gains by selection can be achieved via visual or mass selection.

The estimates of genetic parameters, in general, for the lengths of the hypocotyl and epicotyl in the four experiments showed a situation that is favorable for the identification of superior genotypes. These characters may be useful in the selection of genotypes of different behaviors to contribute to the distinctiveness requirement of the Plant Variety Protection.

The minimum sample size would be the one from which all of the *r* simulations provided values of average and variance within the confidence intervalat 95% probability (CRUZ, 2006b). Thus, the minimum optimal number of plants to be measured to represent the average hypocotyl length was 45, 53, 52 and 45 for BRS Valiosa RR, Água-Marinha RR, UFVS 2010 and NK 7059 RR, respectively, whereas the average length of the epicotyls was 49, 43, 51 and 49 for BRS Valiosa RR, Água-Marinha RR, UFVS 2010 and NK 7059 RR, respectively (Figure 1). For the variance of the length of the hypocotyl, the optimal size of the sample was 60, 64, 58 and 73 plants for BRS Valiosa RR, Água-Marinha RR, UFVS 2010 and NK 7059 RR, respectively; for the variance of the length of the epicotyl, the optimal size was 70, 29, 64 and 70 for BRS Valiosa RR, Água-Marinha RR, UFVS 2010 and NK 7059 RR, respectively (Figure 2).

Table 1. Estimates of the average square of genotypes (QMG), genotypic quadratic component (φ_g), coefficient of environmental variation ($CV_{e\%}$), genotypic variation coefficient ($CV_{g\%}$), $CV_{g\%}/CV_{e\%}$ ($CV_{g\%}/CV_{e\%}$) ratio, and genotypic determination coefficient (H^2) in two additional descriptors (hypocotyl and epicotyl) in the four analyzed experiments, conducted under greenhouse conditions, in Viçosa, Minas Gerais State, 2006-2011.

Experiments	Parameters	Hypocotyl	Epicotyl
	QMG	395.6499**	3341.7815**
	φg	71.5838	655.6388
1	$CV_{e\%}$	18.2839	11.3431
	$CV_{g\%}$	25.1840	36.4231
	$CV_{g\%}/CV_{e\%}$	1.3774	3.2110
	H^2	90.4635	98.0972
	QMG	443.6493**	2863.9230**
	φg	4.3539	28.3400
2	$CV_{e\%}$	11.6280	11.1836
	$CV_{g\%}$	8.4411	10.8847
	$CV_{g\%}/CV_{e\%}$	0.7259	0.9733
	H^2	98.1377	98.9553
	QMG	95.0187**	306.3580**
	φg	16.7656	51.7024
3	$CV_{e\%}$	10.4719	11.7335
	$CV_{g\%}$	12.8177	12.1972
	$CV_{g\%}/CV_{e\%}$	1.2240	1.0395
	H^2	88.2227	84.3824
	QMG	45.8840**	445.9462**
	φg	7.5750	73.1945
4	$CV_{e\%}$	8.0413	13.7596
	$CV_{g\%}$	7.8205	13.1635
	$CV_{g\%}/CV_{e\%}$	0.9725	0.9567
	H^2	82.5456	82.0665

**Significant at1% probability by the F test.

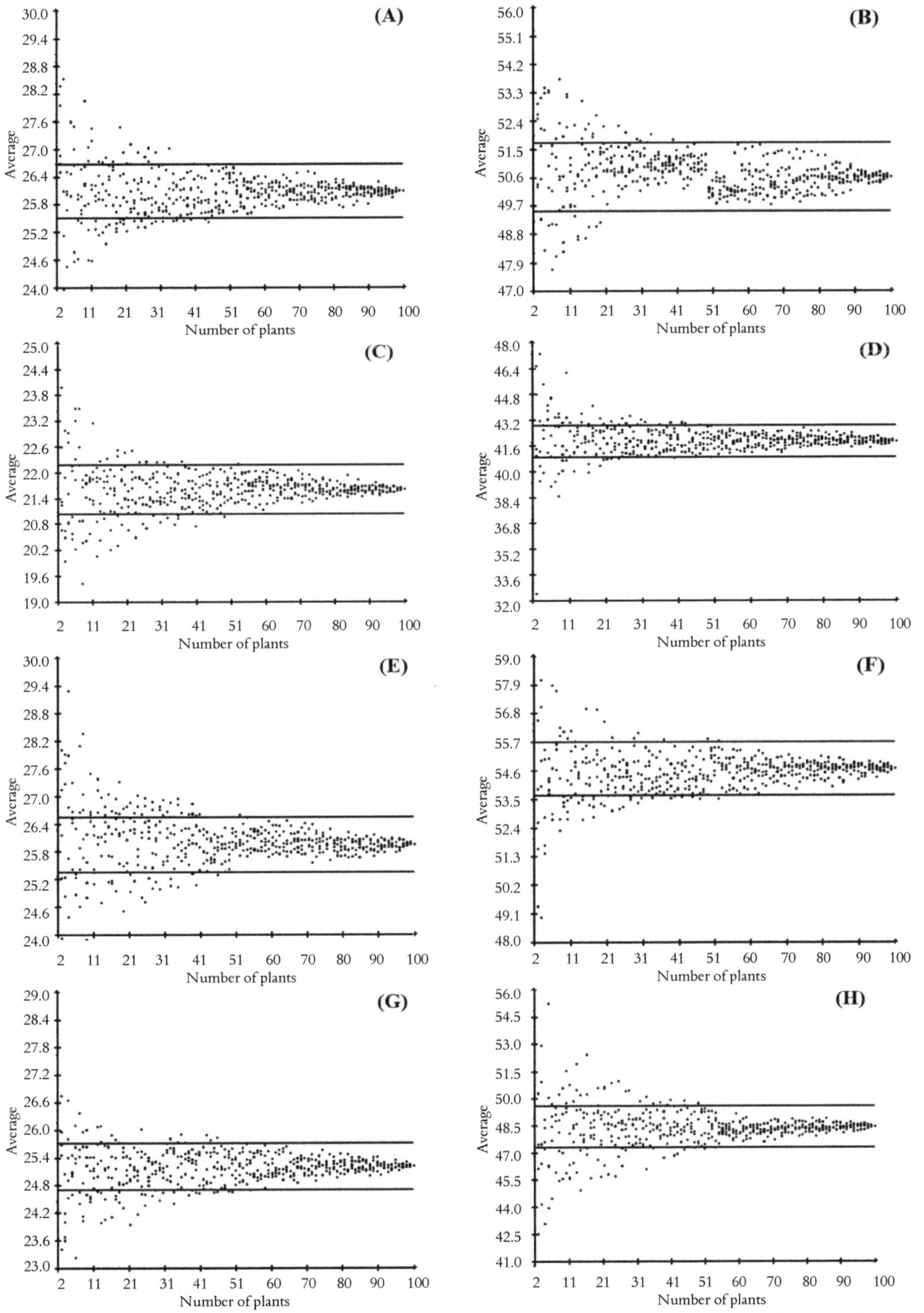

Figure 1. Optimal size of the sample able to represent the average of the length of hypocotyl and epicotyls. (A) Hypocotyl - BRS Valiosa RR, (B) Epicotyl - BRS Valiosa RR, (C) Hypocotyl - Água-Marinha RR, (D) Epicotyl - Água-Marinha RR, (E) Hypocotyl - UFVS 2010, (F) Epicotyl - UFVS 2010, (G) Hypocotyl – NK 7059 RR, and (H) Epicotyl – NK 7059 RR, conducted under greenhouse conditions, in Viçosa, Minas Gerais State, 2006-2011.

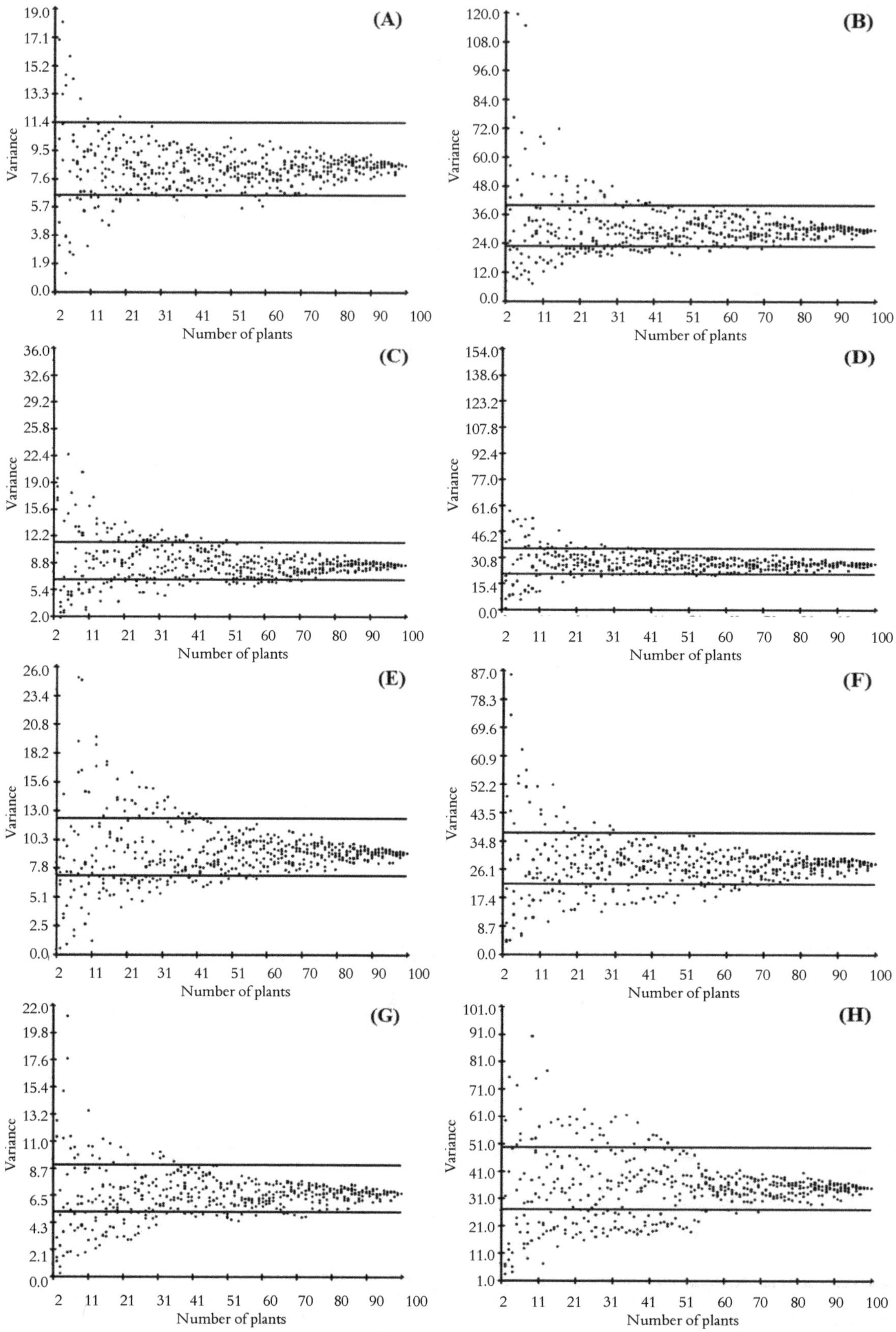

Figure 2. Optimal size of the sample able to represent the variance of the length of the hypocotyl and epicotyl. (A) Hypocotyl - BRS Valiosa RR, (B) Epicotyl - BRS Valiosa RR, (C) Hypocotyl - Água-Marinha RR, (D) Epicotyl - Água-Marinha RR, (E) Hypocotyl - UFVS 2010, (F) Epicotyl - UFVS 2010, (G) Hypocotyl - NK 7059 RR, and (H) Epicotyl - NK 7059 RR, conducted under greenhouse conditions, in Viçosa, Minas Gerais State, 2006-2011.

The optimal size for the average showed small variations for the two variables and for the four genotypes, namely, the samples for the lengths of the hypocotyl and epicotyl varied for eight plants between the genotypes. However, for the optimal minimum number that represents the greatest variance, variations among the genotypes were observed: 15 plants for the length of the hypocotyl and 41 for the length of the epicotyl. The sample size for the average was lower compared with the variance for all of the genotypes. Based on other evaluation methodologies, the variability of the estimate of the sample size among genotypes for the plant height, height of first pod, stem diameter, number of nodes on the stem, number of pods on the stem, number of grains on the stem and grain weight on the stem (ESTEFANEL et al., 1984) and the number of nodes per plant was identified (CARGNELUTTI FILHO et al., 2009).

We used Spearman's correlation (Table 2) and coincidence (Table 3) to compare the variation of the averages of the three data sets (original, 3 classes and 5 classes) after the data conversion. In the correlation between the three data sets in the four experiments, we observed that the estimates were greater than or equal to 0.8000 and 0.8549 for the length of hypocotyl and epicotyl, respectively. The coefficients of coincidence ranged from 81.81 to 100.00% for the hypocotyl length and 85.71 to 100.00% for the epicotyl length in the four experiments.

Table 2. Spearman correlation coefficients among three traits evaluated in two additional descriptors (Hypocotyl - above the diagonal and Epicotyl - below the diagonal) measured in four experiments, conducted under greenhouse conditions, in Viçosa, Minas Gerais State, 2006-2011[1].

	Experiment 1 (45 genotypes)		
	Original	3 Classes	5 Classes
Original		0.8598	0.9782
3 Classes	0.9355		0.8558
5 Classes	0.9731	0.9328	
	Experiment 2 (4 genotypes)		
	Original	3 Classes	5 Classes
Original		0.8000	0.8000
3 Classes	1.0000		1.0000
5 Classes	1.0000	1.0000	
	Experiment 3 (22 genotypes)		
	Original	3 Classes	5 Classes
Original		0.9356	0.9548
3 Classes	0.9514		0.9480
5 Classes	0.9740	0.9469	
	Experiment 4 (14 genotypes)		
	Original	3 Classes	5 Classes
Original		0.9120	0.9824
3 Classes	0.9340		0.8989
5 Classes	0.9516	0.8549	

[1] Estimation of correlation based on the average of the genotypes.

Table 3. Coefficients of coincidence in number and percentage (in parentheses) among three traits analyzed in two additional descriptors of soy measured in four experiments, where the coefficients above the diagonal refer to a higher average and the coefficients below refer to inferior averages, conducted under greenhouse conditions, in Viçosa, Minas Gerais State, 2006-2011[1].

	Experiment 1 (45 genotypes)						
	Hypocotyl				Epicotyl		
	Original	3 Classes	5 Classes		Original	3 Classes	5 Classes
Original	-	20 (90.91)	22 (100.00)	Original	-	20 (90.91)	21 (95.45)
3 Classes	18 (81.81)	-	20 (90.91)	3 Classes	19 (86.36)	-	19 (86.36)
5 Classes	22 (100.00)	18 (81.81)	-	5 Classes	21 (95.45)	19 (86.36)	-
	Experiment 2 (4 genotypes)						
	Hypocotyl				Epicotyl		
	Original	3 Classes	5 Classes		Original	3 Classes	5 Classes
Original	-	2 (100.00)	2 (100.00)	Original	-	2 (100.00)	2 (100.00)
3 Classes	2 (100.00)	-	2 (100.00)	3 Classes	2 (100.00)	-	2 (100.00)
5 Classes	2 (100.00)	2 (100.00)	-	5 Classes	2 (100.00)	2 (100.00)	-
	Experiment 3 (22 genotypes)						
	Hypocotyl				Epicotyl		
	Original	3 Classes	5 Classes		Original	3 Classes	5 Classes
Original	-	10 (90.91)	9 (81.82)	Original	-	10 (90.91)	10 (90.91)
3 Classes	11 (100.00)	-	10 (90.91)	3 Classes	10 (90.91)	-	10 (90.91)
5 Classes	11 (100.00)	11 (100.00)	-	5 Classes	10 (90.91)	10 (90.91)	-
	Experiment 4 (14 genotypes)						
	Hypocotyl				Epicotyl		
	Original	3 Classes	5 Classes		Original	3 Classes	5 Classes
Original	-	6 (85.71)	6 (85.71)	Original	-	6 (85.71)	7 (100.00)
3 Classes	6 (85.71)	-	6 (85.71)	3 Classes	6 (85.71)	-	6 (85.71)
5 Classes	7 (100.00)	6 (85.71)	-	5 Classes	7 (100.00)	6 (85.71)	-

[1] It was considered 50% of the sample for analysis.

The correlation reflects the degree of association between the characters and this knowledge is important because it shows how the selection of a character influences the expression of another (CRUZ et al., 2004). The coefficient of coincidence is calculated by the concordance of occurrence of genotypes in the group formed by the observations from above or below for each pair of characters over a sample of predefined size (CRUZ, 2006a). For this study, it was observed that the data sets showed variation in proportional magnitude and the concordance of the results for both the upper and lower observations showed satisfactory results when the figures were converted into multiple categories with 3 and 5 classes.

Thus, the conversion of the data presented in this study may be considered useful in plant improvement, as the lengths of the hypocotyl and epicotyl showed high estimates of correlation and coincidence coefficients and the possibility of the selection of desirable genotypes. Moreover, the selection based on information from multiple categories (3 or 5 classes) can facilitate the process of evaluating and identifying superior genotypes. However, studies to identify genotype standards are necessary to standardize the methodology of the analysis of the hypocotyl and epicotyl lengths in soybean genotypes.

Conclusion

The magnitude of the genetic parameter estimates characterized the influence of the genetic components in the phenotypic expression of the length of the hypocotyls and epicotyls of different soybean genotypes

The optimal size of the sample, with regard to the length of the hypocotyl, varied among the genotypes and ranged from 45 to 53 for the average and 43 to 51 for the variance. Regarding the length of the epicotyl, the optimal sample size was between 58 and 73 for the average and between 29 and 70 for the variance.

The conversion of quantitative data into multiple categories provided estimates of correlation coefficients greater than 0.80 and coincidence greater than 81% between the converted data and the original data.

Acknowledgements

We thank the National Council of Scientific and Technological Development (CNPq-Brazil) and the Coordination of Improvement of Higher Education Personnel (CAPES) for their financial support of this work.

References

ADAMI, M.; RIZZI, R.; MOREIRA, M. A.; FRIEDRICH, B.; RUDORFF, T.; FERREIRA, C. C. Amostragem probabilística estratificada por pontos para estimar a área cultivada com soja. **Pesquisa Agropecuária Brasileira**, v. 45, n. 6, p. 585-592, 2010.

ANDRADE, F. N.; ROCHA, M. M.; GOMES, R. L. F.; FREIRE FILHO, F. R.; RAMOS, S. R. R. Estimativas de parâmetros genéticos em genótipos de feijão-caupi avaliados para feijão fresco. **Revista Ciência Agronômica**, v. 41, n. 2, p. 253-258, 2010.

ARNHOLD, E.; MILANI, K. F. Rank-ordering coefficients of variation for popping expansion. **Acta Scientiarum. Agronomy**, v. 33, n. 3, p. 527-531, 2011.

BAYS, R.; BAUDET, L.; HENNING, A. S.; LUCCA FILHO, O. A. Recobrimento de sementes de soja com micronutrientes, fungicida e polímero. **Revista Brasileira de Sementes**, v. 29, n. 2, p. 60-67, 2007.

BORGES, V.; FERREIRA, P. V.; SOARES, L.; SANTOS, G. M.; SANTOS, A. M. M. Seleção de clones de batata-doce pelo procedimento REML/BLUP. **Acta Scientiarum. Agronomy**, v. 32, n. 4, p. 643-649, 2010.

CARGNELUTTI FILHO, A.; EVANGELISTA, D. H. R.; GONÇALVES, E. C. P.; STORCK, L. Tamanho da amostra de caracteres de genótipos de soja. **Ciência Rural**, v. 39, n. 4, p. 983-991, 2009.

CARGNELUTTI FILHO, A.; LOPES, S. J.; TOEBE, M.; SILVEIRA, T. R.; SCHWANTES, I. A. Tamanho de amostra para estimação do coeficiente de correlação de Pearson entre caracteres de *Crambe abyssinica*. **Revista Ciência Agronômica**, v. 42, n. 1, p. 149-158, 2011.

CONAB-Companhia Nacional de Abastecimento. **Indicadores da agropecuária**. Brasília: Conab, 2011.

COSTA, J. A.; PIRES, J. L. F.; THOMAS, A. L.; ALBERTON, M. Comprimento e índice de expansão radial do hipocótilo de cultivares de soja. **Ciência Rural**, v. 29, n. 4, p. 609-612, 1999.

COSTA, M. M.; DI MAURO, A. O.; UNÊDA-TREVISOLI, S. H.; ARRIEL, N. H. C.; BÁRBARO, I. M.; SILVEIRA, G. D.; MUNIZ, F. R. S. Heritability estimation in early generations of two-way crosses in soybean. **Bragantia**, v. 67, n. 1, p. 101-108, 2008.

CRUZ, C. D. **Princípios de genética quantitativa**. Viçosa: UFV, 2005.

CRUZ, C. D. **Programa Genes**: estatística experimental e matrizes. Viçosa: UFV, 2006a.

CRUZ, C. D. **Programa Genes**: análise multivariada e simulação. Viçosa: UFV, 2006b.

CRUZ, C. D.; REGAZZI, J. A.; CARNEIRO, P. C. S. **Modelos biométricos aplicados ao melhoramento genético**. 2. ed. Viçosa: UFV, 2004. (v. 1).

DAHER, R. F.; MANDONADO, H.; PEREIRA, A. V.; AMARAL JUNIOR, A. T.; PEREIRA, M. G.; FERREIRA, C. F.; RAMOS, S. R. R.; TARDIN, F. D.; SILVA, M. P. Estimativas de parâmetros genéticos e de coeficiente de repetibilidade de caracteres forrageiros em clones de capimelefante (*Pennisetum purpureum* Schum.). **Acta Scientiarum. Agronomy**, v. 26, n. 4, p. 483-490, 2004.

ESTEFANEL, V.; SACCOL, A. V.; SCHNEIDER, F. M.; BURIOL, G. A.; HELDWEIN, A. B.; MANFRON, P. A. Tamanho da amostra para estimar características agronômicas da soja. **Revista do Centro de Ciências Rurais**, v. 14, n. 3-4, p. 221-229, 1984.

FALCONER, D. S. **Introdução à genética quantitativa**. Viçosa: Imprensa Universitária da UFV, 1987.

FARIAS NETO, J. T.; VELLO, N. A. Avaliação de progênies $F_{4:3}$ e $F_{5:3}$ e estimativas de parâmetros genéticos com ênfase para porcentagem de óleo, produtividade de grãos e óleo em soja. **Ciência e Agrotecnologia**, v. 25, n. 4, p. 812-820, 2001.

FEHR, W. R.; CAVINESS, C. E. **Stages of soybean development**. Ames: Iowa State University of Science and Technology, 1977. (Special Report, 80).

FERRAZ E CAMPOS, S. R.; MACHADO, V. L. S.; VIANA, A. A. N.; AZEVEDO, Z. M. M. Registro e proteção de cultivares. In: SEDIYAMA, T. (Ed.). **Tecnologia de produção e usos da soja**. Londrina: Editora Mecenas, 2009. p. 235-246.

GOMES, R. L. F.; VELLO, N. A.; AZEVEDO FILHO, J. A. Z. Genetic analysis of F_6 and $F_{6:7}$ soybean generations. **Crop Breeding and Applied Biotechnology**, v. 4, n. 1, p. 35-42, 2004.

MARTIN, T. N.; DUTRA, L. M. C.; JAUER, A.; STORCK, L.; ZABOT, L.; UHRY, D.; SANTI, A. L.; STEFANELO, C.; LUCCA FILHO, O. A. Tamanho ótimo de parcela e número de repetições em soja (*Glycine max* (L.) Merrill). **Ciência Rural**, v. 35, n. 2, p. 271-276, 2005.

MAURO, A. O.; SEDIYAMA, T.; SEDIYAMA, C. S. Estimativas de parâmetros genéticos em diferentes tipos de parcelas experimentais em soja. **Pesquisa Agropecuária Brasileira**, v. 30, n. 5, p. 667-672, 1995.

NETO, M. O. V.; BORÉM, A.; SALLES E PORTUGAL, R. Lei de proteção de cultivares. In: BORÉM, A. (Ed.). **Melhoramento de espécies cultivadas**. Viçosa: UFV, 2005. p. 931-960.

NÓBREGA, L. H. P.; VIEIRA, R. D. Avaliação e classificação de cultivares de soja quanto ao comprimento do hipocótilo, sob condições de laboratório e de casa de vegetação. **Revista Brasileira de Sementes**, v. 17, n. 2, p. 160-164, 1995.

NOGUEIRA, A. P. O.; SEDIYAMA, T.; CRUZ, C. D.; REIS, M. S.; PEREIRA, D. G.; JANGARELLI, M. Novas características para diferenciação de cultivares de soja pela análise discriminante. **Ciência Rural**, v. 38, n. 9, p. 2427-2433, 2008.

PEREIRA, W. A.; SÁVIO, F. L.; BORÉM, A.; DIAS, D. C. F. S. Influência da disposição, número e tamanho das sementes no lote de comprimento de plântulas de soja. **Revista Brasileira de Sementes**, v. 31, n. 1, p. 113-121, 2009.

REIS, E. F.; REIS, M. S.; SEDIYAMA, T.; CRUZ, C. D. Estimativas de variâncias e herdabilidades de algumas características primárias e secundárias da produção de grãos em soja (*Glycine max* (L.) Merrill). **Ciência e Agrotecnologia**, v. 26, n. 4, p. 749-761, 2002.

ROCHA, F. H. A.; LOGES, V.; COSTA, A. S.; ARAGÃO, F. A. Z.; SANTOS, V. F. Genetic study with *Heliconia psittacorum* and interspecific hybrids. **Crop Breeding and Applied Biotechnology**, v. 10, n. 4, p. 282-288, 2010.

SEDIYAMA, T.; TEIXEIRA, R. C.; BARROS, H. B. Origem, evolução e importância econômica. In: SEDIYAMA, T. (Ed.). **Tecnologia de produção e usos da soja**. Londrina: Editora Mecenas, 2009. p. 1-5.

SEDIYAMA, T.; TEIXEIRA, R. C.; REIS, M. S. Melhoramento da soja. In: BORÉM, A. (Ed.). **Melhoramento de espécies cultivadas**. Viçosa: UFV, 2005. p. 533-603.

SILVA, A. R.; RÊGO, E. R.; CECON, P. R. Tamanho da amostra para caracterização morfológica de frutos de pimenteira. **Horticultura Brasileira**, v. 29, n. 1, p. 125-129, 2011.

VENCOVSKY, R. Herança quantitativa. In: PATERNIANI, E.; VIEGAS, G. P. (Ed.). **Melhoramento e produção de milho**. Campinas: Editora Fundação Cargil, 1987. p. 135-214.

15

Genetic parameters of iron and zinc concentrations in Andean common bean seeds

Allan Emanoel Mezzomo Zemolin, Nerinéia Dalfollo Ribeiro*, Cleiton Renato Casagrande, Maykon Júnior da Silva and Fernanda Daltrozo Arns

*Departamento de Fitotecnia, Centro de Ciências Rurais, Universidade Federal de Santa Maria, Avenida Roraima, 1000, 97105-900, Santa Maria, Rio Grande do Sul, Brazil. *Author for correspondence. E-mail: nerineia@hotmail.com*

ABSTRACT. The genetic parameter estimates of the iron and zinc concentrations in Andean common bean seeds were obtained using the IAC Boreal × Light Red Kidney and Ouro Branco × Light Red Kidney crosses. The parents and the F_1, F_1 reciprocal, F_2, F_2 reciprocal, and backcross BC_{11} and BC_{12} generations were evaluated in a field experiment that was carried out in the state of Rio Grande do Sul, Brazil. The iron concentration in Andean common bean seeds ranged from 24.70 to 102.40 mg kg^{-1} dry matter (DM), the zinc concentration ranged from 10.73 to 37.50 mg kg^{-1} DM, and no significant maternal effect was observed. The narrow-sense heritability ranged from low (h^2n= 19.04%) to high (h^2n= 63.60%) for the concentrations of iron and zinc, respectively. Hybrid vigor and transgressive segregation were observed for the iron and zinc concentrations in Andean common bean seeds. In the hybrid combination IAC Boreal × Light Red Kidney, it was possible to select recombinants for the iron and zinc biofortification program. From the tested hybrid combinations, recombinants with low iron and zinc concentrations in seeds could be selected to use when the diet needs to be restricted in those minerals.

Keywords: gain with selection, heritability, heterosis, inheritance pattern, maternal effect, *Phaseolus vulgaris*.

Introduction

Beans are significant protein sources and are used daily in human diets in several countries of East Africa and Latin America, representing 65% of the total protein that is consumed and 32% of the total energy (Blair, González, Kimani, & Butare, 2010a). Beans are also important sources of minerals, especially iron, with concentrations ranging from 8.90 to 161.50 mg kg^{-1} dry matter (DM), and zinc, with concentrations ranging from 19.00 to 65.50 mg kg^{-1} DM (Talukder et al., 2010; Tryphone & Nchimbi-Msolla, 2010; Silva, Abreu, Ramalho, & Maia, 2012a), in common bean cultivars. In addition, common bean grains do not require polishing or any other prior mechanical processing and thus can be considered a whole food. Therefore, beans can potentially be used in mineral biofortification programs.

Both high and low iron and zinc concentrations in seeds can be desired. Seeds with high concentrations of these minerals are desired because iron and zinc deficiencies prevail worldwide; affect people of all ages, genders and social classes; and are considered a public health issue. Such deficiencies cause symptoms such as anemia, decreased work capacity, decreased birth weight, immune system changes, dermatitis, diarrhea and others, as described by Gibson (2012) and Abbaspour, Hurrel, and Kelishadi (2014). Therefore, increased iron and zinc concentrations have received significant attention from common bean breeding programs. However, seeds with low concentrations of these minerals are desired by some patients. Patients with a diagnosis of hereditary hemochromatosis are recommended to avoid iron-rich food (Cançado & Chiattone, 2010) due to their increased intestinal absorption of iron, which can lead to iron overload, increasing their risk of fibrosis, functional failure, sclerosis, diabetes and cardiomyopathies (Cançado & Chiattone, 2010; Pietrangelo, 2010). Excess zinc in the human body alters the HDL levels, the immune function and the iron concentration in the body (Guerrero-Romero & Rodríguez-Morán, 2005). Therefore, the selection of common bean cultivars with low iron and zinc concentrations is important for cases in which the diet requires the restriction of these minerals.

To develop common bean cultivars with iron and zinc concentrations in seeds that meet the specific dietary needs of bean consumers, it is necessary to study the genetic parameters of these minerals. The expression of a maternal effect of iron and zinc concentrations in common bean seeds depends on the evaluated gene pool and the tested hybrid combination (Jost, Ribeiro, Cerutti, Poersch, & Maziero, 2009; Rosa et al., 2010; Silva, Abreu, & Ramalho, 2013; Mukamuhirwa, Tusiime, & Mukankusi, 2015; Possobom, Ribeiro, Domingues, & Casagrande, 2015). Similarly, there is a wide range of variation for narrow-sense heritability and predicted gain with selection estimates for iron and zinc concentrations in Middle American and Andean common bean seeds (Cichy, Forster, Grafton, & Hosfield, 2005; Jost et al., 2009; Rosa et al., 2010; Mukamuhirwa et al., 2015; Possobom et al., 2015).

Genetic parameter estimates refer strictly to the analyzed populations, the environmental conditions and the estimation method (Ramalho et al., 2012). For the seed coat combinations of dark red with cream-colored stripes × light red and white × light red, genetic parameter estimates for the iron and zinc concentrations in Andean common bean seeds were not found in the literature. The selection of low iron and zinc concentrations in common bean seeds, aiming to serve a market niche of individuals presenting health problems due to the high ingestion of these minerals, is unprecedented. Thus, the objectives of this study were to obtain genetic parameter estimates for the iron and zinc concentrations in Andean common bean seeds and to select recombinants for mineral biofortification programs, as well as for cases in which diets with restricted iron and zinc are required.

Material and methods

Obtaining segregating generations

The generations that were assessed were obtained from controlled crosses between the Andean common bean cultivars IAC Boreal × Light Red Kidney and Ouro Branco × Light Red Kidney. The parents were selected based on the seed coat colour: IAC Boreal is dark red with cream-colored stripes, Light Red Kidney is light red, and Ouro Branco is white; and due to the contrasting iron and zinc concentrations in the seeds (Ribeiro et al., 2014a).

Crossing blocks were installed in a greenhouse at the Plant Science Department of the Santa Maria Federal University (UFSM), Santa Maria, in the State of Rio Grande do Sul, Brazil (29°42´S lat, 53°49´W long, and 95 m asl). The plants were cultivated in 5 L plastic pots that were filled with a mixture of a typical alitic Argisol Hapludalf, Macplant commercial substrate, carbonized rice husk, and ashes at a volumetric ratio of 3:1:1:1. In each pot, there were two bean plants.

During the summer-fall season of 2013, F_1 (♀ P_1 × ♂ P_2) and F_1 reciprocal (♀ P_2 × ♂ P_1) seeds were obtained for both hybrid combinations using the interlacing method with previous floral bud emasculation. The hybrids and parents were sown in the winter-spring season of 2013 to obtain F_2 (natural selfing of F_1 plants), F_2 reciprocal (natural selfing of F_1 reciprocal plants), and backcross BC_{11} (♀ F_1 × ♂ P_1) and BC_{12} (♀ F_1 × ♂ P_2) seeds. The pods were harvested individually at maturity (R_9), and the seeds were oven dried (65 to 70 °C) to 13% average humidity.

Generations evaluated

The seeds from the F_1, F_1 reciprocal, F_2, F_2 reciprocal, BC_{11} and BC_{12} generations that were obtained for each hybrid combination and from the parents (IAC Boreal, Light Red Kidney and Ouro Branco) were evaluated in an experiment that was performed in a field at the Plant Science Department of UFSM. The soil was classified as typical alitic Argisol, Hapludalf and presented the following chemical composition: pH (H_2O): 6.50; organic matter: 1.80%; P (Mehlich-1): 18.00 mg dm^{-3}; K: 116.00 mg dm^{-3}; Ca^{+2}: 8.70 cmolc dm^{-3}; Mg^{+2}: 4.50 cmolc dm^{-3}; $H^{+}+Al^{3+}$: 1.70 cmolc dm^{-3}; Fe: 2267.40 mg dm^{-3}; Zn: 0.50 mg dm^{-3}; and Cu: 0.90 mg dm^{-3}. The soil was prepared conventionally and fertilized according to the interpretation of soil analysis results. The fertilizers applied were 275 kg ha^{-1} of the 5-20-20 formula (urea: 45% nitrogen; superphosphate: 18% P_2O_5; and potassium chloride: 60% K_2O) at furrow sowing and 67 kg ha^{-1} urea (45% N) at the growth stage of the first trifoliate leaf (V3). Micronutrients were not added to the soil or to the leaves.

The experiment was carried out in two blocks in 1.00-m-long rows that were spaced 0.5 m apart. Sowing was conducted on 17 February 2014, and the number of rows varied with seed availability. In block 1, the generations that were obtained from the cross IAC Boreal × Light Red Kidney were evaluated in different numbers of rows: seven rows for F_1 and F_1 reciprocal, four rows for BC_{11} and BC_{12}, 14 rows for F_2 and F_2 reciprocal, and seven rows for each parent (IAC Boreal, Light Red Kidney and Ouro Branco). In block 2, the generations that were obtained from the cross Ouro Branco × Light Red Kidney were evaluated, with seven rows for F_1 and F_1 reciprocal, five rows for BC_{11} and BC_{12}, 15 rows for F_2 and F_2 reciprocal, and seven rows for each parent. The sowing density was 15 seeds m^{-1} for the parents; 10 seeds m^{-1} for F_1, F_1 reciprocal and backcross; and 8 seeds m^{-1} for F_2 and F_2 reciprocal.

Weed control was carried out by hoeing, and insects were eliminated by applying Engeo™ Pleno (Thiamethoxam + Lambda-cyhalothrin) insecticide at a dose of 125 mL ha^{-1}. Disease control was not carried out. The plants were harvested individually and manually at maturity (R9). The pods were removed manually to avoid contaminating the seeds with metals.

The seeds of each plant were oven dried at 65 to 70°C until reaching 13% average humidity. A random sample of 5 g of seeds was ground in an electrical grinder to produce particles smaller than 1 mm and was not sieved. Part of the raw bean flour that was obtained (0.5 g) was used for the nitric-perchloric digestion process, as described by Ribeiro, Jost, Maziero, Storck, and Rosa (2014b). The iron and zinc concentrations in the seeds were measured in a flame atomic absorption spectrophotometer at wavelengths of 248.30 and 213.90 nm, respectively.

Genetic parameter estimates

The bilateral t test was used to compare the means between the $P_1 \times P_2$, $P_1 \times F_1$, $P_2 \times F_1$ reciprocal, $F_1 \times F_1$ reciprocal and $F_2 \times F_2$ reciprocal contrasts to investigate the maternal effects on the iron and zinc concentrations in common bean seeds for each hybrid combination. The data of the iron and zinc concentrations in F_2 plants were submitted to Shapiro-Wilk testing at the 5% significance level to verify the normality. The number of classes used in the frequency distribution was calculated using the expression $\sqrt{n}$, where n is the number of observations.

The genetic parameters were estimated from the variances of the parents (P_1 and P_2) and the F_1, F_2, BC_{11} and BC_{12} generations. Heritability was estimated in a broad sense ($h_b^2 = \frac{\sigma_g^2}{\sigma_P^2}$) and a narrow sense ($h_n^2 = \frac{\sigma_A^2}{\sigma_P^2}$) according to the backcross method, in which genotypic variance: $\sigma^2_g = \sigma^2_P - \sigma^2_E$; phenotypic variance: $\sigma^2_P = \sigma^2_{F_2}$; additive variance: $\sigma^2_A = 2\sigma^2_{F_2} - (\sigma^2_{BC_{11}} + \sigma^2_{BC_{12}})$; and environmental variance in F_2: $\sigma^2_E = (0.5\,\sigma^2_{F_1} + 0.25\,\sigma^2_{P_1} + 0.25\,\sigma^2_{P_2})$. Heterosis in the F_1 generation was quantified by traditional heterosis ($H\% = \frac{F_1 - P}{P} x\, 100$) and heterobeltiosis ($HT\% = \frac{F_1 - MP}{MP} x\, 100$), where $P = \frac{P_1 + P_2}{2}$, and MP is the better parent.

The prediction of gains with selection was carried out considering a selection of 10% of the F_2 plants with higher iron and zinc concentrations in their seeds for each hybrid combination. The expected gain with selection was estimated using the expression $\Delta G = DS\, x\, h_n^2$ and $\Delta G(\%) = \frac{\Delta G\, x\, 100}{\bar{F_2}}$, where DS = selection differential, as calculated by the difference between the selected F_2 plant means and the general mean of the F_2 plants. Statistical analyses were carried out using the Office Excel and Action spreadsheet and GENES program (Cruz, 2013).

Results and discussion

Genetic variability and maternal effect

In the hybrid combinations of IAC Boreal × Light Red Kidney and Ouro Branco × Light Red Kidney, only contrast between the parents for the zinc concentrations in their seeds was significant (Table 1). For the iron concentration in seeds, the $P_1 \times P_2$ contrast was not significant for any of the hybrid combinations, indicating no genetic variability between the parents, in contrast with the observations of Ribeiro et al. (2014a) for the IAC Boreal, Light Red Kidney and Ouro Branco cultivars. In this case, the differences in the soil pH, soil iron content and amount of precipitation during the growing season explain the variation that was observed in the iron concentrations of common bean seeds (Moraghan, Padilha, Etchevers, Grafton, & Acosta-Gallegos, 2002; Silva, Abreu, Ramalho, & Corrêa, 2012b; Petry, Boy, Wirth, & Hurrell, 2015; Possobom et al., 2015).

Because the parents that were used in the controlled crosses were contrasting regarding the zinc concentration in their seeds, it was possible to obtain recombinants with genetic variability, with zinc concentrations ranging from 10.73 to 37.50 mg kg^{-1} DM (Figure 1C and D). However, for iron, the range observed in the F_2 generation was greater, from 24.70 to 102.40 mg kg^{-1} DM (Figure 1A and B). Therefore, even when using convergent crosses for the iron concentration, a wide segregation was observed for this mineral in the F_2 generation, comparable to that observed previously by Jost et al. (2009) in segregants obtained from crosses between the common bean cultivars Diamante Negro and IAPAR 44, which did not present significant differences in the iron concentrations in the seeds. From the hybrid combinations that were tested in the present study, the genetic variability for the iron and zinc concentrations in Andean common bean seeds was obtained, enabling the selection of recombinants in early generations.

The $F_1 \times F_1$ reciprocal contrast was not significant for the iron and zinc concentrations in Andean common bean seeds in the crosses between IAC Boreal × Light Red Kidney and Ouro Branco × Light Red Kidney (Table 1). This result indicates the absence of a maternal effect and that the seeds of the F_1 generation represent fertilization between the parents in both hybrid combinations. Our results agree with those of Possobom et al. (2015) and Rosa et al. (2010), who reported non-significant maternal effects for iron and zinc concentrations, respectively, in hybrid seeds of common bean.

The expression of maternal effect for the iron and zinc concentrations in common bean seeds seems to be related to the distribution of these minerals in the fractions of the seed. Iron and zinc are minerals that, depending on the cultivar, can be concentrated in higher quantities in the seed coat or embryo in common bean seeds and can even occur in similar proportions in the seed fractions (Moraghan et al., 2002; Ribeiro et al., 2012; Possobom et al., 2015). Possobom et al. (2015) observed that the parents CNFP 10104 and CHC 01-175 accumulated from 54.61 to 67.92% of the total iron in the seed coat of common bean seeds and observed a significant maternal effect of this mineral in the hybrids. In the other evaluated hybrid combination (Cal 96 × Hooter), the authors did not find any maternal effect expression, which was verified by the greater accumulation of iron, from 69.40 to 73.44%, in the seed embryo. Therefore, the selection of superior common bean recombinants for iron concentration should begin in seeds of the F_3 generation if iron accumulates more in the seed coat or in seeds of the F_2 generation if iron is more concentrated in the seed embryo (Jost et al., 2009; Possobom et al., 2015).

The locations of iron and zinc in the different fractions of common bean seeds will directly affect the bioavailability of these minerals because common bean seed embryos contain significant amounts of phytates, which inhibit iron and zinc absorption by the human body (Ariza-Nieto, Blair, Welch, & Glahn, 2007). Thus, as the results of this study suggest, if the accumulation of iron and zinc is higher in the embryos of common bean seeds, it is expected that these minerals will be less bioavailable. However, Ramírez-Cárdenas, Leonel, Costa, and Reis (2010) observed that the grains of cooked beans from the Ouro Branco cultivar presented high zinc and phytate concentrations and a greater zinc bioavailability among other evaluated cultivars. Therefore, it is possible to select common bean cultivars with high micromineral and phytate concentrations without compromising the bioavailability of these minerals.

Table 1. Number of plants (NP); mean values of the iron and zinc concentrations in common bean seeds; respective standard deviation obtained in parents (P_1 and P_2) and in the F_1, F_1 reciprocal (F_1r), F_2, F_2 reciprocal (F_2r) and backcross (BC_{11} and BC_{12}) generations in the crosses IAC Boreal (IAC B, P_1) × Light Red Kidney (LRK, P_2) and Ouro Branco (OB, P_1) × Light Red Kidney (LRK, P_2); and the probability as determined by the *t* test (p value) for the contrasts $P_1 \times P_2$, $P_1 \times F_1$; $P_2 \times F_1$ reciprocal; $F_1 \times F_1$ reciprocal; and $F_2 \times F_2$ reciprocal.

Generation	Iron (mg kg^{-1} dry matter)				Zinc (mg kg^{-1} dry matter)			
	IAC B × LRK		OB × LRK		IAC B × LRK		OB × LRK	
	NP	Mean	NP	Mean	NP	Mean	NP	Mean
P_1	24	71.86 ± 9.65	16	49.33 ± 7.33	25	22.47 ± 2.92	24	18.13 ± 2.38
P_2	21	69.03 ± 10.02	16	56.21 ± 7.49	25	16.98 ± 2.68	24	16.39 ± 2.02
F_1	17	77.94 ± 9.74	11	60.96 ± 7.31	17	24.62 ± 2.35	15	19.63 ± 1.76
F_1r	19	74.69 ± 11.20	12	59.70 ± 8.76	20	22.63 ± 3.00	16	20.17 ± 2.59
F_2	55	53.88 ± 17.65	56	52.86 ± 10.11	55	20.98 ± 4.00	56	19.78 ± 3.39
F_2r	54	65.08 ± 15.01	50	50.50 ± 11.22	56	21.97 ± 4.50	50	18.71 ± 4.15
BC_{11}	12	68.20 ± 12.12	20	52.52 ± 9.45	12	21.28 ± 4.02	17	19.83 ± 2.73
BC_{12}	20	76.48 ± 17.72	20	47.94 ± 10.78	20	23.91 ± 3.83	17	20.11 ± 3.48
Probability (%)								
$P_1 \times P_2$		65.68		1.29		<0.01		0.87
$P_1 \times F_1$		5.22		0.05		1.47		4.00
$P_2 \times F_1r$		9.62		26.55		<0.01		<0.01
$F_1 \times F_1r$		63.42		71.28		1.59		50.10
$F_2 \times F_2r$		0.07		25.56		22.60		14.28

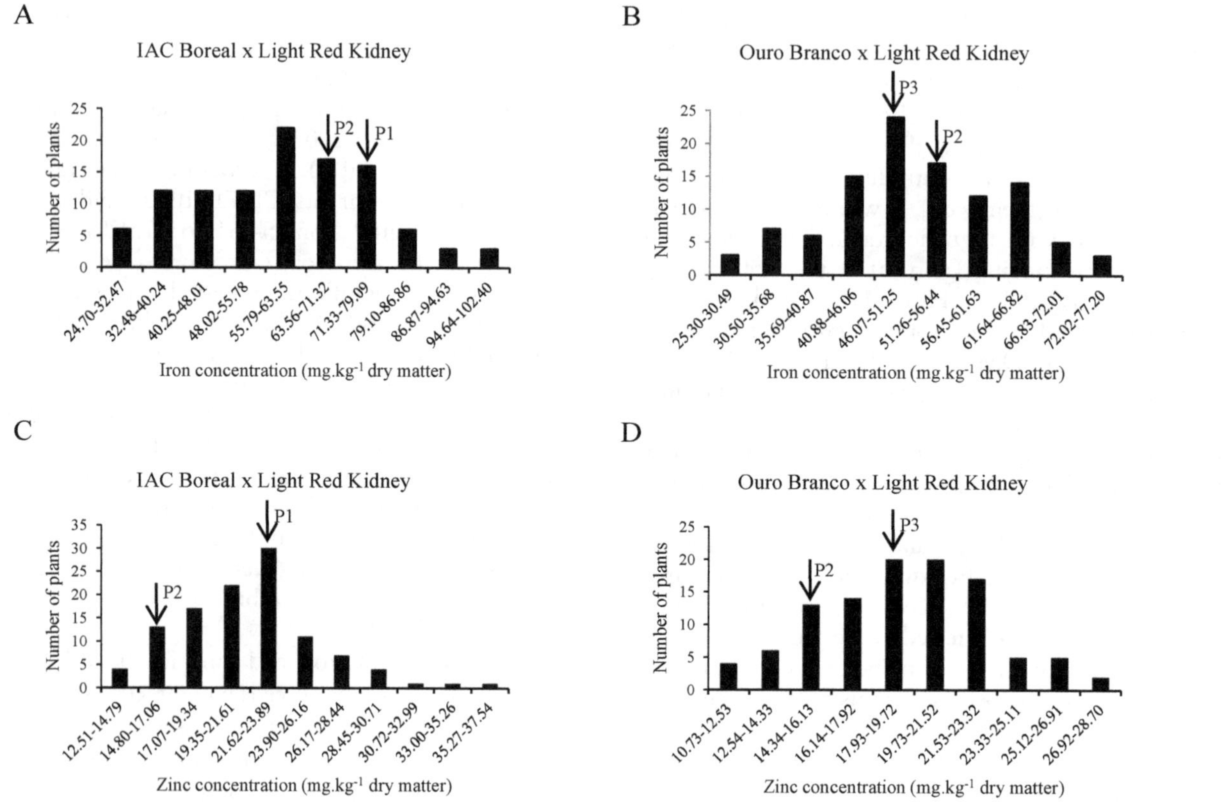

Figure 1. Frequency distribution of the F_2 plants that were obtained from the cross IAC Boreal × Light Red Kidney for iron (A) and zinc (C) concentrations and cross Ouro Branco × Light Red Kidney for iron (B) and zinc (D) concentrations in common bean seeds during the dry season of 2014.

** The arrows indicate the mean values that were obtained from the parents: P1 = IAC Boreal, P2 = Light Red Kidney, and P3 = Ouro Branco.

Inheritance pattern, heritability and heterosis

The iron and zinc concentrations in Andean common bean seeds of the F_2 generation presented a continuous, nearly normal distribution, indicating quantitative inheritance (Figure 1A, B, C and D). A continuous distribution of the iron and zinc concentrations in common bean seeds was previously described in early (Possobom et al., 2015) and advanced (Blair, Astudilo, Grusak, Graham, & Beebe, 2009; Cichy, Caldas, Snapp, & Blair, 2009; Blair et al., 2010b; Ribeiro et al., 2014b; Teixeira, Lima, Abreu, & Ramalho, 2015) generations in recombinants that were obtained from the crosses Andean × Andean, Andean × Middle American and Middle American × Middle American. These results affirm that the iron and zinc concentrations in common bean present quantitative inheritance in populations of low and high homozygosity, regardless of the gene pool, and that they oppose the initial hypothesis of Cichy et al. (2005) regarding the monogenic control of the zinc concentration in common bean seeds.

The iron concentration in Andean common bean seeds presented broad-sense heritability estimates of intermediate (h^2b= 52.36%; Ouro Branco × Light Red Kidney) to high (h^2b= 67.83%; IAC Boreal × Light Red Kidney) magnitude (Table 2), indicating that the genotypic variance was greater than the phenotype variance. These values are comparable to the broad-sense heritability estimates of iron concentrations in Andean common bean seeds (Possobom et al., 2015) and Middle American common bean seeds (Jost et al., 2009; Teixeira et al., 2015). The narrow-sense heritability estimates varied from 19.04 to 45.29% in the crosses of Ouro Branco × Light Red Kidney and IAC Boreal × Light Red Kidney, respectively, suggesting difficulties in the fixation of this character with the advance of generations. Narrow-sense heritability estimates of similar magnitude were observed previously by Ribeiro et al. (2014b) and Possobom et al. (2015) for the iron concentrations in common bean seeds; however, these estimates were inferior to those obtained by Jost et al. (2009). This result is supported by the fact that heritability estimates refer strictly to the populations assessed, the environmental conditions and the estimation method and are not constant or fixed values (Ramalho et al., 2012).

Table 2. Genetic parameter estimates and prediction of gains with selection for the iron and zinc concentrations in common bean seeds that were obtained from the crosses IAC Boreal (IACB) × Light Red Kidney (LRK) and Ouro Branco (OB) × Light Red Kidney (LRK) during the dry season of 2014.

Genetic parameters	Iron (mg kg^{-1} DM)		Zinc (mg kg^{-1} DM)	
	IACB × LRK	OB × LRK	IACB × LRK	OB × LRK
P_1 variance	93.08	53.75	8.54	5.67
P_2 variance	100.39	56.10	7.20	4.08
F_1 variance	94.95	53.41	5.50	3.09
Phenotypic variance (σ^2_P)	297.92	113.69	18.25	14.35
Environment variance (σ^2_E)	95.85	54.16	6.68	3.99
Genotypic variance (σ^2_G)	202.07	59.52	11.56	10.36
Additive variance (σ^2_A)	134.93	21.64	5.66	9.12
Broad-sense heritability (h^2_b)	67.83	52.36	63.36	72.22
Narrow-sense heritability (h^2_n)	45.29	19.04	31.04	63.60
Traditional heterosis (H%)	10.63	15.54	24.84	13.75
Heterobeltiosis (HT%)	8.46	8.46	9.59	8.28
Minimum value in parents	52.20	36.80	12.10	12.99
Maximum value in parents	94.90	65.70	27.30	22.68
Minimum value in F_2	24.70	25.30	12.50	10.73
Maximum value in F_2	102.40	77.20	37.50	28.71
Original mean in F_2	59.43	51.75	21.48	19.27
Selection for high values				
Mean of selected plants	91.49	69.90	30.09	26.16
Selection differential (DS)	32.06	18.15	8.61	6.88
Gain with selection (ΔG)	14.52	3.45	2.67	4.38
Gain with selection (ΔG%)	24.43	6.68	12.44	22.72
Predicted mean after the first selection cycle	73.95	55.20	24.15	23.65
Selection for low values				
Mean of selected plants	30.72	32.21	15.11	12.54
Selection differential (DS)	-28.71	-19.54	-6.37	-6.73
Gain with selection (ΔG)	-13.00	-3.72	-1.98	-4.28
Gain with selection (ΔG%)	-21.88	-7.19	-9.21	-22.22
Predicted mean after the first selection cycle	46.43	48.03	19.50	14.99

The broad-sense heritability of the zinc concentrations of Andean common bean seeds had a high magnitude (h^2b= 63.36 to 72.22%), and the narrow-sense heritability had an intermediate (h^2n= 31.04%; IAC Boreal × Light Red Kidney) to high (h^2n= 63.60%; Ouro Branco × Light Red Kidney) magnitude. The variation range that was obtained for the broad-sense and narrow-sense heritability estimates of zinc concentrations was similar to that found for Middle American common bean seeds by Rosa et al. (2010) and was inferior to the estimates reported by Cichy et al. (2005). In this study, a high narrow-sense heritability was observed for the zinc concentration in the cross Ouro Branco × Light Red Kidney, indicating the predominance of additive variance. In this case, the fixation of favorable alleles can be verified in the next generation (Borém & Miranda, 2007), favoring the selection of the recombinants obtained from this hybrid combination.

The values of traditional heterosis and heterobeltiosis varied from 8.28 to 24.84%, indicating that it was possible to obtain common bean hybrids with iron and zinc concentrations in seeds superior to the average and maximum iron and zinc concentrations of the parent seeds (Table 2). The existence of hybrid vigor was previously described for the iron and zinc concentrations in common bean seeds (Jost et al., 2009; Rosa et al., 2010; Silva et al., 2013). The development of common bean hybrid cultivars is not usual due to the low efficiency that is obtained in controlled crosses, the non-existence of a male sterility mechanism and the low use of seeds for cropping. Therefore, segregating generations will lead to homozygosis, and it is expected that heterosis will be reduced by half in each self-fertilized generation.

Selection for high iron and zinc concentrations in seeds

In the tested hybrid combinations, transgressive segregation was observed for the iron and zinc concentrations in the seeds (Figure 1, Table 2). Recombinants with higher and lower iron and zinc concentrations in seeds than in the parents were obtained from both hybrid combinations, similar to the results reported for other bi-parental crosses that have been carried out in common bean (Jost et al., 2009; Rosa et al., 2010; Silva et al., 2013; Ribeiro et al., 2014b; Mukamuhirwa et al., 2015; Possobom et al., 2015).

In the cross IAC Boreal × Light Red Kidney, three F_2 plants were obtained with an iron concentration in seeds greater than 95 mg kg^{-1} DM (Figure 1A), which is considered high for common bean inbred lines (Ribeiro, Domingues, Zemolim, & Possobom, 2013). The selection of recombinants with high iron concentrations in seeds is promising in this cross due to the high broad-sense heritability estimates (h^2b= 67.83%) and predicted gain with selection (ΔG%= 24.43%) and due to the intermediate narrow-sense heritability (h^2n= 45.29%). The development of common bean cultivars that are biofortified for iron meets the demand for foods that can be used to fight against anemia. Iron deficiency also affects cognitive development, immunity and work capacity (Abbaspour et al., 2014).

We obtained three recombinants with a zinc concentrations in seeds greater than 31 mg kg^{-1} DM, which is considered high according to the stratification presented by Cichy et al. (2005) and Tryphone and Nchimbi-Msolla (2010), in the cross IAC Boreal × Light Red Kidney. The selection and advance of these populations will result in the development of common bean cultivars that are biofortified for zinc, which can contribute to immune system improvement, dermatitis treatment, bone maturation and taste sensitivity, which are all affected by zinc deficiency in the body (Gibson, 2012).

Selection for low iron and zinc concentrations in seeds

Recombinants with an iron concentration lower than 42 mg kg^{-1} DM were obtained in both hybrid combinations (Table 2). This value characterizes low iron for common bean lines according to the classification presented by Tryphone and Nchimbi-Msolla (2010). However, the selection of recombinants with a low iron concentration in seeds is more promising in the cross IAC Boreal × Light Red Kidney due to the lowest value of predicted gain with selection (ΔG%= -21.88%).

The selection of common bean lines with a low iron concentration in seeds is important for the diet of patients with hereditary hemochromatosis. These patients have an increased intestinal absorption of iron, which can lead to iron overload, causing cellular and tissue injury, fibrosis, functional failure, sclerosis, diabetes and cardiomyopathies (Cançado & Chiattone, 2010; Pietrangelo, 2010). Therefore, these patients are recommended to avoid eating iron-rich food (Cançado & Chiattone, 2010).

In the cross Ouro Branco × Light Red Kidney, 104 F_2 plants were obtained with a zinc concentration in seeds of less than 26 mg kg^{-1} DM (Figure 1D), which is considered low for common bean inbred lines (Tryphone and Nchimbi-Msolla, 2010). In this cross, we also observed a higher predicted gain with selection for a low zinc concentration (ΔG%= -22.22%). The development of common bean cultivars with a low zinc concentration serves a market niche for individuals seeking to normalize their levels of HDL in the blood or for individuals with altered immune function and iron concentration due to zinc toxicity (Guerrero-Romero & Rodriguez-Morán, 2005).

Conclusion

In Andean common bean seeds in the F_2 generation, the iron concentration ranges from 24.70 to 102.40 mg kg^{-1} dry matter (DM), the zinc concentration ranges from 10.73 to 37.50 mg kg^{-1} DM, and no maternal effect was observed.

Narrow-sense heritability estimates is from low (h^2n= 19.04%) to high (h^2n= 63.60%) magnitude for the iron and zinc concentrations.

There is a hybrid vigor for the iron and zinc concentrations in seeds of Andean common bean and transgressive segregation was observed.

Recombinants with low and high iron and zinc concentrations in seeds can be selected from the hybrid combinations assessed.

Acknowledgements

The authors would like to thank the National Council of Technological and Scientific Development (CNPq) for the financial aid and scholarships granted. To the Coordination for the Improvement of Higher Education Personnel (Capes) for the scholarship granted.

References

Abbaspour, N., Hurrell R., & Kelishadi R. (2014). Review on iron and its importance for human health. *Journal of Research in Medical Sciences*, *19*(2), 164-174.

Ariza-Nieto, M., Blair, M. W., Welch, R. M., & Glahn, P. (2007). Screening of iron bioavailability patterns in eight bean (*Phaseolus vulgaris* L.) genotypes using the caco-2 cell in vitro model. *Journal of Agricultural and Food Chemistry*, *55*(19), 7950-7956. doi: 10.1021/jf070023y

Blair, M. W., Astudilo, C., Grusak, M. A., Graham, R., & Beebe, S. E. (2009). Inheritance of seed iron and zinc concentrations in common bean (*Phaseolus vulgaris* L.). *Molecular Breeding*, *23*(2), 197-207. doi: 10.1007/s11032-008-9225-z

Blair, M. W., González, L. F., Kimani, P. M., & Butare, L. (2010a). Genetic diversity, inter-gene pool introgression and nutritional quality of common beans (*Phaseolus vulgaris* L.) from Central Africa. *Theoretical and Applied Genetics*, *121*(2), 237-248. doi: 10.1007/s00122-010-1305-x

Blair, M. W., Medina, J. I., Astudilo, C., Rengifo, J., Beebe, S. E., Machado, G., & Graham, R. (2010b). QTL for seed iron and zinc concentration and content in a Mesoamerican common bean (*Phaseolus vulgaris* L.) population. *Theoretical and Applied Genetics*, *121*(6), 1059-1070. doi: 10.1007/s00122-010-1371-0

Borém, A., & Miranda, G. V. (2007). *Melhoramento de plantas* (4a ed.). Viçosa, MG: UFV.

Cançado, R. D., & Chiattone, C. S. (2010). Visão atual da hemocromatose hereditária. *Revista Brasileira de Hematologia e Hemoterapia*, *32*(6), 469-475. doi: 10.1590 /S1516-84842010000600011

Cichy, K. A., Forster, S., Grafton, K. F., & Hosfield, G. L. (2005). Inheritance of seed zinc accumulation in navy bean. *Crop Science*, *45*(1), 864-870. doi: 10.2135/cropsci2004.0104

Cichy, K. A., Caldas, G. V., Snapp, S. S., & Blair, M. W. (2009). QTL analysis of seed iron, zinc, and phosphorus levels in an Andean bean population. *Crop Science*, *49*(5), 1742-1750. doi: 10.2135/cropsci2008. 10.0605

Cruz, C. D. (2013). Genes - a software package for analysis in experimental statistics and quantitative genetics. *Acta Scientiarum Agronomy*, *35*(3), 271-276. doi: 10.4025/actasciagron.v35i3.21251

Gibson, R. (2012). Zinc deficiency and human health: etiology, health consequences, and future solutions. *Plant and Soil*, *361*(1), 291-299. doi: 10.1007/s11104-012-1209-4

Guerrero-Romero, F., & Rodríguez-Morán, M., (2005). Complementary therapies for diabets: the case for chromium, magnesium, and antioxidants. *Archives of Medical Research*, *36*(3), 250-257. doi: 10.1016/j.arcmed. 2005.01.004

Jost, E., Ribeiro, N. D., Cerutti, T., Poersch, N. L., & Maziero, S. M. (2009). Potencial de aumento do teor de ferro em grãos de feijão por melhoramento genético. *Bragantia*, *68*(1), 35-42. doi: 10.1590/50006-8705

Moraghan, J. T., Padilla, J., Etchevers, J. D., Grafton, K., & Acosta-Gallegos, J. A. (2002). Iron accumulation in seed of common bean. *Plant and Soil*, *246*(2), 175-183. doi: 10.1023/A:1020616026728

Mukamuhirwa, F., Tusiime, G., & Mukankusi, M. C. (2015). Inheritance of high iron and zinc concentration in selected bean varieties. *Euphytica*, *205*(2), 369-380. doi: 10. 1007/s10681-015-1385-4

Petry, N., Boy, E., Wirth, J. P., & Hurrell, R. F. (2015). The potential of the common bean (*Phaseolus vulgaris*) as a vehicle for iron biofortification. *Nutrients*, 7(2), 1144-1173. doi: 10.3390/nu7021144

Pietrangelo, A. (2010). Hereditary hemochromatosis: pathogenesis, diagnosis and treatment. *Gastroenterology*, *139*(2), 393-408. doi: 10.1053/j.gastro.2010.06.013

Possobom, M. T. D. F., Ribeiro, N. D., Domingues, L. da S., & Casagrande, C. R. (2015). Genetic control of iron concentration in Mesoamerican and Andean common bean seeds. *Pesquisa Agropecuária Brasileira*, *50*(5), 383-391. doi: 10.1590/S0100-204X201500050 0005

Ramalho, M. A. P., dos Santos, J. B., Pinto, C. A. B. P., Souza E. A., Gonçalves, F. M. A., & Souza, J. C. (2012). *Genética na agropecuária* (5a ed.). Lavras, MG: UFLA.

Ramírez-Cárdenas, L., Leonel, A. J., Costa, N. M. B., & Reis, F. P. (2010). Zinc biovailability in different beans as affected by cultivar type and cooking conditions. *Food Research International*, *43*(2), 573-581. doi: 10. 1016/j.foodres.2009.07.023

Ribeiro, N. D., Maziero, S. M., Prigol, M., Nogueira, C. W., Rosa, D. P., & Possobom, M. T. D. F. (2012). Mineral concentrations in the embryo and seed coat of common bean cultivars. *Journal of Food Composition and Analysis*, *26*(1-2), 89-95. doi: 10.1016/j.jfca.2012.03.003

Ribeiro, N. D., Domingues, L. da S., Zemolin, A. E. M., & Possobom, M. T. D. F. (2013). Selection of common bean lines with high agronomic performance and high calcium and iron concentrations. *Pesquisa Agropecuária Brasileira*, *48*(10), 1368-1375. doi: 10.1590/ S0100-204X2013001000008

Ribeiro, N. D., Rodrigues, J. A., Prigol, M., Nogueira, C. W., Storck, L., & Gruhn, E. M. (2014a). Evaluation of special grains bean lines for grain yield, cooking time and mineral concentrations. *Crop Breeding and Applied Biotechnology*, *14*(1), 15-22. doi: 10.1590/S1984-70332014000100003

Ribeiro, N. D., Jost, E., Maziero, S. M., Storck, L., & Rosa, D. P. (2014b). Selection of common bean lines with high grain yield and high calcium and iron concentrations. *Revista Ceres*, *61*(1), 77-83. doi: 10.15 90/S0034-737X2014000100010

Rosa, S. S., Ribeiro, N. D., Jost, E., Reiniger, L. R. S., Rosa, D. P., Cerutti, T., & Possobom, M. T. D. F. (2010). Potential for increasing the zinc content in common bean using genetic improvement. *Euphytica*, *175*(2), 207-213. doi: 10.1007/s10681-010-0163-6

Silva, C. A., Abreu, A. F. B., Ramalho, M. A. P., & Maia, L. G. S. (2012a). Chemical composition as related to seed color of common bean. *Crop Breeding and Applied Biotechnology*, *12*(2), 132-137. doi: 10.1590/S1984-70332012000200006

Silva, C. A., Abreu, A. F. B., Ramalho, M. A. P., & Corrêa, A. D. (2012b). Interaction genotype by season and its influence on the identification of beans with high content of zinc and iron. *Bragantia*, *71*(3), 336-341. doi: 10.1590 /S0006-87052012005000037

Silva, C. A., Abreu, A. F. B., & Ramalho, M. A. P. (2013). Genetic control of zinc and iron concentration in common bean seeds. *African Journal of Agricultural Research*, *8*(11), 1001-1008. doi: 10.5897/AJAR12.2172

Teixeira, R. K. S., Lima, D. C., Abreu, A. F. B., & Ramalho, M. A. P. (2015). Implications of early selection for grain colour on iron and zinc content and productivity of common bean. *Plant Breeding*, *134*(2), 193-196. doi: 10. 1111/pbr.12241

Talukder, Z. I., Anderson, E., Miklas, P. N., Blair, M. W., Osorno, J., Dilawari, M., & Hossain, K. G. (2010). Genetic diversity and selection of genotypes to enhance Zn and Fe content in common bean. *Canadian Journal of Plant Science*, *90*(1), 49-60. doi: 10. 4141/CJPS09096

Tryphone, G. M., & Nchimbi-Msolla, S. (2010). Diversity of common bean (*Phaseolus vulgaris* L.) genotypes in iron and zinc contents under screenhouse conditions. *Journal of Agricultural Research*, *5*(8), 738-747. doi: 10. 5897/AJAR10.304

16

Evaluation of common bean accesses with multi-category variables

Amanda Nogueira de Albuquerque*, Marco Antônio Aparecido Barelli, Leonarda Grillo Neves, Vinícius Ribeiro Arantes and Kleberson Luiz Miranda da Silva

*Departamento de Agronomia, Universidade do Estado do Mato Grosso, Av. Tancredo Neves, 1095, Cavalhada II, 78200-000, Cáceres, Mato Grosso, Brazil. *Author for correspondence. E-mail: amanda_vha@hotmail.com*

ABSTRACT. The morphologic traits of the common bean genotype (*Phaseolus vulgaris* L.) in Cáceres MT, Brazil, are evaluated to optimize genotypes for future improvement breeding cultures. Sixteen genotypes were evaluated through 33 morphological characters. The experiment was carried out in randomized blocks with four replicates at the Mato Grosso Company for Research and Extension (EMPAER) in Cáceres, Mato Grosso State. A dendrogram was produced by Unweighted Pair Group Method with Arithmetic Mean (UPGMA) and Plan projection. Results showed the occurrence of genetic variability among the genotypes and the dendrogram revealed the formation of two main groups, one comprised genotypes of the black group and the other comprised genotypes of the Cores and Carioca group, placed side by side. This fact may also be observed by a projection on a bi-dimension plane. The most divergent materials were LP 98-122 and Jalo Precoce, whereas the most similar genotypes were IPR Chopin and BRS Grafite, IPR Chopin and Graúna, IPR Chopin and LP 98-122. Genotypes in the experiment have a possible genetic variability of being used in genetic breeding programs.

Keywords: *Phaseolus vulgaris*, morphological description, UPGMA, genetic divergence.

Introduction

Common beans are an important protein source in human diet in developing countries of tropical and subtropical regions. In Brazil, the common bean is one of the basic components of the population's diet, especially for the economically disadvantaged section of society (WANDER, 2005). Due to their nutritional and therapeutic properties, the common bean is a highly desirable component of diets since it combats hunger and malnutrition (AIDAR; YOKOYAMA, 2003).

According to Silva and Costa (2003), common bean species are distributed worldwide and have been cultivated in tropical and temperate zones of the northern and southern hemispheres.

Vieira (1988) reports that common bean culture is very widespread across Brazil, especially in homesteads. This is due to diverse preferences of consumers and farmers, but mainly because of the environmental conditions in which they exploit the culture. According to the same author, the enormous genetic variability existing in the national germplasm is beyond the innumerable introductions that are being carried through. The availability of ample variability is essential for successful improvement programs practically across all the relevant economic characteristics.

The genetic diversity of plants is fundamental for survival and at the same time allows the development of different varieties with their distinct levels of adaptation within the available environment necessary for culture (FUKUDA, 1996).

The genetic variability in the common bean germplasm used in family agriculture is essential as a survival strategy for small farmers. In fact, they employ suitable materials within their economic and agricultural-ecologic conditions. The efficiency of conservation and the exploration of this variability increases when this is duly characterized (RODRIGUES et al., 2002).

The genetic variability of the common bean has been preserved, evaluated, organized and made available to the scientific community. This fact brings about high genetic profits for culture and for Brazilian agriculture. The knowledge and exploration of this variability favors continued research on the culture and guarantees the support of the competitive and productive Brazilian agricultural system (FONSECA; SILVA, 2005).

The determination of existing genetic divergence in the germplasm bank facilitates the characterization and identification of duplicates. This also opens up the possibility of identifying genitors of interest in the attainment of hybrids with bigger potential heterotic effects. The later will be used in improvement programs (SUDRÉ et al., 2006).

Querol (1993) reported that the botanical characterization of germplasm identified and differentiated the phenotypes of the varieties, preventing the duplication of accesses within the collection. They thus used descriptions with high heritability, easy identification and expressions in all environments. The agronomical describers were of low heritability, even desirable to the producer. They had to visualize adaptation of the material and its productive potential, placing in evidence promising genotypes and their use in improvement programs.

Morphological describers include all characteristics that easily identify and differentiate them among the accesses of culture in the field. They generally have high heritability and they are expressed in all environments. The convergence of the morphological describers with the descriptions of molecular biochemists constitutes a valuable tool for the identification of duplicate accesses in the collection (FUKUDA; GUEVARA, 1998).

Cruz and Carneiro (2003) report that multi-categorical characteristics are commonly used in vegetal genetic improvement, especially those related to morphology and plant structure. Moreover, they are of great interest to the commercialized product quality. The value of the multi-categorical characteristics, when organized on an established scale, may be analyzed as a quantitative, changeable and discrete variable.

The above authors have, consequently, attributed numerical values to each category. Thus, the use of traditional measures for dissimilation was not appropriate, which was also the case in the Euclidean distance, as it was not yet possible to affirm that the individuals of a pair had close values (CRUZ; CARNEIRO, 2003).

In the case of multi-categorical variables the similarity index is used, in which a determined value is expressed in a percentage of coincidences, taking into consideration some analyzed characteristics. Thus, obtaining the dissimilarity from these indexes for posterior use is an accomplishment of the grouping analyses (CRUZ; CARNEIRO, 2003).

According to Cruz and Carneiro (2003), the prior analysis of dissimilarity grouping and multi-categorical data are presented as a viable alternative to evaluate the divergence among genotypes.

Daros et al. (2002) used multi-categorical variables in the morphological characterization of 14 sweet potato accesses of the germplasm collection from the Universidade Estadual do Norte Fluminense (UENF), state of Rio de Janeiro, Brazil. Trial was undertaken with 20 characteristics of the aerial part and roots, and mainly evidenced the occurrence of variability among the accesses proportionate to the pubescence characteristics of the apex rames, the inferior ribbing pigmentation leaves and roots.

Bertan et al. (2006) used changeable multi-categorical experiments with wheat until the blooming period. They demonstrated the existence of dissimilarity among the genotypes of the studied characteristics.

Sudré et al. (2006) reported that the accomplishment of the collection of multi-categorical data was a practical alternative because it was economical and faster to compare with quantitative or molecular data. However, each type of data has its singular importance and becomes interesting in the wider study of the collection of germplasm, providing greater support to researches and to data bank collection.

Current research evaluated suitable morphologic characteristics to cultivate the common bean adapted to Mato Grosso's conditions and furthermore to estimate the genetic divergence, through multi-categorical data, propitiating the identification of genitors which will be able to produce hybrids in programs of local improvement.

Material and methods

Research was carried in an experimental area belonging to the Mato Grosso Company for Research and Extension (EMPAER) in Cáceres, Mato Grosso State, Brazil (16°09`20.03" S and 57°37`49.15" N; altitude 118 meters), during the crop season in July 2008.

Table 1 show the sixteen common bean accesses used. They were obtained from the collection of the Agronomic Institute of Paraná (IAPAR), Agronomic Institute of Campinas (IAC), and the National Center of Researches in Rice and Common Bean (CNPaf/EMBRAPA).

Treatments were set out in completely randomized blocks, with four replications, in which the experimental parcels were made up of 5 rows of 5 meters in length, with 50 cm spacing between the rows and 35 cm between plants. The cultural treatments used during the experiment had been necessary for culture and manual harvest. They were made when 90% of the string beans were dried, separating the plants of each parcel and each block and labeling them.

Table 1. Genotypes of beans during harvest 2007/08, commercial group, and institution origin.

Order	Genotypes	Group	Origin
1	IAC Jabola	Colors	IAC
2	Mouro	Black	IAC
3	BRS Valente	Black	EMBRAPA
4	BRS Horizonte	Carioca	EMBRAPA
5	BRS Grafite	Black	EMBRAPA
6	Uirapuru	Black	IAPAR
7	Carioca magnífico	Carioca	IAC
8	Goiano Precoce	Colors	IAC
9	Jalo Precoce	Colors	IAC
10	Carioca Silmar	Carioca	IAC
11	IPR Chopin	Black	IAPAR
12	Juruti	Carioca	IAPAR
13	Graúna	Black	IAPAR
14	LP 98-122	Black	IAPAR
15	Diamante Negro	Black	IAPAR
16	Unknown	Colors	IAC

The evaluated characteristics had been affected through morphological describers for *Phaseolus vulgaris* L., standardized according to International Plant Genetic Resources Institute (IPGRI, 2001). The following characteristics were used: bracteole form, vegetation habit, standard color, wings color, string bean color, string bean transversal section, string bean bending, string bean suture, string bean color at the physiological maturation, wall of the string bean staple fibers, string bean standard tegument, seeds tegument darker color, seeds brightness, seeds form, hypocotyls pigmentation, emergent cotyledon color, color tonality from the chlorophyll leaf, anthocyanin in the leaf, leaf form, leaves persistence, lodging, floral size, bracteoles button size, bracteole/chalice length relation, chalice/bracteole color, wings opening, stiletto/keel relation, string bean position in the plant, string bean apex position and dry string bean color and seeds ribbing.

Data were analyzed following the methodology of multi-categorical variables, according to Cruz and Carneiro (2003), using the formula below:

$$d_{ii'} = \frac{D}{C + D}$$

in which:

$d_{ii'}$: dissimilarity, taking into consideration a set of multi-categorical variables;

D: category agreement;

C: category discord.

To avoid possible problems caused by indetermination in conditions where the coefficient is zero, the inverse of the similarity coefficient was increased by a unit (CRUZ; CARNEIRO, 2003).

UPGMA was the method for average grouping among groups. A dendrogram was therefore constructed from the genetic distances among the evaluated materials. The measures of genetic dissimilarity had also been projected in a bi-dimensional plan to allow the best visualization of the relative position accesses (CRUZ; CARNEIRO, 2003).

The statistical analyses had been carried through with genes software, version 2007 (CRUZ, 2006).

Results and discussion

An analysis of Table 2 and Figure 1 together shows that the most divergent genotypes were Jalo Precoce and LP 98-122, with 21 different characteristics. They had the lowest distance, with 29 characteristics in common (dii´ = 52.8%) and 3 not common characteristics, or rather, BRS Graphite, IPR Chopin and Graúna (bracteoles different forms, leaves persistence and seed forms) and IPR Chopin and LP 98-122 (different for the floral button size, bracteolas form and leaves persistence).

Table 2. Measures of genetic dissimilarity between the genotypes of beans, in relation to 32 characteristics with the multi-categorical variables used as bases.

Genotypes[1]	1	2	3	4	5	6	7	8	9	10	11	12	13	14	15	16
1	0.50	0.70	0.71	0.65	0.68	0.65	0.63	0.68	0.68	0.70	0.66	0.64	0.70	0.68	0.68	0.67
2		0.50	0.65	0.61	0.60	0.56	0.59	0.66	0.65	0.67	0.59	0.64	0.59	0.59	0.62	0.70
3			0.50	0.75	0.58	0.62	0.70	0.70	0.73	0.75	0.59	0.68	0.59	0.58	0.58	0.71
4				0.50	0.65	0.63	0.55	0.65	0.63	0.58	0.63	0.59	0.66	0.65	0.68	0.67
5					0.50	0.53	0.63	0.75	0.77	0.66	0.52	0.65	0.53	0.56	0.56	0.65
6						0.50	0.60	0.70	0.70	0.68	0.54	0.62	0.53	0.57	0.57	0.65
7							0.50	0.62	0.63	0.64	0.62	0.60	0.66	0.66	0.66	0.63
8								0.50	0.57	0.68	0.75	0.66	0.75	0.75	0.73	0.70
9									0.50	0.62	0.75	0.63	0.77	0.80	0.71	0.65
10										0.50	0.65	0.61	0.68	0.66	0.70	0.65
11											0.50	0.62	0.52	0.52	0.54	0.68
12												0.50	0.65	0.63	0.66	0.69
13													0.50	0.53	0.54	0.70
14														0.50	0.56	0.73
15															0.50	0.63
16																0.50

[1]1 = IAC Jabola, 2 = Mouro, 3 = BRS Valente, 4 = BRS Horizonte, 5 = BRS Grafite, 6 = Uirapuru, 7 = Carioca Magnífico, 8 = Goiano Precoce, 9 = Jalo Precoce, 10 = Carioca Silmar, 11 = IPR Chopin, 12 = Juruti, 13 = Graúna, 14 = LP 98 122, 15 = Diamante Negro, 16 = Unknown.

The 3 combinations presented the genotypes IRP Chopin in common. Within the combinations, the genotypes BRS Graphite, IRP Graúna and LP 98-122 produced low values of dissimilarity (5 and 13 = 0.54; 5 and 14 = 0.56), which indicated kinship between them or common genitors during the new cultivation process. Even the most similar characteristics had some genetic diversity; sometimes they did not have the minimum distance value, or rather, 52.83% and not 50%.

According to Maluf and Ferreira (1983), the measures of genetic dissimilarity were used as the predictive hybrid values for initiating the genetic improvement program. This is a useful tool because instead of choosing randomly the genitors for the hybrids, they may be chosen by a study on genetic diversity, carrying through and limiting the crossings only with the most promising combinations. The process turns out to be more efficient.

The dendrogram used in the UPGMA method (Figure 1) is taken as a measurement of dissimilarity in the distance based on multi-categorical variables, where the divergence inside each group is restricted. When carrying out this type of genetic distance analysis, the accesses with lesser distances had been grouped into 2 main groups.

The first group was subdivided into 3 sub-groups: subgroup Ia, containing genotypes IRP Chopin, Graúna and LP 98-122; subgroup Ib, with BRS Graphite, Uirapuru and Diamante Negro genotypes; subgroup Ic containing the Mouro and BRS genotypes. The second group also was divided into 3 sub-groups: subgroup IIa contained the genotypes Goiano Precoce and Jalo Precoce; the subgroup IIb contained the genotypes BRS Horizonte, Carioca Magnífico, Carioca Silmar and Juruti, and the subgroup IIc that contained genotypes IAC Jabola and Unknown.

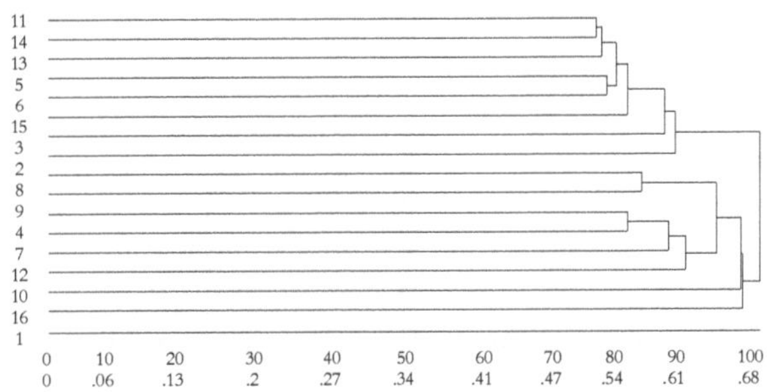

Figure 1. Representative dendrogram of genetic divergence among 32 bean genotypes (*Phaseolus vulgaris* L.), obtained by the method of grouping Media Liaison between groups (UPGMA), using dii´, as a measure of dissimilarity.
Genotypes: 1 = IAC Jabola, 2 = Mouro, 3 = BRS Valente, 4 = BRS Horizonte, 5 = BRS Grafite, 6 = Uirapuru, 7 = Carioca Magnífico, 8 = Goiano Precoce, 9 = Jalo Precoce, 10 = Carioca Silmar, 11 = IPR Chopin, 12 = Juruti, 13 = Graúna, 14 = LP 98 122, 15 = Diamante Negro, 16 = Unknown.

The formatted material group belonging to the Black commercial group was isolated from group II because it contained materials pertaining to the commercial groups Color and Carioca.

Results by the multi-categorical variable analysis of the bean accesses of the experiment demonstrate the existence of genetic variability between the cultivars. In the case of the 32 evaluated characteristics, only anthocyanin in the leaf, relation stiletto/keel, the string suture bean, staple fibers wall in the string bean, seeds ribbing tonality and the chlorophyll color in the leaf did not have any detected variability. All the evaluated accesses had the same value for the mentioned characteristics.

A considerable genetic similarity between BRS Graphite of the subgroup and IPR Chopin of the subgroup Ia could be verified between the subgroup Ia and Ib. Within the subgroup Ia, all the materials (IPR Chopin, LP 98-122 and Graúna) were genetically very similar, as also occurred with regard to the functions of the characteristics under analysis.

When groups I and II were compared, the differences in morphological characteristics among the genotypes were highlighted, determining that the two groups had hypocotyls pigmentation, emergent cotyledon, standard color, wings color, string bean color, and darker seeds color.

When analyzing the projection of the distance in the plan (Figure 2), the genotypes partition into two distinct groups in which the Jalo Precoce genotype of group I and LP98-122 of group II are most dissimilar, confirms the results evidenced by the dendrogram.

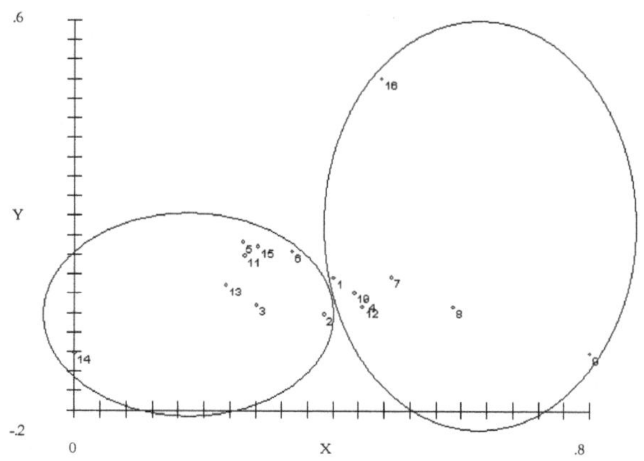

Figure 2. Projection of the distance in the plan, considering the 16 genotypes of *Phaseolus vulgaris* L.

Morphological diversity is an important tool to identify the studied genotypes, to differentiate them from the ones that contain some similar characteristics and to identify the possible duplicates in germplasm banks or in situ, where there might be equal genotypes with different names or different accesses that received the same name. Moreover, it precociously allows the election of ancestors for the improvement of future programs, as registered by Bertan et al. (2006).

Sudré et al. (2006) report that the use of multi-categorical data is a practical, economic and fast alternative when compared to quantitative and molecular data. However, each has a singular importance. It is, moreover, preferable that a germplasm collection was widely studied to improve ample supported researches and the formation of databases. The collection of the multi-categorical characterization and the divergence study based on quantitative data are a viable alternative to start the study of germplasm banks with few financial resources and a little man power.

The gauging of the genetic divergence between the genotypes used in the experiment is practical and economically faster to demonstrate which ones have the most and the least similar materials. It may be used to initiate an improvement program using the materials with greater genetic diversity. It is thus possible to obtain materials with the desired characteristics.

When materials that present greater dissimilarity are used, the farmer gains time and maximizes efforts. In fact, such combinations are highly promising. When genetically similar materials are used, they generate lineages with a low genetic diversity, reducing the possibility for genetic profits when these combinations are made. Since information on genetic dissimilarity among materials is an excellent tool, it may be an asset in improvement programs coupled to a higher efficiency in germplasm banks in the conservation of materials. Consequently, the number of duplicate samples is reduced.

Conclusion

Results show that the morphological characteristics evaluated enabled the identification of genotypes Jalo Precoce and LP 98-122 as the most divergent; identified three combinations as the most similar, IPR Chopin, making links between the genotypes BRS Grafite, and Graúna and LP 98-122; the commercial group Black was isolated by groups of commercial Color and Carioca; the access IPR Chopin, Graúna and LP 98-122 were less divergent.

Acknowledgements

The author would like to thank the Foundation for the Support of Research of the State of Mato Grosso (Fapemat), and Mato Grossense Compant for Research Technique and Rural Extension (Empaer).

References

AIDAR, H.; YOKOYAMA, L. P. **Cultivo do feijoeiro comum**. Goiás: Embrapa Arroz e Feijão, 2003. (Sistema de produção, n. 2).

BERTAN, I.; CARVALHO, F. I. F.; OLIVEIRA, A. C.; VIEIRA, E. A.; HARTWIG, I.; SILVA, J. A. G.; SHIMIDT, D. A. M.; VALÉRIO, I. P.; BUSATO, C. C.; RIBEIRO, G. Comparação de métodos de agrupamento na representação da distância morfológica entre genótipos de trigo. **Revista Brasileira de Agrociência**, v. 12, n. 3, p. 279-286, 2006.

CRUZ, C. D. **Programa genes**: análise multivariada e simulação. Viçosa: UFV, 2006.

CRUZ, C. D.; CARNEIRO, P. C. S. **Modelos biométricos aplicados ao melhoramento genético**. Viçosa: UFV, 2003.

DAROS, M.; AMARAL JÚNIOR, A. T.; PEREIRA, T. N. S.; LEAL, N. R.; FREITAS, S. P.; SEDIYAMA, T. Caracterização morfológica de acesso de batata doce. **Horticultura Brasileira**, v. 20, n. 1, p. 43-47, 2002.

FONSECA, J. R.; SILVA, H. T. **Banco ativo de germoplasma de feijão (*Phaseolus vulgaris* L.)**. Santo Antônio de Goiás: Embrapa Arroz e Feijão, 2005.

FUKUDA, W. M. G. **Banco de germoplasma de mandioca**: manejo, conservação e caracterização. Cruz das Almas: Embrapa/CNPMF, 1996.

FUKUDA, E. M. G.; GUEVARA, C. L. **Descritores morfológicos e agronômicos para caracterização de mandioca (*Manihot esculenta*, Crantz)**. Cruz das Almas: Embrapa-CNPMF, 1998.

IPGRI-International Plant Genetic Resources Institute. **Descritores para *Phaseolus vulgaris***. Rome, 2001.

MALUF, W. R.; FERREIRA, P. Genetics divergence in tomatoes and its relationship with heterosis. **Revista Brasileira de Genética**, v. 6, n. 3, p. 453-460, 1983.

QUEROL, D. **Recursos genéticos, nosso tesouro**: abordagem técnica. Rio de Janeiro: AS-PTA, 1993.

RODRIGUES, L. S.; ANTUNES, I. F.; TEIXEIRA, M. G.; SILVA, J. B. Divergência genética entre cultivares locais e cultivares melhoradas de feijão. **Pesquisa Agropecuária Brasileira**, v. 37, n. 9, p. 1275-1284, 2002.

SILVA, H. T.; COSTA, A. O. **Caracterização botânica de espécies silvestres do gênero *Phaseolus* L. Leguminosae)**. Santo Antônio de Goiás: Embrapa Arroz e Feijão, 2003.

SUDRÉ, C. P.; CRUZ, C. D.; RODRIGUES, R.; RIVA, E. M.; AMARAL JÚNIOR, A. T.; SILVA, D. J. H.; PEREIRA, T. N. S. Variáveis multicategóricas na determinação da divergência genética entre acessos de pimenta e pimentão. **Horticultura Brasileira**, v. 24, n. 1, p. 88-93, 2006.

VIEIRA, C. Perspectivas da cultura do feijão e de outras leguminosas de grão no país e no mundo. In: ZIMMERMANN, M. J. O.; ROCHA, M.; YAMADA, T. (Ed.). **Cultura do feijoeiro**. Piracicaba: ABPPF/Potafós, 1988. p. 3-20.

WANDER, A. E. **Cultivo do feijão irrigado na região noroeste de Minas Gerais**. Goiás: Embrapa Arroz e Feijão, 2005. (Sistemas de produção, n. 5).

17

Heterosis and combining ability of seed physiological quality traits of single cross *vs.* three-way sorghum hybrids

Marisol Galicia-Juárez[1], Leopoldo Ernesto Mendoza-Onofre[1*], Víctor Arturo González-Hernández[1], María Eugenia Cisneros-López[2], Ignacio Benítez-Riquelme[1] and Leobigildo Córdova-Téllez[1]

*[1]Colegio de Postgraduados, Campus Montecillo, Estado de México, México. [2]Campo Experimental Río Bravo, Instituto Nacional de Investigaciones Forestales, Agrícolas y Pecuarias, km 61, Carretera Matamoros-Reynosa, Cx. Postal 88900, Río Bravo, Tams, México. *Author for correspondence. E-mail: leopoldo@colpos.mx*

ABSTRACT. The aim of this research was to compare genetic parameters for traits related with seed germination and seedling vigour of single cross (SCH) *vs.* three-way (TWH) sorghum hybrids. The study was conducted in a sand bed under a greenhouse; 67 genotypes (four pairs of A- and B- lines, four R- lines, 13 SCH and 42 TWH) were allocated in a randomized complete blocks experimental design with three replications. Statistical analysis included ANOVA's and Student "t" tests, while Tukey test ($p \leq 0.05$) was applied for mean comparisons. The R-lines showed better physiological seed quality attributes than the A- and B- lines. The seed quality and heterosis values of SCH *vs.* that of TWH did not show significant differences ($p > 0.05$) for any variable; however, heterobeltiosis of SCH was higher than that of TWY for normal seedlings, dry weight of plumule and seedling emergence rate. The A2- line and R14 restorer line showed the highest GCA values in both types of hybrids. No differences ($p \leq 0.05$) were found between maternal and paternal effects of the A- and B- lines involved in the male sterile cross (female parent) of the TWH, except for plumule length in Line 5.

Keywords: *Sorghum bicolor* L. Moench, maternal effects, seedling emergence, sorghum lines, top-cross tests.

Introduction

In commercial seed production of single cross hybrids of grain sorghum [*Sorghum bicolor* (L.) Moench] two lines are involved: a male-sterile line (A- line, female), and a male fertile line (R- line, restorer) (León-Velasco, Mendoza-Onofre, Castillo-González, Cervantes-Santana, & Martínez-Garza, 2009a). In the three-way sorghum hybrids three lines are used since the female parent is a male-sterile cross of an A- line and a B- non isogenic line, and the male parent is an R- line. In both hybrid types, the progenie may express a higher plant vigour than the parents, an effect named heterosis, which is a relevant genetic trait frequently measured and used in plant breeding.

Mean heterosis (called as heterosis hereafter) is calculated based on the performance of the F1 hybrid with regard to the average performance of both parental lines (Kumar, Reedy, Ramaiah, &

Sharma, 2011); the high parent heterosis (called as heterobeltiosis hereafter) refers to the F1 hybrid behaviour with respect to the best parental line (Hayes & Rooney, 2014). Heterosis brings gains in grain yield, earlier flowering, greater number of leaves, taller plants, more tillers and panicles per plant, heavier seeds, more seeds per panicle (Acquaah, 2012), a faster seedling emergence rate, more vigourous seedlings and a higher plumule dry weight (Yu & Tuinstra, 2001; Cisneros-López, Mendoza-Onofre, Mora-Aguilera, Córdova-Téllez, & Livera-Muñoz, 2007; León-Velasco et al., 2009a). Compared to the three-way experimental sorghum hybrids, the single cross hybrids show a higher heterosis for grain yield (Rooney, 2004).

The three-way sorghum hybrids are not commercially grown in México (Montes-García, Williams-Alanís, Moreno-Gallegos, Cisneros-López, & Pecina-Quintero, 2012) nor anywhere else. However, in corn (*Zea mays* L.) these hybrids are highly profitable for seed producing companies due to the high heterosis obtained for seed yield in the female single cross (female parent of this type of hybrid), as contrasted to the lower seed yield harvested from a female pure-line (female parent of the single cross hybrid) (Beck, 2002).

Several A-, B- and R- sorghum lines adapted to the Mexican Highlands have been developed and single cross hybrids have been evaluated under field conditions for different purposes (Osuna-Ortega et al., 2000; Cisneros-López et al., 2007; 2010; Cisneros-López, Mendoza-Onofre, & González-Hernández, 2012; León-Velasco et al., 2009a). In this region (mean annual rainfall, 655 mm; average daily temperature, 16.3°C; 2,250 m of altitude) seedling growth of experimental conventional (single cross) sorghum hybrids is slow at the beginning of the growing season. In seedling emergence tests conducted in sand beds, seed size has not been the main responsable factor of this slow sorghum seedling growth (Cisneros-López et al., 2007; Valadez-Gutiérrez et al., 2007). On the other hand, the physiological seed quality traits of three-way sorghum hybrids are unknown. Therefore, it is necessary to compare these traits in three-way hybrids *vs.* single cross hybrids, assuming that in the three-way hybrids the single cross female parent seeds will have better physiological attributes than those produced by the pure female line.

A genetic trait closely related to heterosis is the combining ability of the parents. The general combining ability (GCA) is the capacity of an individual or a population of individuals to combine with others, and it is estimated by the average performance in their hybrid combinations, whereas the specific combining ability (SCA) is the particular performance of two parental lines in a hybrid combination. These two responses are useful to select parental lines of sorghum hybrids (Kenga, Alabi, & Gupta, 2004; Solanki, Patel, Patel, & Desai, 2007; León-Velasco, Mendoza-Onofre, Castillo-González, Cervantes-Santana, & Martínez-Garza, 2009b).

The diallel crosses are widely used for measuring the GCA and SCA effects of agronomic traits in plant breeding programs. The top-cross test that involves crosses of corn lines with a tester, is another option to estimate GCA (Bradshaw, 2016). In three-way corn hybrids the more common tester is a single cross hybrid or a pure-line (Beck, 2002), whereas in single cross hybrids the elite lines are used as testers (Hallauer, Carena, & Miranda Filho, 2010). According to Castañón, Jeffers, Hidalgo, & Tosquy (1998), the top-cross test has successfully substituted the diallel crosses because it reduces pollination costs.

The aim of this research was to compare the magnitude of the heterosis, heterobeltiosis, and the general and specific combinig ability on seed physiological quality traits of single cross hybrids *vs.* three-way hybrids of sorghum, as well as maternal *vs.* paternal effects of single AxB crosses for those traits.

Material and methods

Seed of isogenic and non isogenic single crosses AxB was increased in the spring-summer growing season of year 2012, according to a diallel squeme (Figure 1). Four pairs of A- and B- lines were included, so sixteen AxB male-sterile crosses were obtained. Four of these crosses correspond to isogenic AxB crosses; it means these are seed increments of the A- line (the female parental line of single cross hybrids). The other twelve are non isogenic crosses that correspond to the female parental single cross of the three-way hybrids. In addition, in this diallel we can also differentiate maternal *vs.* paternal effects of the isogenic A- and B- lines. For example, the average of A1xB2, A1xB3 and A1xB5 crosses will correspond to the performance of the A1- line as a female parent (maternal effects), and the average of A2xB1, A3xB1 and A5xB1 crosses will correspond to the performance of the B1- line (known as the maintainer line, isogenic of the A1- line) as a male parent (paternal effects).

Seed from the single cross hybrids and the three-way hibrids was increased in the 2013 growing season, by hand made pollinations between the sixteen AxB crosses and four R- lines (R14, R17, R19, and R25).

Male sterile lines	Maintainer lines			
	B1	B2	B3	B5
A1	A1xB1	**A1xB2**	**A1xB3**	**A1xB5**
A2	*A2xB1*	A2xB2	A2xB3	A2xB5
A3	*A3xB1*	A3xB2	A3xB3	A3xB5
A5	*A5xB1*	A5xB2	A5xB3	A5xB5

Figure 1. Diallel scheme of isogenic and non isogenic AxB crosses. Isogenic crosses (boxes in white); non isogenic crosses (boxes in grey). Non isogenic crosses to analize maternal effects of the A1 line (bold letters) and paternal effects of the B1 line (italics letters).

Seedling emergence test

This experiment was established in June 2014, including 67 genotypes (four pairs of A- and B- lines; four R- lines; 13 out of the expected 16 single cross hybrids, and 42 out of the 48 possible three-way hybrids), according to seed availability. Seed was sown in sand beds within a greenhouse. The experimental units were rows of 50 cm length, 5 cm apart. In each row 25 seeds were placed every 2 cm at 5 cm depth. Treatments were allocated in a randomized blocks design with three replications.

Variables measured in each plot were:

Seedling emergence rate (SER): Seedlings were counted daily (as soon as the plumule became visible) until no more seedlings emerged in three consecutive days. Then SER was estimated with the equation of Maguire (1962): $SER = \Sigma^{n}_{i} (Xi/Ni)$, where Xi = number of emerged seedlings on day i, and Ni = number of days after sowing including day i.

Normal seedlings (NS), abnormal seedlings (AS) and dead seeds (DS) were also counted at the end of the essay, as specified by International Seed Testing Association (ISTA, 1999) rules, and its respective percentages were calculated. Aditionally, plumule length (PL) and plumule dry weight (PDW) from 10 normal seedlings were measured; PL was the length (cm) from the base of the stem to the apex of the youngest leaf; and PDW (mg per seedling) was recorded after seedlings were dried in an oven at 75°C for 72h.

Heterosis [(Performance of the F1/Average performance of the parental lines: two in single cross hybrids and three in three-way hybrids) x 100] and heterobeltiosis [(Performance of the F1/Performance of the best parent) x 100] (Hayes & Rooney, 2014) were calculated for each hybrid. Data of the fertile lines (B and R) was used in these calculations to assure a full seed set of the panicles.

Estimates of GCA and SCA in both types of hybrids were obtained considering top-cross tests. In these tests, the term "lines" were assigned to each A-line in the analysis of single cross hybrids, and to the non isogenic AxB crosses in the three-way hybrids. The term "tester" was assigned to the R- lines. Therefore, the GCA of lines and testers were estimated through their respective averages; the SCA corresponded to the interaction of line x tester.

Maternal effects *vs.* paternal effects were compared for each AxB non isogenic crosses participating in the three-way hybrids. Maternal effects correspond to hybrids in which the version A of the line acted as female parent (direct cross), while paternal effects were those hybrids in which the maintainer B-line participated as the male parent (reciprocal cross) (Figure 1) of the male sterile single crosses. Therefore, the statistical contrast between maternal *vs.* paternal effects of Line 1, would correspond to the average performance of A1xB2, A1xB3, A1xB5 crosses (direct crosses, maternal effects) *vs.* the average performance of A2xB1, A3xB1 and A5xB1 crosses (reciprocal crosses, paternal effects), crossed to each R-line.

Statistical analyses

The statistical package SAS (2002) was used. Prior to the statistical analyses, an angular transformation [arcosin $(Yi)^{1/2}$] to the percentage data (Yi) was applied but results are shown in percentages. Mean comparison tests were applied (Tukey, $p \leq 0.05$) among genotype groups (A-, B-, R- lines, single cross hybrids and three-way hybrids). Since each hybrid group was represented by different number of observations (because there was not enought seed in some hybrids), PROC GLM of SAS (2002) was used to carry out the ANOVA's analysis. Statistical contrasts between the averages of the heterosis estimates of the 13 single cross hybrids *vs.* that of the 42 three-way hybrids were made by the Student "t" base on an unpaired design. To compare the differences of maternal effects *vs.* paternal effects the Student "t" test was also performed.

Results and discussion

Comparisons of groups of genotypes

There were significant differences ($p \leq 0.05$) among genotypes for all the genotypes groups (grouped according to their role in a sorghum breeding program), for all seed physiological quality traits, except for abnormal seedlings (Table 1).

Among parental lines, the male R-lines group showed much better seed physiological quality attributes than the A- and B- line groups; *i.e.*, higher percentage of normal seedlings, faster seedling emergence, as well as longer and heavier seedling plumules (Table 2). Most interestingly, the seed from the R- lines group was as good as the seed from the single cross and three-way hybrids regarding sed physiological quality traits.

Instead, the A- and B- inbred lines groups showed similar seed quality due to their isogenic nature. Under field conditions these lines are "3-dwarf" types (Brown, Rooney, Franks, & Kresovich, 2008) although our R- lines are usually more vigouous and reach higher stem size than isolines at harvest (León-Velasco et al., 2009a). There were no differences in seed physiological quality traits between the single cross hybrids and the three-way hybrids. These results clearly show that morphological differences in plant vigour can be detected as early as the seedling growth stage.

Cisneros-López et al. (2007) compared also seed germination and seedling emergence rate of single cross hybrids and their parental lines under field (same location as the present study), and laboratory conditions. They did not find either any significant difference between the hybrids and the R- lines, for percentages of normal seedlings, abnormal seedlings and dead seeds, plumule length and seedlings emergence rate. Yu &Tuinstra (2001) found similar results for variables related with seedling emergence rate in sorghum. That is, the *per se* seed performance of our group of R- lines is as good as that of seed obtained from hybrids. The absence of significant differences between the means of both groups of hybrids also suggest that the seed physiological quality traits do not rely on the female parents involved (pure-line for single cross hybrids or AxB non isogenic crosses for three-way hybrids), since all our male parents (R- lines) were the same in both types of hybrids.

Heterosis

Table 3 shows three features on the heterosis and heterobeltiosis results: (i) There were not significant differences ($p > 0.05$) for heterosis between single cross and three-way hybrids, in any of the four seed quality traits; (ii) Heterosis had greater values than those of heterobeltiosis for all traits in both types of hybrids; (iii) Single cross hybrids had higher heterobeltiosis than three-way hybrids for normal seedlings, plumule dry weight and seedling emergence rate.

Heterotic effects are expected to be maximized in single crosses depending upon the genetic diversity of the parental lines and the homocigous level of these lines. Contrary to cross pollinated species such as maize, sorghum is a self pollinated specie in which homocigosity can be reached without an important decrease in vigour. The high heterosis values obtained here for the single cross hybrids (from 126 and 166 % depending upon to the trait involved) indicate that in this group of parental lines exist genetic divergence between the A- lines and the R- lines.

Table 1. Square means and significance of the source of variation of the groups of genotypes for seed physiological quality traits.

Source of variation	DF	Normal seedlings	Abnormal seedlings	Dead seeds	Plumule length	Plumule dry weight	Seddlings emergence rate
Groups	4	1337.76**	47.17ns	1267.43**	30.12**	334.11**	8.43**
Replications	2	97.62ns	821.93ns	87.90ns	38.53**	146.59**	0.41ns
Error	194	85.88	70.70	79.19	1.40	12.59	0.51
VC (%)		17	62	28	13	26	28

DF: Degrees of freedom; VC: variation coefficient; **: Significance of the F test ($p \leq 0.01$): ns: not significant.

Table 2. Means of seed physiological quality traits for each group of sorghum genotypes.

Genotype group	Normal Seedlings (%)	Abnormal Seedlings (%)	Dead Seeds (%)	Plumule Length (cm)	Plumule dry weight (mg)	Seed e mergence rate (Nd^{-1})
A- lines (n=4)	45 b	9 a	44 b	7.78 b	8.5 b	1.76 b
B- lines (n=4)	40 b	6 a	52 b	6.24 c	6.1 b	1.42 b
R- lines (n=4)	73 a	5 a	19 a	8.75 ab	13.7 a	2.93 a
Single cross hybrids (n=13)	74 a	6 a	20 a	9.55 a	16.5 a	3.03 a
Three-way hybrids (n=42)	64 a	5 a	28 a	9.08 a	14.0 a	2.50 a

A: Male sterile lines; B: Maintainer lines; R: Restorer lines; (n): Number of genotypes involved in the average; Values with the same letter in each column are not significant different (Tukey, $p \leq 0.05$).

Table 3. Values of heterosis and heterobeltiosis for seed physiological quality traits for single cross and three-way sorghum hybrids.

Type of hybrids	Normal seedlings	Plumule length	Plumule dry weight	Seedling emergence rate
		Heterosis		
Single cross hybrids (n=13)	126 a	126 a	166 a	137 a
Three-way hybrids (n=42)	125 a	132 a	164 a	130 a
		Heterobeltiosis		
Single cross hybrids (n=13)	100 a	111 a	126 a	104 a
Three-way hybrids (n=42)	88 b	105 a	105 b	86 b

n: Number of genotypes involved in the average; values with the same letter in each column, are not significant different (Student "*t*" test, $p \leq 0.05$).

After evaluating parental lines with similar pedigrees under field conditions, Cisneros-López et al. (2007) found lower heterosis values (normal seedlings, 90%; plumule length, 98% and seedling emergence rate, 107%) for single cross hybrids than in our study. These differences in heterosis might be due to differences in environmental conditions, more adverse in the field for seed germination and seedlings emergence rate with respect to the sand bed test done under greenhouse conditions in our study. Heterosis in biomass production at early stages of development has also been observed under chilling temperatures in sorghum (Windpassinger et al., 2016) as well as in classical molecular biology studies on *Arabidopsis* (Meyer, Törjék, Becher, & Altmann, 2004).

According to Márquez (1988), heterosis will decrease as the number of parental lines involved in a hybrid increases (two in a single cross hybrid, three in a three-way hybrid or four in a double cross hybrid), probably because it is less likely that all pairs of genes will combine for favourable characters. Our results indicate that in this group of parental lines this phenomenon do not occur, as three-way crosses and single crosses has similar heterosis values (Table 2). This response may be related to a similar proportion of allelic additively expressed genes associated with heterosis in the early stage of development of both types of hybrids.

The heterobeltiosis values were of lower magnitude that those of heterosis in both types of hybrids. Cisneros-López et al. (2007) also found that heterobeltiosis values of the group of single cross hybrids were lower that the heterosis values (normal seedlings, 89 *vs.* 90%; plumule length, 94 *vs.* 98%; seed emergence rate, 92 *vs.* 107%). However, heterobeltiosis data represents greater agronomical importance than heterosis as the first indicates the superiority of the hybrid compared to the best parental line

General and specific combining ability

Lines and testers in single cross hybrids

No significant difference ($p > 0.05$) were found in the GCA of lines and testers for most of the seed quality measured traits, but line A2- tended to be better than the other females A- lines. There were not significant differences among testers either, except for seedling emergence rate of R 14 and R19 male lines (Table 4). Regarding the specific combining ability (SCA), the best combination for normal seedlings and seedlings emergence rate was the A2xR14 single cross hybrid.

Lines and testers in three-way hybrids

Similar to the single cross hybrids results, in the three-way hibrids the A and B versions of Line 2, and R14 among testers, were the only ones with high values of GCA for normal seedlings, plumule dry weight and sedling emergence rate (Table 5). Regarding SCA estimates, the superiority of R14 tester is evident, because when it is crossed to any female single cross, seeds of the three-way hybrids produce more normal seedlings, which can emerge at a faster rate (Table 5).

Table 4. Estimated values of general and specific combining ability (GCA and SCA) for seed physiological quality traits of female and male lines in single cross hybrids.

Female lines	Testers (male lines)				
	R14	R17	R19	R25	GCA
	Normal seedlings (%)				
A1	67 ab	--	--	79 ab	73 AB
A2	85 a	77 ab	83 a	81 ab	82 A
A3	82 a	65 ab	64 ab	--	70 AB
A5	72 ab	47 ab	74 ab	69 ab	66 B
GCA	77 W	63 W	74 W	76 W	
	Plumule length (cm)				
A1	8.53 a	--	--	9.40 a	8.97 A
A2	9.80 a	10.50 a	9.93 a	9.67 a	9.98 A
A3	10.07 a	9.40 a	9.70 a	--	9.72 A
A5	8.70 a	9.10 a	9.50 a	9.80 a	9.27 A
GCA	9.27 W	9.67 W	9.71 W	9.62 W	
	Plumule dry weight (mg)				
A1	12.4 a	--	--	16.4 a	14.4 A
A2	18.1 a	18.2 a	20.6 a	17.7 a	18.7 A
A3	18.0 a	15.1 a	19.1 a	--	17.4 A
A5	12.8 a	14.6 a	15.4 a	16.3 a	14.8 A
GCA	15.3 W	16.0 W	18.4 W	16.8 W	
	Seedlings emergence rate (N d^{-1})				
A1	2.62 ce	--	--	3.38 ad	3.00 B
A2	4.14 a	3.04 bd	3.72 ab	3.69 ac	3.65 A
A3	3.37 ad	2.71 be	2.40 de	--	2.83 B
A5	2.70 be	1.93 e	2.68 be	3.06 ad	2.59 B
GCA	3.21 W	2.56 X	2.93 WX	3.38 W	

GCA values with the same capital letter, or SCA values with the same lower case letter in each trait, are not significant different (Tukey, $p \leq 0.05$); --: hybrid not included in test.

Table 5. Estimated values of general and specific combining ability (GCA and SCA) for seed physiological quality traits of female and male lines in three-way hybrids.

Female lines	Testers (male lines)				GCA
	R14	R17	R19	R25	
	Normal seedlings (%)				
A1/B1	80 a	48 e	61 ce	61 ce	63 A
A2/B2	72 ac	55 de	68 ad	73 ac	67 A
A3/B3	78 ab	58 ce	54 de	65 bd	64 A
A5/B5	69 ad	53 de	63 ce	60 ce	61 A
GCA	75 W	53 Y	62 X	65 X	
	Plumule length (cm)				
A1/B1	9.47 a	9.66 a	8.64 a	8.41 a	9.05 A
A2/B2	8.98 a	9.59 a	9.20 a	8.87 a	9.16 A
A3/B3	9.65 a	9.71 a	8.72 a	8.77 a	9.21 A
A5/B5	9.28 a	9.04 a	8.82 a	8.56 a	8.93 A
GCA	9.35 WX	9.50 W	8.84 XY	8.65 Y	
	Plumule dry weight (mg)				
A1/B1	15.1 a	14.8 a	13.7 a	12.3 a	14.0 A
A2/B2	14.1 a	14.9 a	16.1 a	13.1 a	14.6 A
A3/B3	16.0 a	15.3 a	13.2 a	13.5 a	14.5 A
A5/B5	14.9 a	12.4 a	14.0 a	11.8 a	13.3 A
GCA	15.0 W	14.3 WX	14.3 WX	12.7 X	
	Seedlings emergence rate (Nd^{-1})				
A1/B1	3.36 a	1.67 g	2.27 dg	2.37 dg	2.42 AB
A2/B2	3.12 ac	2.10 eg	2.57 ce	2.90 ad	2.67 A
A3/B3	3.31 ab	2.25 dg	1.97 eg	2.64 be	2.54 AB
A5/B5	2.82 ad	1.85 fg	2.21 dg	2.49 cf	2.34 B
GCA	3.15 W	1.96 Z	2.25 Y	2.60 X	

GCA values with the same capital letter, or SCA values with the same lower case letter in each trait, are not significant different (Tukey, $p \leq 0.05$).

Maternal *vs.* paternal effects in the non isogenic female cross of three-way hybrids

There were not significant differences between maternal effects (the A- versions) *vs.* paternal effects (the B- versions) of each pair of A/B lines, except for the plumule length of Line 5. Three-way hybrids in which the A- version was involved showed longer plumule (9.22 cm) than those hybrids in which the female crosses included the B- version (8.61 cm). These results indicate that there are not nuclear and mitochondrial gene interactions in the seedling growh response of our A/B elite lines.

Conclusion

There were significant differences among genotypes for physiological quality traits of sorghum seeds. The R- lines group showed better seed quality traits than A- and B- lines groups. Single cross hybrids and three-way hybrids showed similar seed quality and heterosis values. The A2-line and the R14 restorer line stand out as the best candidates to be used as parental lines for seed physiological quality traits. There were not significant differences between maternal *vs.* paternal effects in physiological quality traits of sorghum seeds. Morphological differences in plant vigour can be detected in the seedling growth stage.

Acknowledgements

The authors gratefully acknowledge the scholarship granted to the first author by the National Council of Science and Technology of México (CONACYT).

References

Acquaah, G. (2012). *Principles of Plant Genetics and Breeding* (2nd ed.). Chapter: 34. Breeding Sorghum. Oxford, UK: John Wiley & Sons, Ltd.; Wiley Online Library.

Beck, D. L. (2002). *Management of hybrid maize seed production.* El Batan, Mexico: CIMMYT.

Bradshaw, J. E. (2016). *Plant Breeding, Past, Present and future.* Edinburg, UK: Spring.

Brown, P. J., Rooney, W. L., Franks, C., & Kresovich, S. (2008). Efficient mapping of plant height quantitative trait loci in a sorghum association population with introgressed dwarfing genes. *Genetics, 180*(1), 629-637.

Castañón, G., Jeffers, D., Hidalgo, H., & Tosquy, H. (1998). Prueba de mestizos de maíz en el estado de Veracruz, México. *Agronomía Mesoamericana, 9*(2), 89-96.

Cisneros-López, M., Mendoza-Onofre, L. E., Mora-Aguilera, G., Córdova-Téllez, L., & Livera-Muñoz, M. (2007). Cold tolerant sorghum hybrids and parental lines. I: Seed quality and its effects on seedling establishment. *Agrociencia, 41*(1), 45-55.

Cisneros-López, M. E., Mendoza-Onofre, L. E., Zavaleta-Mancera, H. A., González-Hernández, V. A., Mora-Aguilera, G., Córdova-Téllez, L., & Hernández-Martínez, M. (2010). Pollen–pistil interaction, pistil histology and seed production in A x B grain sorghum crosses under chilling field temperatures. *Journal of Agricultural Science, 148*(1), 73-82.

Cisneros-López, Ma. E., Mendoza-Onofre, L. E., & González-Hernández, V. A. (2012). Male parent effects on stigma receptivity and seed set of sorghum A-lines under chilling field temperatures. *Plant Breeding, 131*(1), 88-93.

Hallauer, A. R., Carena, M. J., & Miranda Filho, J. B. (2010). *Quantitative genetics in maize breeding.* New York, US: Springer.

Hayes, C. M., & Rooney, W. L. (2014). Agronomic performance and heterosis of specialty grain sorghum hybrids with a black pericarp. *Euphytica, 196*(3), 459-466.

International Seed Testing Association [ISTA]. (1999). *International rules for seed testing.* Zurich, SW: Int. Seed Testing Ass.

Kenga, R., Alabi, S. O., & Gupta, S. C. (2004). Combining ability studies in tropical sorghum (*Sorghum bicolor* (L.) Moench). *Field Crops Research, 88*(2-3), 251-260.

Kumar, A. A., Reedy, B. V. S., Ramaiah, B., & Sharma, R. (2011). Heterosis in white-grained grain mold resistant sorghum hybrids. *Journal of SAT Agricultural Research 9 (1), 1-6.*

León-Velasco, H., Mendoza-Onofre, L. E., Castillo-González, F., Cervantes-Santana, T., & Martínez-Garza, A. (2009a). Evaluation of two generations of cold tolerant sorghum hybrids and parental lines. I: Genetic variability and adaptability. *Agrociencia, 43*(5), 483-496.

León-Velasco, H., Mendoza-Onofre, L. E., Castillo-González, F., Cervantes-Santana, T., & Martínez-Garza, A. (2009b). Evaluation of two generations of cold tolerant sorghum hybrids and parental lines. II: Combining ability, heterosis and heterobeltiosis. *Agrociencia, 43*(6), 609-623.

Maguire, J. D. (1962). Speed of germination: Aid in selection and evaluation for seedling emergence and vigor. *Crop Science, 2*(2), 176-177.

Márquez, S. F. (1988). *Genotecnia vegetal. métodos, teoría, resultados. Tomo II.* México, DF: AGT Editor, S. A.

Meyer, C. R., Törjék, O., Becher M., & Altmann, T. (2004). Heterosis of biomass production in Arabidopsis. Establishment during early development. *Plant Physiology, 134*(4), 1813-1823.

Montes-García, N., Williams-Alanís, H., Moreno-Gallegos, T., Cisneros-López, M. E., & Pecina-Quintero, V. (2012). 'RB Paloma', a white sorghum variety for grain and forage production. *Revista Fitotecnia Mexicana, 35*(2), 185-187.

Osuna-Ortega, J., Mendoza-Onofre, L. E., González-Hernández, V. A., Castillo-González, F., Mendoza-Castillo, M. C., & Williams-Alanís, H. (2000). Potential of cold tolerant germplasm in the adaptation and adaptability of sorghum in México. I: High Valleys. *Agrociencia, 34*(5), 561-572.

Rooney, W. L. (2004). Sorghum improvement-integrating traditional and new technology to produced improved genotypes. *Advances in Agronomy, 83*, 38-96.

SAS Institute. (2002). *SAS/SAT. Version 9.00.* Cary, NC.: SAS.

Solanki, B. G., Patel, D. M., Patel, P. B., & Desai, R. T. (2007). Combining ability in sorghum [*Sorghum bicolor* (L.) Moench] for yield and its attributing traits (II). *Crop Research (Hisar), 33*(1/3), 187-191.

Valadez-Gutiérrez, J., Mendoza-Onofre, L. E., Córdova-Téllez, L., Vaquera-Huerta, H., Mendoza-Castillo, Ma. C., & García-de los Santos, G. (2007). Seed sizes, invigorization substances, and vigor tests in cold tolerant sorghums. *Agrociencia, 41*(2), 169-179.

Windpassinger, S., Friedt, W., Deppé, I., Wener, C., Snowdon, R., & Wittkop, B. (2016). Towards enhancement of early-stage chilling tolerance and root developmente in sorghum F1 hybrids. *Journal of Agronomy and Crop Science.* doi: 10.1111/jac.12171. Retrieved on Feb. 18, 2016 from http://online library.wiley.com/doi/10.1111/jac.12171/epdf

Yu, J., & Tuinstra, M. R. (2001). Genetic analysis of seedling growth under cold temperature stress in grain sorghum. *Crop Science, 41*(5), 1438-1443.

18

Evaluation of soybean cultivars on the embryogenic and organogenic potential

Leandra Regina Texeira[1,2*], Alessandro de Lucca e Braccini[1], Bruno Grimaldo Martinho Churata[3], Elisa Serra Negra Vieira[2], Polyana Kelly Martins[2] and Ivan Schuster[2]

*[1]Programa de Pós-graduação em Genética e Melhoramento, Departamento de Agronomia, Centro de Ciências Agrárias, Universidade Estadual de Maringá, Av. Colombo, 5790, 87020-900, Maringá, Paraná, Brazil. [2]Cooperativa Central de Pesquisa Agrícola, BR 467, km 98, Cx. Postal 301, 85813-450, Cascavel, Paraná, Brazil. [3]Departamento de Estatística, Universidade Federal do Paraná, Curitiba, Paraná, Brazil. *Author for correspondence. E-mail: lrtexeira@gmail.com*

ABSTRACT. The first step in plant transformation using somatic embryogenesis or organogenesis for plant regeneration *in vitro* has been the identification of genotypes with the best totipotency. Thus, the objective of this experiment was to evaluate the totipotency of ten cultivars and one soybean line from the Coodetec Gene Bank. The somatic embryos developed from immature cotyledons measuring from 4 to 6 mm in length, and the organogenesis was evaluated from mature soybean seeds. The cultivars CD 201, CD 216 and CD 220 were the most promising genotypes to produce somatic embryos, whereas the cultivar CD 217 had the highest quantity of differentiated explants, which represents the best regeneration efficiency for participating in organogenic systems. In the experiment with somatic embryogenesis, time to plant regeneration was ten months, but the plants were regenerated after only four months in the experiment with organogenesis.

Keywords: *Glycine max*, somatic embryogenesis, organogenesis.

Introduction

Parallel to the traditional plant breeding, methods for enhancing the genetic variability has been developed and applied to all major crops. The establishment of these protocols is essential for increasing the number of superior genotypes. In soybean (*Glycine max* (L.) Merrill), for example, the cultivation of transgenic cultivars has been possible because of the several protocols in plant transformation.

Successful plant transformation consists of inserting DNA into target cells that must be capable of integrating this exogenous DNA, and these transformed cells must have effective participation in the germination tissue.

Complete transformation, however, requires a mechanism to develop the adult plants, and this mechanism requires *in vitro* protocols to regenerate these plants (PARROT; CLEMENTE, 2004). In soybean, fertile plants have been regenerated from cell or tissue cultures by using somatic embryogenesis or organogenesis mechanisms.

The main constraints to regenerate the transformed explants have been their levels of nutritional and hormonal requirements (DUSI, 1998). The continuous refinement of these protocols has also been necessary when new genes are isolated, characterized and available for testing because numerous variables have to be adapted to the conditions of every laboratory. In the somatic embryogenesis of soybean plants, for example, research reports have showed the genotypic influence from donor plants (KOMATSUDA; OHYAMA, 1988; KOMATSUDA et al., 1992), type of explant, growth regulators (HARTWECK et al., 1988; BUCHHEIM et al., 1989), solidifying agents (FU; HUANG et al., 1995) and pH (KOMATSUDA; KO, 1990).

Most of the soybean genotypes with embryogenic potential are not yet well adapted to the field conditions, and they are easily overcome by newly elite cultivars. In transgenic breeding programs, therefore, gene transference from genotypes with embryogenic potential to these elite cultivars is necessary, and this transference has been made by plant crosses and selection.

Some reports have showed that plant regeneration by organogenesis depends less on the genotypes and, therefore, the organogenesis has been used in the routines of several laboratories (OLHOFT et al., 2003; OPABODE, 2006; PAZ et al., 2004; SHAN et al., 2005). Organogenesis, however, is a system much more complex because of the interactive effects from explant sources, growing media and environmental factors (GEORGE, 1993; JOY IV; THORPE, 1999; PIERIK, 1997). The organogenesis also depends on exogenous growth regulators as the auxins and cytokinins as well as tissue responses to hormonal changes during the cultivation period (SUGIYAMA, 1999).

Thus, the objective of this experiment was to select soybean cultivars and a line with potential to regenerate plants *in vitro* using somatic embryogenesis and organogenesis, and evaluate the regenerative efficiency of these genotypes.

Material and methods

Genetic material

The embryogenic and organogenic potential of nine soybean cultivars and one line from the Coodetec Gene Bank (CD 201, CD 202, CD 206, CD 211, CD 216, CD 217, CD 218, CD 220, CD 221 and the line CD 98-2644) were evaluated at the Tissue Culture Laboratory of the Cooperativa Central de Pesquisa Agrícola (Coodetec), Cascavel, Paraná State, Brazil.

Somatic embryogenesis

Soybean plants were cultivated under greenhouse conditions with photoperiod of 16 hours to obtain the cotyledonary explants. Cotyledons measuring from 4 to 6 mm in length were collected at the R_5 stage from healthy pods of vigorous plants. First, these pods were disinfested using 70% ethanol for 5 min. and 10% sodium hypochlorite (v v^{-1}) for 20 min. Thereafter, they were rinsed in distilled and deionized water to remove the excess of the sodium hypochlorite. The pods were opened to excise the cotyledon pairs which were abaxially faced on the culture medium.

Every genotype had ten cotyledons cultivated in Petri dishes, 90 x 15 mm (SANTARÉM et al., 1997). These Petri dishes were filled with 25 mL of the medium MSD40 (FINER; NAGASAWA, 1988) prepared with MS salts at half strength (MURASHIGE; SKOOG, 1962), B5 vitamin (GAMBORG et al., 1968), inositol (100 mg L^{-1}), 3% sucrose, 0.2% Phytagel, 180 μM 2,4-D and pH 7.0. Ten Petri dishes per genotype were incubated at 28 ± 1°C for 30 days under the photoperiod of 16h and light irradiance of 25-30 μmol m^{-2} s^{-1} (SANTARÉM et al., 1997). The explants were subcultivated for more 15 days upon the development of the primary globular embryos. Thereafter, these globular embryos were isolated from regions close to the original explants and transferred onto the medium MSD20 to develop and multiply the embryogenic clusters. This growing medium was prepared with the MS (MURASHIGE; SKOOG, 1962) at half strength (WRIGHT, 1991), B5 vitamin (GAMBORG et al., 1968), inositol (100 mg L^{-1}), 3% sucrose, 0.2% Phytagel, 90 μM 2,4-D, and pH 5.8. The subcultures were also maintained in similar conditions during the induction period. During these steps, the embryogenic tissues were selected to choose embryos in globular stage. The proliferation period was extended for 6-8 weeks.

Embryos in the globular stage were cultivated in hystodifferentiation medium (MSM6AC) to obtain the cotyledonary embryos. This growing medium was prepared with MS salts (MURASHIGE; SKOOG, 1962), B5 vitamin (GAMBORG et al., 1968), 6% maltose, 0.5% active charcoal, 0.2% Phytagel and pH 5.8 (BAILEY et al., 1993). In the cotyledonary stage, these embryos were transferred to the medium MSM6 (activated-free charcoal MSM6AC) until the physiological maturity (FINER; McMULLEN, 1991) when they were desecated. The desecation process consisted of set up several embryos on Petri dishes which were sealed with plastic film for two days. Thereafter, these embryos were transferred to the germination and conversion medium MSO composed of MS salts (MURASHIGE; SKOOG, 1962), B5 vitamin (GAMBORG et al., 1968), 3% sucrose, 0.2% Phytagel and pH 5.8. In this phase, the photoperiod was 23 hours to avoid premature flowering. As soon as the hypocotyl and radicle were visually detected, they were transferred to Magenta boxes (Sigma) with medium MSO for seedling elongation and development.

Acclimatization

Seedlings with well-developed radicular systems were transplanted into plastic containers filled with vermiculite and transferred into a growth chamber where they stayed for 20 days under the previous conditions of temperature and photoperiod. In this growth chamber they were protected by plastic film. Soon after, these seedlings were transplanted into plastic pots filled with soil for more 10 days. Thereafter, the plants were transferred to greenhouse conditions, and the photoperiod was reduced to 16h for seed production.

Experimental design

The experimental design was completely randomized with ten replications. Every experimental unit had ten cotyledons with their inner side on the growing medium. These data were reported by descriptive statistics.

Organogenesis

In this experiment, mature seeds were used as explants. First, the seeds were disinfested in 70% ethanol for 5 min., incubated in 2.5% sodium hypochloride (v v^{-1}) plus 0.01% Tween 20 for 20 min., rinsed four times in distilled, deionized and autoclaved water, and dried on sterilized filter papers. Next, these seeds were distributed on Petri dishes (90 x 15 mm) containing growing medium (GM) prepared with B5 salts, B5 vitamin (GAMBORG et al., 1968), 3% sucrose , 3 mM MES (2-N-morfolino-ethanosulfonic acid), 0.7% Phytagel and pH 5.8. Finally, the seeds were maintained at 27 ± 1°C, photoperiod of 16 hours, and irradiance of 24-36 µmol m^{-2} s^{-1} for 4 days.

Every cultivar had 50 normal seedlings collected from 80 germinated seeds. The explants were obtained by the method of Olhoft et al. (2003) after removing the seed tegument and excising the terminal end of hypocotyls below the cotyledonary node. Thus, two explants were obtained from every seed through the cotyledon separation along the remained hypocotyl. One hundred explants from every cultivar was set up on the induction medium (SIM) prepared with B5 salts (GAMBORG et al., 1968), B5 vitamin (GAMBORG et al., 1968), 3% sucrose, 3 mM MES, 1.67 mg L^{-1} BAP (6-benzilaminopurina), 0.6% Phytagel and pH 5.6. The Petri dishes were incubated under photoperiod of 16 hours and light irradiance of 24-36 µmol m^{-2} s^{-1} during two weeks for stimulating plantlets induction. Thereafter, the terminal end of the hypocotyl was eliminated, and the explants with newly developed plantlets were transferred to Petri dishes with fresh growing medium for more two weeks. Afterward, the cotyledons were removed and the plantlets transferred to the elongating medium (SEM). This medium consisted of MS salt (MURASHIGE; SKOOG, 1962), B5 vitamin (GAMBORG et al., 1968), 3% sucrose, 3 mM MES, 0.5 mg L^{-1} giberellic acid, 50 mg L^{-1} asparagine, 0.1 mg L^{-1} IAA, 1 mg L^{-1} zeatin, 0.7% Phytagel and pH 5.6. These media were subcultivated two times every 15 days. Individual plantlets with 3 cm in length were transferred into rooting medium (RM) with B5 salt (GAMBORG et al., 1968) at half strength, 3% sucrose, 3 mM MES, 1 mg L^{-1} AIB (indol-3-butíric acid), 0.7% Phytagel and pH 5.6.

Acclimatization

Seedlings with well-developed radicular systems were transferred to plastic containers filled with vermiculite, covered with plastic bags, and maintained in the growth chamber under the same conditions of temperature and photoperiod for 20 days. Next, they were transplanted to plastic containers filled with soil and grown for more 10 days. Thereafter, the plants were transferred to greenhouse conditions.

Experimental design

The experimental design was completely randomized with three replications of 100 explants per cultivar. Data from plantlets induction, number of plantlets elongating, number of elongated plantlets, rooting and plant regeneration were reported by their means.

Results and discussion

Embryogenesis

The present protocol permitted that somatic embryos could be differentiated in the inner surface of immature cotyledons during only four weeks (Figure 1). Concentration of 2,4-D at 40 mg L^{-1} has been efficient for inducing the development of somatic embryos in soybean cultivars (BUCHHEIM et al., 1989). This efficiency, however, was highly discrepant among these cultivars. In six weeks and after two periods of sub-cultivations on the medium MSD40, the number of embryos per cotyledon ranged from 0.12 to 2.30 (Figure 2). The cultivar CD 201 developed an average of 2.3, CD 202 developed 0.98, CD 216 developed 0.86 and CD 220 developed 0.94 somatic embryos. Otherwise, CD 217 with 0.13 and CD 218 with 0.12 somatic embryos had small embryogenic capacity. Similarly, variability in the somatic embryogenesis of soybean genotypes was reported by Parrot et al. (1989), Komatsuda and Ko (1990) and Bonacin et al. (2000).

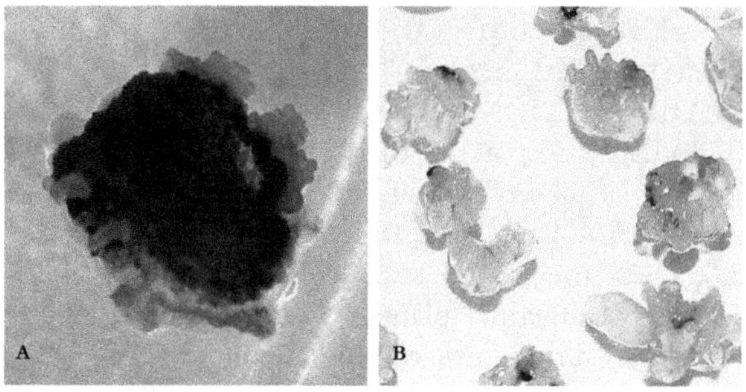

Figure 1. Clusters of somatic embryos from immature soybean cotyledons: A=Immature cotyledon cultivated for six weeks on MSD40 growing medium; B=Isolated somatic embryos on the MSD20 growing medium.

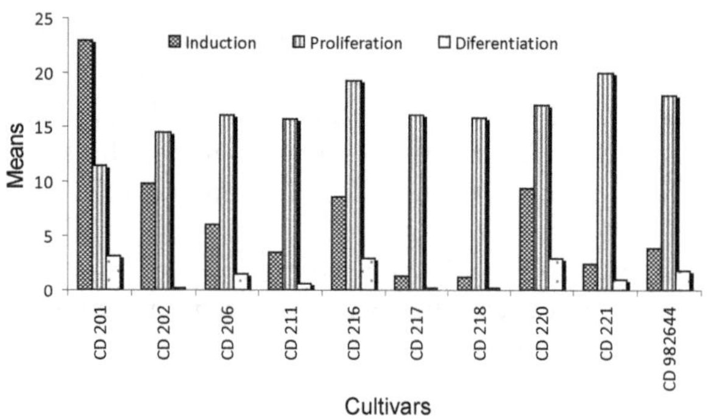

Figure 2. Average numbers of induction, proliferation and differentiation somatic embryos.

Every cultivar started the proliferation phase with different number of somatic embryos because all the globular embryos were inoculated on the medium MSD20. At the beginning of this phase, the somatic embryos developed from 0.87 to 2.0 globular embryos (Figure 3). In terms of similar number of inoculated embryos, the cultivar CD 216 developed 1.93 and the CD 221 developed 2.0 embryos. Although the CD 201 had the highest embryo production in the induction phase, it had the lowest number of embryos per inoculated embryo in the proliferation phase. These responses are similar to reports from Gesteira (2002) who found that the cultivar CD 201 was among the best cultivars in the induction phase, but the lower frequency in the proliferation phase. Weak and positive correlation between these phases has also been observed in some research reports (BAILEY et al., 1993, SIMMONDS; DONALDSON, 2000).

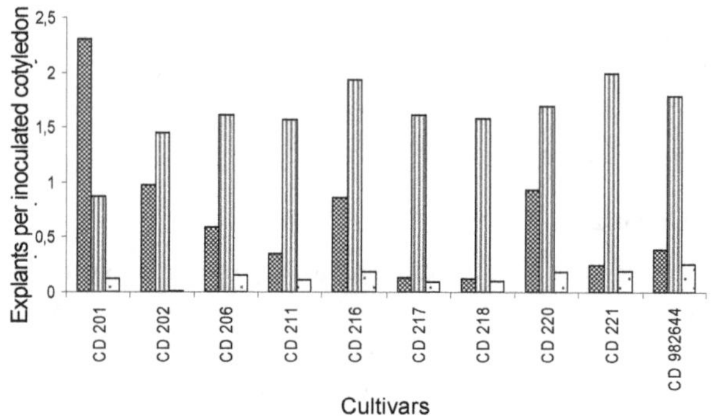

Figure 3. Number of embryos from inoculated embryos in several phases of somatic embryogenesis in soybean cultivars. Number of embryos per inoculated cotyledon; . Number of somatic globular embryos per inoculated embryo at the beginning of the proliferation; Differentiated embryos per inoculated embryo

In the present experiment, these soybean cultivars had marked influence on the induction of somatic embryogenesis as previously described through different protocols (BAILEY et al., 1993; KOMATSUDA; OHYAMA, 1988; KOMATSUDA et al., 1992; MEURER et al., 2001; RANCH et al., 1985; SANTARÉM et al., 1997; SHOEMAKER et al., 1991; SIMMONDS; DONALDSON, 2000).

In the differentiation phase, the growing medium without 2,4-D effectively inhibited the development of somatic embryos and allowed the differentiation of tissues. Morphological abnormalities were avoided by the excessive exposure of these embryos to the activated charcoal, therefore, reducing the residual effect of 2,4-D. Similar to auxins, other compounds from cultivated tissues have also inhibitory effects on the growth and development of these embryos (BUCHHEIM et al., 1989).

The somatic embryos obeyed the same pattern of development of the zygotic embryos transferred to the hystodifferentiation growing medium (Figure 4). After the globular stage, they differentiated into cordiforme, torpedo and cotyledonary stages. Evidently, the presence of abnormal embryos was detected during this stage.

The differentiation phase was less efficient than the induction and proliferation. Every somatic embryo produced from 0.014 to 0.25 differentiated embryos. The best genotype was the line CD 98-

2644, but the indices of differentiated explants from the cultivars CD 206, CD 216, CD 220 and CD 221 were above 10.0% unlike the CD 202, CD 211, CD 217 and CD 218 that had the lowest indices.

Figure 4. Somatic embryos of soybean cultivars from the end of hystodifferentiation phase growing on the medium MS6MAC.

Mature cotyledonary embryos were characterized by their color changes from green to creamy yellow, and time to physiological maturity was recorded between four and six weeks. Full maturation is unnecessary to obtain plants from somatic embryos, but it is required for reaching high levels of plant regeneration (BUCHHEIM et al., 1989). The regenerative potential from physiologically mature cotyledonary embryos that germinated on the medium MSO is showed in the Figure 5A. Desiccation is the best way to enhance the embryo germination which is generally low (LIU et al., 1994) although they are morphologically similar to the zygotic embryos (LIU et al., 1994).

Different capacity of regeneration was detected among the cultivars. The cultivar CD 201 with 82% had the best response, but the CD 220 had 58% and the CD 216 had the regenerative capacity of 51% which is a satisfactory level.

Loganathan (2010) studied the plant regeneration from somatic embryos of soybean and reported different responses from several media and acclimatization conditions under laboratory and greenhouse conditions. Although most of the embryos from the regulator-free growth medium developed the roots and shoots, some of them were incapable of overcoming the desiccation process. They showed some dark and necrotic appearance and did not germinate. The explanation is the long period of treatment with 2,4-D which may have affected the embryogenic potential.

Figure 5. (A) Stems and well-developed radicles from the conversion and regeneration medium (MS0); (B) regenerated soybean seedling from the medium MSO; (C) acclimatization of regenerated plants *in vitro*; (D) plant acclimatization in the greenhouse.

In terms of total explants per immature embryo inoculated at the beginning of induction phase, the best responses were obtained from CD 201, CD 216 and CD 220, which are the preferential cultivars for somatic embryogenesis. The cultivar CD 201 highlighted in the present experiment because of its high capacity of induction, unlike CD 216 and CD 220 that had similar performance during the induction, proliferation, and differentiation phases. The best cultivar performance in some phase not necessarily is the best in another one because the variability depends on cultivar responses.

In soybean, some studies have demonstrated the effects of maturation groups on the induction of somatic embryogenesis. Shoemaker et al. (1991), Tian et al. (1994) and Tomlin et al. (2002) reported that soybean genotypes from early maturation groups had high embryogenic responses than cultivars from late maturation groups. In part, these responses are in agreement with the present results because the early cultivar CD 216 had similar performance in the induction, proliferation and differentiation phases. Furthermore, Ko et al. (2004) observed high embryogenic responses from genotypes classified in the intermediate maturation group. The present responses are also in agreement with these reports because the CD 201 and CD 220, which were efficient during the embryogenesis induction, are clustered in a semi-precocious group of maturation. Otherwise, Ranch et al. (1985) and

Bailey et al. (1993) did not find any relationship between maturation groups and somatic embryogenesis.

The present results can help plant breeders to select soybean cultivars and they show some alternatives that could be used to improve the efficiency of the somatic embryogenesis. An example is the protocol adjustment to permit the proliferation of CD 201 or induce the embryo development of the cultivars CD 216 and CD 220. Some cultivars as the CD 206 and CD 221 that differentiated few explants could be recommended for successful somatic embryogenesis if the induction could be increased. Others, as CD 211, CD 217 and CD 218 should be avoided because they were less efficient in two of three regeneration phases.

Organogenesis

All the cultivars had organogenic capacity of plant regeneration, and the plants developed in less than 90 days (Figure 6).

Figure 6. Organogenesis from soybean cotyledonary nodes: (A) Plantlets induction; (B) isolated plantlets from cotyledons in the beginning of the elongation phase; (C) elongated plantlets; (D) elongated plantlets and isolate in rooting medium; (E) acclimatization of regenerated plants.

From every cultivar, however, different responses were detected in the number of induced plantlets. The explanation for these findings may rest with differences in the seed quality of the explant sources or the cultivar performance *in vitro*. Similarly, these cultivars had different responses in number of elongate plantlets, number of seedling with roots and regenerated plants (Figure 7). The present cultivars had differences in the number of elongated shoots per plantlet, but the CD 217 was the most responsive (Figure 8).

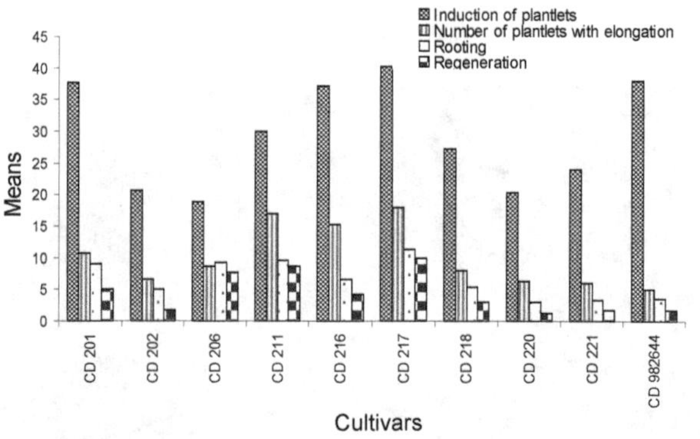

Figure 7. Number of differentiated structures during the organogenesis.

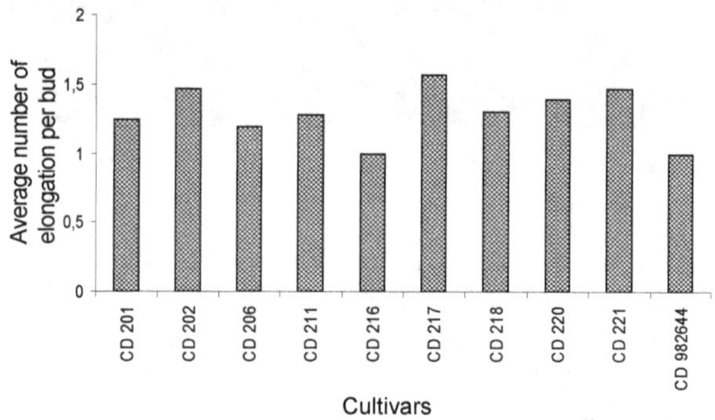

Figure 8. Number of elongations obtained by shoots that produced elongations

Elongated plantlets with the average length from 3 and 4 cm were isolated and transferred to test tubes with rooting media amended with 1 mg L^{-1} AIB where they remained for 15-20 days. These elongated plantlets were positively responsive to rooting. Again, the CD 217 was the best cultivar, and it was followed by those cultivars from intermediate maturation group as the CD 201, CD 202, CD 206, CD 218, CD 220 and CD 221.

As acclimatization of plants is a critical phase in tissue culture, only the plantlets with well-developed roots were acclimatized. Even so there were losses. For example, leaf necrosis followed by seedling death was observed between five and seven days after the transplanting into the vermiculite, but the cultivar CD 217 was still considered satisfactory.

Somatic embryogenesis and organogenesis periods

Cultivars with high frequency of embryogenesis or organogenesis must be selected to participate in genetic programs of plant transformation. However, some aspects must still be considered before choosing the regeneration pathway. One relevant factor is time to plant development. In somatic embryogenesis, time to full development was ten months, and more 45 days was necessary to obtain the explants. Otherwise, the organogenesis is faster because time to completely developed soybean plants is only four months. Another factor is the level of plant regeneration, which was higher in plants submitted to organogenesis (88%) than in those submitted to embryogenesis (82%). In oat plants, Bered et al. (1998) reported the preferential regeneration through organogenesis.

Conclusion

The efficiency of somatic embryogenesis and organogenesis is affected by the plant genotypes. The regeneration of the soybean cultivars CD 201, CD 216 and CD 220 by somatic embryogenesis is more efficient. The regeneration of the cultivar CD 217 through organogenesis is more efficient. The embryogenic and organogenic systems can be used efficiently in the genetic transformation of soybean plants.

References

BAILEY, M. A.; BOERMA, H. R.; PARROT, W. A. Genotype effects on proliferative embryogenesis and regeneration of soybean. **In Vitro Cellular and Developmental Biology**, v. 29, n. 3, p. 102-108, 1993.

BERED, F.; SERENO, M. J. C. M.; CARVALHO, F. I. F.; LANGE, C. E.; HANDEL, C. L.; DORNELES, A. L. C. Regeneração de plantas de aveia a partir de calos embriogênicos e organogênicos. **Pesquisa Agropecuária Brasileira**, v. 33, n. 11, p. 1827-1833, 1998.

BONACIN, G. A.; MAURO, A. O.; OLIVEIRA, R. C.; PERECIN, D. Induction of somatic embryogenesis in soybean: physicochemical factors influencing the development of somatic embryos. **Genetics and Molecular Biology**, v. 23, n. 4, p. 865-868, 2000.

BUCHHEIM, J. A.; COLBURN, S. M.; RANCH, J. P. Maturation of soybean somatic embryos and the transition to plantled growth. **Plant Physiology**, v. 89, n. 3, p. 768-775, 1989.

DUSI, D. M. A. Transformação Genética de Plantas In: TORRES, A.; CALDAS, L. S.; BUSO, J. A. (Ed.). **Cultura de tecidos e transformação genética de plantas**. Brasília: Embrapa-SPI/Embrapa CNPH, 1998. v. 2. p. 679-717.

FINER, J. J.; McMULLEN, M. D. Transformation of soybean via particle bombardment of embryogenic suspension culture tissue. **In Vitro Cellular and Developmental Biology**, v. 27, n. 4, p. 175-182, 1991.

FINER, J. J.; NAGASAWA, A. Development of an embryogenic suspension cultura of soybean [*Glycine max* (L.) Merrill]. **Plant Cell, Tissue and Organ Culture**, v. 15, n. 2, p. 125-136, 1988.

FU, J.; HUANG, B. Involvement of antioxidants and lipid peroxidation in the adaptation of two cool-season grasses to localized drought stress. **Environmental and Experimental Botany**, v. 45, n. 2, p. 151-158, 1968.

GAMBORG, O. L.; MILLER, R. A.; OJIMA, K. Nutrient requirement of suspension cultures of soybean root cells. **Experimental Cell Research**, v. 50, n. 1, p. 151-158, 1968.

GEORGE, E. F. **Plant propagation by tissue culture**: the technology. 2nd. ed. London: Exegetics, 1993.

GESTEIRA, A. S.; OTONI, W. C.; BARROS, E. G.; MOREIRA, M. A. RAPD – based detection of genomic instability in soybean plants derived from somatic embryogenesis. **Plant Breeding**, v. 121, n. 3, p. 269-271, 2002.

HARTWECK, L. M.; LAZZERI, P. A.; CUI, D. Auxin orientation effects and histological analysis of soybean somatic embryogenesis. **In Vitro Cellular and Developmental Biology-Plant**, v. 24, n. 8, p. 821-828, 1988.

JOY IV, R. W.; THORPE, T. A. Shoot morphogenesis: structure, physiology, biochemistry and molecular biology. In: SOH, W. Y.; BHOJWANI, S. S. (Ed.). **Morphogenesis in plant tissue cultures**. London: Kluwer Academic, 1999. p. 171-214.

KO, T. S.; RANDALL, L. N.; KORBAN, S. S. Screening multiple soybean cultivars (MG to MG VIII) for somatic embryogenesis following *Agrobacterium*-meddiated transformation of immature cotyledons. **Crop Science**, v. 44, n. 5, p. 1825-1831, 2004.

KOMATSUDA, T.; KO, S. W. Screening of soybean [*Glycine max* L. Merrill] genotypes for somatic embryo production from immature embryo. **Japanese Journal of Breeding**, v. 40, n. 2, p. 249-251, 1990.

KOMATSUDA, T.; OHYAMA, K. Genotypes of high competence for somatic embryogenesis and plant regeneration in soybean *Glycine max*. **Theoretical and Applied Genetics**, v. 75, n. 5, p. 695-700, 1988.

KOMATSUDA, T.; LEE, W.; OKA, S. Maturation and germination of somatic embryo in soybeans *Glycine gracilis* Skvortz and *Glycine max* L. Merr. **Plant Cell, Tissue and Organ Culture**, v. 28, n. 1, p. 103-113, 1992.

LIU, W.; HILDEBRAND, D. F.; MOORE, P. J.; COLLINS, G. B. Expression of desiccation-induced and lipoxygenase genes during the transition from maturation to the germination phases in soybean somatic embryos. **Planta**, v. 194, n. 1, p. 69-76, 1994.

LOGANATHAN, M.; MARUTHASALAM, S.; SHIU, L. Y.; LIEN, W. C.; HSU, W. H.; LEE, P. F.; YU, C. W.; LIN, C. H. Regeneration of soybean [*Glycine max* L. Merrill] throught direct somatic embryogenesis from the immature embryonic shoot tip. **In Vitro Cellular and Developmental Biology**, v. 47, n. 3, p. 265-273, 2010.

MEURER, C. A.; DINKINS, R. D.; REDMOND, C. T.; McALLISTER, K. P.; TUCKER, D. T.; WALKER, D. R.; PARROTT, W. A.; TRICK, H. N.; ESSIG, J. S.; FRANTZ, H. M.; FINER, J. J.; COLLINS, G. B. Embryogenic response of multiple soybean [*Glycine max* L. Merrill] cultivars across three locations. **In Vitro Cellular and Developmental Biology-Plant**, v. 37, n. 1, p. 62-67, 2001.

MURASHIGE, T.; SKOOG, F. A revised medium for rapid growth and bioassays with tobacco tissue cultures. **Plant Physiology**, v. 15, n. 3, p. 473-479, 1962.

OLHOFT, P. M.; FLAGEL, L. E.; DONOVAN, C. M.; SOMERS, A. D. Efficient soybean transformation using hygromycin B selection in the cotyledonary-node method. **Planta**, v. 216, n. 5, p. 723-735, 2003.

OPABODE, J. T. *Agrobacterium*-mediated transformation of plants: emerging factors that influence efficiency. **Biotechnology and Molecular Biology Review**, v. 1, n. 1, p. 12-20, 2006.

PARROT, W. A.; CLEMENTE, T. E. Transgenic soybean. In: BOERMA, H. R.; SPECHT, J. E. (Ed.). **Soybeans**: improvement, production, and uses. 3rd ed. Madison: ASA/CSSA/SSSA, 2004. p. 635-667.

PARROT, W. A.; WILLIAMS, E. G.; HIDERBRANS, D. F.; COLLINS, G. B. Effect of genotype on somatic embryogenesis from immature cotyledons of soybean. **Plant Cell, Tissue and Organ Culture**, v. 16, n. 1, p. 15-21, 1989.

PAZ, M. M.; SHOU, H.; GUO, Z.; ZHANG, Z.; BANERJEE, A. K.; WANG, K. Assessment of conditions affecting *Agrobacterium*-mediated soybean transformation using the cotyledonary node explant. **Euphytica**, v. 136, n. 2, p. 167-179, 2004.

PIERIK, R. L. M. **In vitro culture of higher plants**. 4th ed. Dordrecht: Kluwer Academic, 1997.

RANCH, J. P.; OGLESBY, L.; ZIELINSKI, A. C. Plant Regeneration from embryo-derived tissue cultures of soybeans. **In Vitro Cellular and Developmental Biology**, v. 21, n. 11, p. 653-658, 1985.

SANTARÉM, E. R.; PELISSIER, B.; FINER, J. J. Effect of explant orientation, pH, solidifying agent and wounding on initiation of soybean somatic embryos. **In Vitro Cellular and Developmental Biology**, v. 33, n. 1, p. 13-19, 1997.

SHAN, Z.; RAEMAKERS, K.; TZITZIKAS, E. N.; MA, Z.; VISSER, R. G. F. Development of a highly efficient, repetitive system of organogenesis in soybean (*Glycine max* L. Merrill). **Plant Cell Reports**, v. 24, n. 9, p. 507-512, 2005.

SHOEMAKER, R. C.; AMBERGER, L. A.; PALMER, L. O.; RANCH, J. P. Effect of 2,4-dichlorophenoxyacetis acid concentration on somatic embryogenesis and heritable variation on soybean [*Glycine max* (L.) Merrill]. **In Vitro Cellular and Developmental Biology**, v. 27, n. 2, p. 84-88, 1991.

SIMMONDS, D. H., DONALDSON, P. A. Genotype screening for proliferative embryogenesis and biolistc transformation of short-season soybean genotypes. **Plant Cell Reports**, v. 19, n. 5, p. 485-490, 2000.

SUGIYAMA, M. Organogenesis in vitro. **Current Opinion in Plant Biology**, v. 2, n. 1, p. 61-64, 1999.

TIAN, L. N.; BROWN, D. C. W.; VOLDENG, H.; WEBB, J. In vitro response and pedigree analysis for somatic embryogenesis of long-day photoperiod adapted soybean. **Plant Cell, Tissue and Organ Culture**, v. 36, n. 1, p. 269-273, 1994.

TOMLIN, E. S.; BRANCH, S. R.; CHAMBERLAIN, D.; GABE, H.; WRIGHT, M. S.; STEWART JR., C. N. Screening of soybean, *Glycine max* L. Merril lines for somatic embryo induction and maturation capability from immature cotyledons. **In Vitro Cellular and Developmental Biology**, v. 38, n. 6, p. 543-248, 2002.

WRIGHT, M. S.; LAUNIS M. S.; NOVITZKY K. L.; DUESING, R. J. H.; HARMS, C. T. A simple method for the recovery of multiple fertile plants from individual somatic of soybean [*Glycine max* L. Merrill]. **In Vitro Cellular and Developmental Biology**, v. 27, n. 3, p. 153-157, 1991.

19

Genetic dissimilarity among sweet potato genotypes using morphological and molecular descriptors

Elisângela Knoblauch Viega de Andrade[1], Valter Carvalho de Andrade Júnior[1], Marcelo Luiz de Laia[2], José Sebastião Cunha Fernandes[1], Altino Junior Mendes Oliveira[1], and Alcinei Mistico Azevedo[3*]

*[1]Departamento de Agronomia, Universidade Federal dos Vales do Jequitinhonha e Mucuri, Campus JK, Rodovia MGT-367, km 583, 5000, Alto da Jacuba, Cx. Postal 3037, 39100-000, Diamantina, Minas Gerais, Brazil. [2]Departamento de Engenharia Florestal, Universidade Federal dos Vales do Jequitinhonha e Mucuri, Diamantina, Minas Gerais, Brazil. [3]Universidade Federal de Minas Gerais, Instituto de Ciências agrárias, Montes Claros, Minas Gerais, Brasil. *Author for correspondence. E-mail: alcineimistico@hotmail.com*

ABSTRACT. This study aimed to evaluate the genetic dissimilarity among sweet potato genotypes using morphological and molecular descriptors. The experiment was conducted in the Olericulture Sector at Federal University of Jequitinhonha and Mucuri Valleys (UFVJM) and evaluated 60 sweet potato genotypes. For morphological characterization, 24 descriptors were used. For molecular characterization, 11 microsatellite primers specific for sweet potatoes were used, obtaining 210 polymorphic bands. Morphological and molecular diversity was obtained by dissimilarity matrices based on the coefficient of simple matching and the Jaccard index for morphological and molecular data, respectively. From these matrices, dendrograms were built. There is a large amount of genetic variability among sweet potato genotypes of the germplasm bank at UFVJM based on morphological and molecular characterizations. There was no duplicate suspicion or strong association between morphological and molecular analyses. Divergent accessions have been identified by molecular and morphological analyses, which can be used as parents in breeding programmes to produce progenies with high genetic variability.

Keywords: *Ipomoea batatas* (L.) Lam., morphological markers, microsatellites, genetic diversity, germplasm bank.

Introduction

Sweet potato (*Ipomoea batatas* L. Lam.) is a tuberous vegetable of economic and social importance, grown mainly by small farmers and used as a staple food for the world's poorest regions (Roesler, Gomes, Moro, Kummer, & Cereda, 2008). Since it has high levels of carbohydrates, sugars, minerals, and vitamins A, C, and B complex, it can be used for multiple purposes (Azevedo, Andrade Júnior, Fernandes, Pedros, & Oliveira, 2015). Vale do Jequitinhonha is among the poorest regions in Brazil. It is diversified and has a poorly developed agriculture due to an arid climate. Consequently, sweet potato is among the most common crops grown in family farms in this region, due to its production ruggedness, resistance to pests and diseases, low demand in soil fertility and drought tolerance (Monteiro, 2007; Roesler et al., 2008).

Yield of sweet potato marketable roots can easily reach 45.4 t ha^{-1} (Andrade Júnior et al., 2009; Azevedo et al., 2014); however, in Vale do Jequitinhonha, this productivity is much lower. Thus, obtaining agronomically superior cultivars adapted to Vale do Jequitinhonha through genetic improvement can be an important strategy in the region and throughout Brazil. Thus, a germplasm bank was created in the Federal University of Jequitinhonha and Mucuri Valleys (*Universidade Federal dos Vales do Jequitinhonha and Mucuri* - UFVJM). The bank contained more than 80 sweet potato accessions, mostly collected in rural properties in the Vale do Jequitinhonha region.

In any germplasm bank, an important step towards genetic improvement is to study dissimilarity among accessions; this enables the identification of possible duplicates and hybridizations that can generate high genetic variability in the offspring. Dissimilarity studies may be made by multivariate analysis based on morphological and/or molecular markers (Cruz, Regazzi, & Carneiro, 2012).

Morphological markers allow for easy discrimination between sweet potato accessions (Chávez et al., 2006; Veasey et al., 2007; Moulin et al., 2012). Similarly, genetic divergence can be evaluated by means of molecular markers, enabling germplasm characterization (Souza, Carvalho, Martins, Guedes, & Oliveira, 2008), phylogenetic analysis, DNA fingerprinting, gene link detection with mono and polygenic characters, identification of varieties, gene introgression, and indirect selection of agronomic traits. The effectiveness of different types of labels is confirmed by the wide use of these techniques in in genetic storage studies and for improving plants (Souza, 2015).

Simple sequence repeat (SSR) microsatellites are molecular markers that facilitate rapid differentiation between related individuals because they are abundant and evenly distributed throughout the genome. In addition, due to their high polymorphism rate, codominant inheritance and multiallelic nature, SSR markers are highly reproducibility and require small amounts of DNA (Oliveira, Pádua, Zucchi, Vencovsky, & Vieira, 2006; Oliveira & Silva, 2008). Several studies with sweet potato culture using SSR markers have observed high genetic diversity among accessions, including Veasey et al. (2008) and Koussao et al. (2014).

When considering the socioeconomic importance of sweet potatoes, it is essential to characterize the genotypes in germplasm banks for genetic dissimilarity in order to eliminate duplicates and guide further crosses in breeding programmes. Therefore, the objective of this work was to evaluate genetic dissimilarity among sweet potato genotypes of the germplasm bank of UFVJM through morphological and molecular descriptors.

Material and methods

The experiment was conducted in the Olericulture Sector of the Federal University of Jequitinhonha and Mucuri Valleys, in JK Campus, in Diamantina city, Minas Gerais State, Brazil (18°12'01"S, 43°34'20"W and 1,400-m altitude). Sixty sweet potato genotypes were assessed, which belong to the germplasm bank of UFVJM. They are named UFVJM (01, 02, 03, 04, 05, 06, 07, 08, 09, 10, 13, 14, 16, 17, 19, 20, 21, 22, 23, 24, 25, 26, 27, 28, 29, 30, 31, 33, 34, 35, 37, 38, 39, 41, 42, 43, 44, 45, 46, 48, 49, 50, 51, 52, 54, 55, 56, 57, and 58), Arruba, Cariru Vermelha, Brazlândia-Branca, Cambraia, Espanhola, Licuri, Palmas, Princesa, Tomba Carro 1, Tomba Carro 2, and Coquinha. Sweet potato branches containing three to five buds were planted in expanded polystyrene trays containing commercial substrate and were kept for 35 days in a greenhouse. Subsequently, seedlings containing eight nodes were transplanted to the field, with the burial of four nodes in prepared areas. The planting of sweet potato seedlings was performed on December 18, 2012.

In the field, the area was plowed and harrowed, followed by the incorporation of fertilizers (20, 90, and 90 kg ha^{-1} of N, P_2O_5 and K_2O, respectively). Subsequently, another plowing was performed to create windrows with heights between 30 cm and 40 cm. Plots consisted of a row that was 2.4-m long, with spaces of 1.0 metres between planting rows and 0.3 metres between plants, totalling eight plants per section. Culture management was performed according to Filgueira (2008).

At 90 days after planting (DAP), the morphological characteristics of plant shoots were evaluated; and at 150 DAP, roots were assessed. Evaluations were made according to the descriptors and scoring scales that were recommended by the International Board for Plant Genetic Resources (IBPGR), which were designed by Huamán (1991) and are listed in Table 1.

Table 1. Scoring scale related to the evaluated morphological descriptors in 60 accessions of sweet potato (*Ipomoea batatas*). Source: Huamán (1991). UFVJM, Diamantina, Minas Gerais State, 2016.

Characteristics	Score
1. Predominant foliage colour	1-Green; 3-green with a few purple stains; 4-green with many purple stains; 5- green with many dark purple stains; 6-predominantly purple; 7-predominantly dark purple; 8-completely purple; 9-completely dark purple.
2. Secondary foliage colour	0-Absent; 1-green base; 2-green apex; 3-green nodes; 4-reddish base; 5-reddish apex; 6-reddish nodes; 7-other.
3. Foliage apex pubescence	0-Absence of hairiness; 3- thin hairiness; 5- moderate; 7-dense; 9- very dense.
4. Principal foliage length	3-Erect (<75 cm); 5-semi-erect (75-150 cm); 7-scattered (150-250 cm); 9-extremely scattered (>250 cm).
5. Leaf lobule types	0-Absence of lobules; 1-very superficial lobules; 3-superficial lobules; 5-moderate; 7-deep; 9-very deep.
6. Number of leaf lobules	0- Absence of lobules- round and smooth leaf; 1-absence of lateral lobules; 3-three lobules; 5-five lobules; 7-seven lobules; 9-nine lobules.
7. Central lobule format	0-Absent; 1-jagged; 2-triangular; 3-semi-circular; 4-semi-elliptical; 5-elliptical; 6-lanceolate; 7-oblong and lanceolate; 8-linear and wide; 9-linear and narrow.
8. Pigmentation of veins	1-Yellow; 2-green; 3-purple stains in principal base of veins; 4-purple stains in many veins; 5-principal vein predominantly purple; 6-principal vein completely purple; 7-partially purple veins; 8-veins totally purple; 9-lower surface and veins completely purple.
9. Petiole length	1-Very short (<10 cm); 3-short (10-20 cm); 5-intermediate (21-30 cm); 7-long (31-40 cm); 9-very long (>40 cm).
10. Petiole pigmentation	1-Green; 2-green with purple base; 3-green with purple leaf insertion; 4-green with purple in both extremities; 5-green with purple stains in all petiole; 6-green with purple stripes; 7-purple with green leaf insertion; 8-some green and some purple petioles; 9-completely of almost purple.
11. Mature leaf colour	1-Greenish yellow; 2-green; 3-green with purple ends; 4-greyish green; 5-green with purple veins; 6-smooth purple; 7-predominantly purple; 8-green upper surface and purple lower surface; 9-purple in both surfaces.
12. Immature leaf	1-Greenish yellow; 2-green; 3-green with purple ends; 4-greyish green; 5-green with purple veins; 6-smooth purple; 7-predominantly purple; 8-green upper surface and purple lower surface; 9-purple in both surfaces.
13. Mature leaf size	3-Small (<8 cm); 5-average (8-15 cm); 7-large (16-25 cm); 9-very large (>25 cm).
14. Mature leaf general format	1-Round; 2-reniform; 3-cordate; 4-triangular; 5-lanceolate; 6-lobed; 7-almost separated.
15. Internode diameter	1-Very thin (<4 mm); 3-thin (4-6 mm); 5-intermediate (7-9 mm); 7-thick (10-12 mm); 9-very thick (>12 mm).
16. Internode length	1-Very short (<3 cm); 3-short (3-5 cm); 5-intermediate (6-9 cm); 7-long (10-12 cm); 9-very long (>12 cm).
17. Flaws in root surface	0-Absent; 1-alligator skin; 2-prominent veins; 3-shallow horizontal constrictions; 4-deep horizontal constrictions; 5-shallow longitudinal slots; 6-deep longitudinal slots; 7-constrictions and deep cracks; 8-other.
18. Root format	1-Round; 2-round and elliptical; 3-elliptical; 4-oval; 5-oboval; 6-oblong; 7-oblong and long; 8-long and elliptical; 9-irregular and long.
19. Predominant colour of root bark	1-White; 2-cream; 3-yellow; 4-orangish; 5-orangish brown 6-pinkish; 7-red; 8-reddish purple; 9-dark purple.
20. Root bark colour intensity	1-Pale; 2-intermediate; 3-dark.
21. Secondary colour of root bark	0-Absent; 1-white; 2-cream; 3-yellow; 4-orangish; 5-orangish brown; 6-pinkish; 7-red; 8-reddish purple; 9-dark purple.
22. Pulp predominant colour	1-White; 2-light cream; 3-dark cream; 4-light yellow; 5-dark yellow; 6-light orange; 7-orangish brown; 8-dark orange; 9-strongly purple.
23. Secondary pulp colour	0-Absent; 1-white; 2-cream; 3-yellow; 4-orangish brown; 5-pinkish; 6-red; 7-reddish purple; 8-purple; 9-dark purple.
24. Distribution of secondary pulp colour	0-Absent; 1-ring close to the bark; 2-large ring close to the bark; 3-intense sparse stains; 4-thin ring in the pulp; 5-large ring in the pulp; 6-ring and other stains in the pulp; 7-stains in longitudinal sections; 8-stains covering the surface almost completely; 9-stains covering the hole pulp.

Extraction of genetic material (DNA) was performed according to the protocol proposed by Veasey et al., (2008) with some modifications. DNA samples were analysed for the quality and quantity that were required for molecular procedures. Each sample was subjected to electrophoresis on 0.8% agarose gel in 1x TAE buffer under a constant voltage of 100 μV for 90 minutes. A size and weight molecular standard (1-kb DNA Ladder, BioLabs Inc.) was used with the samples and served as a standard to estimate the size (in base pairs) and weight (in nanograms that was present in each sample. After the run, the gel was examined under ultraviolet light, and the image was scanned with LpixImage (Loccus Biotecnologia). PCR reactions were performed according to the protocol presented by Veasey et al. (2008). Thirty-two oligonucleotide primers were used; Jarret and Bowen (1994), Buteler, Jarret, and La Bonte (1999), describe them. Initially, ten sweet potato accessions with different morphological characteristics were examined in order to choose those with good amplification, high resolution, reproducibility and polymorphisms. The optimal annealing temperature was 50°C for amplification using all selected oligonucleotides.

PCR amplifications were performed in an Eppendorf thermocycler, model Mastercycler Gradiente, under the following conditions: 94°C for 3 minutes, followed by 40 cycles at 94°C for 45 sec., 50°C for 30 sec., and 72°C for 1 minute. At the end of 40 cycles, an additional step at 72°C for 1 minute was used followed by a 4°C hold

Products resulting from amplification were separated by electrophoresis on 6% polyacrylamide gels in a TBS running buffer under a constant voltage of 80 V for 4 hours in a vertical vessel (20x20 cm). Then, 7 μL of PCR product and 2 μL of loading buffer were loaded in each well. A size marker (50-pb Ladder, Biolabs Inc.) was used as standard. Amplified fragments were detected by staining with silver nitrate.

For multicategoric data (Table 1), a dissimilarity matrix was obtained based on a coefficient of simple matching. Sweet potato is a hexaploid species (Azevedo et al., 2015); there are many different band patterns. From the presence and absence of each DNA fragment, the dissimilarity matrix was designed with a Jaccard similarity coefficient (Jaccard, 1901). For multicategoric, molecular and combined data, a cutoff equal to the average dissimilarity was performed. The R program (R Core Team, 2014) and ade4 package (Dray & Dufour, 2007) were used for the estimates of dissimilarity matrices. To combine molecular and multicategoric data, an algebraic sum of these two matrices was calculated (Alves et al., 2013), generating a new matrix for joint analysis. The correlation between matrices and their significance were estimated by Mantel test with 1,000 permutations using package ade4 with 'mantel.rtest' function in R software (R Core Team, 2014). To obtain the dendrogram, the unweighted pair group method with arithmetic mean (UPGMA) grouping method was used, with 'hclust' function.

Results and discussion

The grades of the 24 evaluated descriptors are presented in Table 2. A large variability among accessions was observed, which was not expected given the large number of phenotypic classes with a high percentage of hits in each category. In studies of genetic dissimilarity, the evaluation of the largest possible number of morphological descriptors is indicated, in order to obtain a more complete dissimilarity analysis. Variables that showed little morphological variability were the secondary colour of the branches (76.7% absent of colour); the pubescence of the branch tip (98.33% absent); the mature leaf (90% green-coloured); the size of the mature leaf (93.3% with medium size between 8 to 15 cm); the secondary colour of the root bark (98.33% absent of colour); the secondary colour of the root pulp (78.3% absent of colour) and the distribution of the pulp colour (76.7% without ring formation) (Table 2).

Table 2. Scores of morphological descriptors used in the characterization of 60 accessions of sweet potato. UFVJM, Diamantina-MG, 2016.

Genotypes	Morphological descriptors [1]																							
	1	2	3	4	5	6	7	8	9	10	11	12	13	14	15	16	17	18	19	20	21	22	23	24
Arruba	1	0	0	6	5	5	6	7	2	5	3	5	5	3	3	3	1	9	5	1	0	1	0	0
Braz.-Branca	2	0	0	3	0	1	1	8	5	6	2	5	5	3	5	7	1	4	2	1	0	1	0	0
Cambraia	3	0	0	5	7	5	6	7	2	6	4	3	5	5	3	3	0	3	2	1	0	1	0	0
C. Vermelha	9	0	0	5	7	9	6	5	2	6	9	5	5	5	3	3	1	3	2	1	0	1	0	0
Coquinha	1	0	0	6	5	5	6	7	2	1	1	5	5	5	3	3	0	4	5	1	0	2	0	0
Espanhola	1	0	0	6	5	7	6	6	2	6	3	3	5	3	1	3	0	5	2	1	0	1	0	0
Licuri	1	0	0	6	7	5	6	4	2	1	4	5	7	5	1	3	0	7	5	1	0	1	0	0
Palmas	1	0	3	4	1	1	2	3	2	2	3	5	5	3	3	5	0	7	6	2	0	2	0	0
Princesa	9	4	0	7	9	5	9	6	2	3	9	3	5	3	3	7	0	8	2	1	0	1	0	0
T Carro 1	7	5	0	5	7	5	5	7	2	1	8	3	5	3	3	7	0	8	2	1	0	1	0	0
T Carro 2	1	0	0	4	1	1	2	3	2	1	3	3	5	3	1	5	1	6	2	1	0	9	6	1
UFVJM-01	1	0	0	6	7	5	4	5	2	1	1	5	7	5	3	5	1	8	5	1	0	1	0	0
UFVJM-02	1	0	0	6	7	5	4	7	2	1	4	3	5	3	1	3	0	9	2	1	0	1	0	0
UFVJM-03	3	4	0	6	5	5	4	5	2	5	3	5	5	3	3	3	0	8	2	1	0	1	0	0
UFVJM-04	3	0	0	6	7	1	6	7	2	2	4	5	5	3	1	3	1	8	2	1	0	1	0	0
UFVJM-05	1	0	0	4	3	5	1	3	2	1	1	5	5	3	1	5	1	8	2	2	0	9	8	1
UFVJM-06	9	0	0	6	5	7	4	6	2	1	9	3	5	3	3	3	1	4	4	1	0	2	0	0
UFVJM-07	1	0	0	3	1	5	4	4	2	1	3	5	5	5	3	3	0	8	5	1	0	1	0	0
UFVJM-08	5	5	0	6	5	9	4	7	5	3	4	5	5	3	3	5	1	5	5	1	0	1	0	0
UFVJM-09	1	0	0	4	1	5	3	4	2	2	3	5	5	3	3	3	0	8	2	2	0	9	8	1
UFVJM-10	1	0	0	4	1	5	1	5	2	1	3	5	5	3	3	3	1	5	6	1	0	2	0	0
UFVJM-13	1	0	0	4	1	1	2	5	2	1	3	5	5	5	3	3	2	3	6	1	0	2	0	0
UFVJM-14	1	0	0	4	1	5	1	8	2	1	3	5	7	3	3	5	1	3	2	1	0	4	0	0
UFVJM-16	3	0	0	6	5	5	4	8	2	5	3	5	5	3	3	3	1	5	5	1	0	1	0	0
UFVJM-17	1	0	0	4	1	5	3	4	2	1	3	5	5	3	3	3	0	4	6	1	0	2	0	0
UFVJM-19	3	0	0	4	1	1	1	3	2	1	1	3	5	3	1	5	0	4	2	2	0	9	8	1
UFVJM-20	3	0	0	6	7	7	6	8	2	5	3	3	5	3	3	3	4	9	2	1	0	1	0	0
UFVJM-21	9	5	0	6	7	5	6	5	7	8	3	7	5	3	5	7	0	9	2	1	0	1	0	0
UFVJM-22	1	0	0	5	3	3	4	5	2	1	4	3	5	3	3	3	1	4	2	2	0	4	0	0
UFVJM-23	1	0	0	5	7	5	6	7	2	1	4	5	5	3	1	3	1	9	5	1	0	1	0	0
UFVJM-24	1	4	0	6	5	5	4	8	2	7	2	5	5	3	3	5	2	5	5	1	0	1	0	0
UFVJM-25	1	0	0	3	1	5	1	4	2	2	1	3	5	5	3	3	0	9	6	1	0	2	0	0
UFVJM-26	1	0	0	4	1	3	1	5	2	2	1	7	5	5	3	3	2	5	6	1	0	2	0	0
UFVJM-27	1	0	0	4	1	5	1	3	2	2	1	3	5	5	3	3	0	8	6	1	0	2	0	0
UFVJM-28	1	0	0	3	1	0	1	8	2	6	4	3	5	3	5	7	0	9	2	1	0	1	0	0
UFVJM-29	1	0	0	3	1	3	2	5	2	1	2	5	5	5	3	3	0	8	6	1	0	2	0	0
UFVJM-30	1	0	0	4	0	1	1	3	2	1	1	3	5	3	3	5	3	3	2	2	0	9	0	1
UFVJM-31	1	0	0	4	0	1	1	2	2	2	1	3	5	3	3	5	0	4	2	2	0	9	8	1

Continue...

Continuation...

Genotypes	Morphological descriptors [1]																							
	1	2	3	4	5	6	7	8	9	10	11	12	13	14	15	16	17	18	19	20	21	22	23	24
UFVJM-33	1	0	0	4	1	1	2	4	2	2	1	5	5	5	3	3	0	8	6	2	0	1	0	0
UFVJM-34	1	2	0	4	1	5	2	5	1	1	3	7	5	5	3	3	0	3	5	1	0	1	0	0
UFVJM-35	1	0	0	4	1	5	2	7	2	2	3	5	5	5	3	3	0	8	6	2	0	2	0	0
UFVJM-37	1	0	0	4	1	5	2	3	2	2	1	5	5	3	3	5	0	5	2	2	0	9	8	1
UFVJM-38	1	0	0	4	1	1	2	3	2	1	2	5	5	3	1	5	0	4	2	2	0	9	8	1
UFVJM-39	1	0	0	4	1	1	1	3	2	1	1	3	5	3	3	3	1	5	2	2	0	9	8	1
UFVJM-41	4	0	0	4	0	1	2	8	2	2	3	3	5	5	5	7	2	6	2	1	0	1	0	0
UFVJM-42	1	0	0	4	1	1	2	3	2	2	1	5	5	3	3	3	0	5	2	2	0	9	8	1
UFVJM-43	1	0	0	4	1	3	1	7	2	3	1	5	5	5	3	3	4	3	6	1	0	1	0	0
UFVJM-44	1	0	0	4	1	1	2	8	2	5	3	5	5	5	5	5	0	3	2	1	0	1	0	0
UFVJM-45	1	0	0	3	1	5	3	4	2	1	3	3	5	5	1	3	1	2	6	1	0	2	0	0
UFVJM-46	1	0	0	4	1	3	1	3	2	1	3	3	5	3	1	3	1	8	2	2	0	9	8	1
UFVJM-48	3	5	0	4	1	1	1	8	2	5	8	5	5	5	1	3	0	8	2	1	0	1	0	0
UFVJM-49	1	0	0	6	3	5	2	3	2	2	1	3	5	5	3	3	1	9	6	2	0	1	0	0
UFVJM-50	6	1	0	6	3	3	4	8	2	5	9	3	5	3	3	3	1	8	8	3	0	9	8	1
UFVJM-51	1	0	0	6	3	3	2	2	2	2	1	3	5	3	1	3	3	8	6	2	0	1	0	0
UFVJM-52	4	4	0	6	5	5	2	6	2	5	4	3	5	3	1	3	1	4	2	1	0	2	0	0
UFVJM-54	1	2	0	6	7	5	5	3	2	5	1	3	5	3	3	3	1	3	2	1	0	1	0	0
UFVJM-55	8	0	0	3	1	1	4	8	5	5	9	5	7	5	1	3	3	5	8	3	0	9	8	1
UFVJM-56	1	2	0	4	1	5	4	3	2	2	1	5	5	3	3	3	3	7	5	1	3	0	0	0
UFVJM-57	9	6	0	6	3	5	4	8	5	5	9	3	5	5	3	5	0	9	8	3	0	9	8	1
UFVJM-58	5	2	0	4	0	1	1	7	2	2	5	3	5	3	1	3	1	7	2	2	0	7	0	0

[1]Morphological descriptors: 1-predominant foliage color; 2-secondary color of foliage; 3-pubescence point in foliage apex; 4-mature leaf general format; 5-foliage lobule types; 6-number of foliage numbers; 7-central lobule format; 8-pigmentation of veins; 9-mature leaf color; 10-immature leaf color; 11-petiole pigmentation; 12-petiole length; 13-mature leaf size; 14-internode diameter; 15-internode length; 16-principal foliage length; 17-root surface flaws; 18-format of roots; 19-predominant color of root bark; 20-intensity of root bark color; 21-secondary color of bark; 22-predominant color of pulp; 23-secondary color of pulp; 24-pulp color distribution. The description of each score is shown in the Material and Methods section.

Molecular characterization involved a screening with 32 primers (Jarret & Bowen, 1994; Buteler, Jarret, & La Bonte, 1999), of which 11 were selected and evaluated (Table 3). These primers were chosen because they had a high average number of evident bands and because they are polymorphic and present a high reproducibility. A total of 210 polymorphic bands were obtained with the 11 selected primers. The average number of polymorphic bands per primer was 19.09. The primer that had the most polymorphic bands was IBSSR02, which amplified 28 bands, while the least polymorphic primer was Ib-255, which amplified 10 polymorphic bands. Moulin et al. (2012) studied sweet potato and performed a screening with 32 primers, from which 18 were selected and evaluated.

Genetic dissimilarity estimated from multicategoric data ranged from 0.08 to 0.92, averaging 0.52 (Table 4). The most divergent genotypes was the pair of UFVJM-55 accessions and Tomba Carro 1; UFVJM-37 and UFVJM-21 were considered the most similar with 0.08 distance (Table 4). Similarly, Veasey et al. (2007) observed variations of 0.12 to 1 in the morphological characterization of sweet potato plants demonstrating large diversity in the traditional variety group in Vale do Ribeira. Longer distances obtained from molecular data ranged from 0.14 to 1.00 with a dissimilarity average of 0.76, indicating high genetic diversity. Through this analysis, it was determined that UFVJM-42 and Espanhola accessions and UFVJM-07 and UFVJM-44 accessions with a distance equal to 1.00 were the farthest. Moreover, the genotype pairs of UFVJM-26 and UFVJM-27 were the closest to each other, with a distance equal to 0.14 (Figure 1).

Table 3. List of SSR primers selected for DNA amplification of sweet potato genotypes (*Ipomoea batatas*) with its base sequences. UFVJM, Diamantina, Minas Gerais State, 2016.

Primer (SSR)	Sequence (5' → 3')	Number of Polymorphic Bands	Percentage of Polymorphic bands
Ib-242	F: GCGGAACGGACGAGAAAA R: ATGGCAGAGTGAAAATGGAACA	21	10,0
Ib-255	F: TGGGCATTCTCATATTTTGCT R: GCCACTCCAACAGCACATAA	10	4,8
Ib-316	F: CAAACGCACAACGCTGTC R: CGCGTCCCGCTTATTTAAC	14	6,7
Ib-318	F: AGAACGCATGGGCATTGA R: CCCACCGTGTAAGGAAATCA	22	10,5
IBSSR02	F: GTAACTGAAGTAGTTGGCGAGG R: CACTGTGGCTCAATGTAGGG	28	13,3
IBSSR14	F:CAAAACTCAGACGACCCATGATG R: CCCCGAGGTATTTATATCACACCC	26	12,4
IBSSR15	F: TGGAGCAGATGTTCCTGGAC R: GATTTGAAGCAGTTCATCC	16	7,6
IBSSR18	F: GATCTTGAATTAGCCCAC R: AGATGGATGACCGTATGC	19	9,0
IBSSR19	F: GCGAATCAAGTCTTTTGTCCAC R: GGGACTGTCCTTTGGGTATG	19	9,0
IBSSR21	F: AAACAACCAACGGGTCTTTGC R: CTCTAGGGTCGCCATAAAAATCAC	17	8,1
IBSSR27	F: GTGTTTATCACATCGTTTTCTG R: GGCTCGTACAATTTTCAAAG	18	8,6
Total		210	

*F = Forward; R = Revers

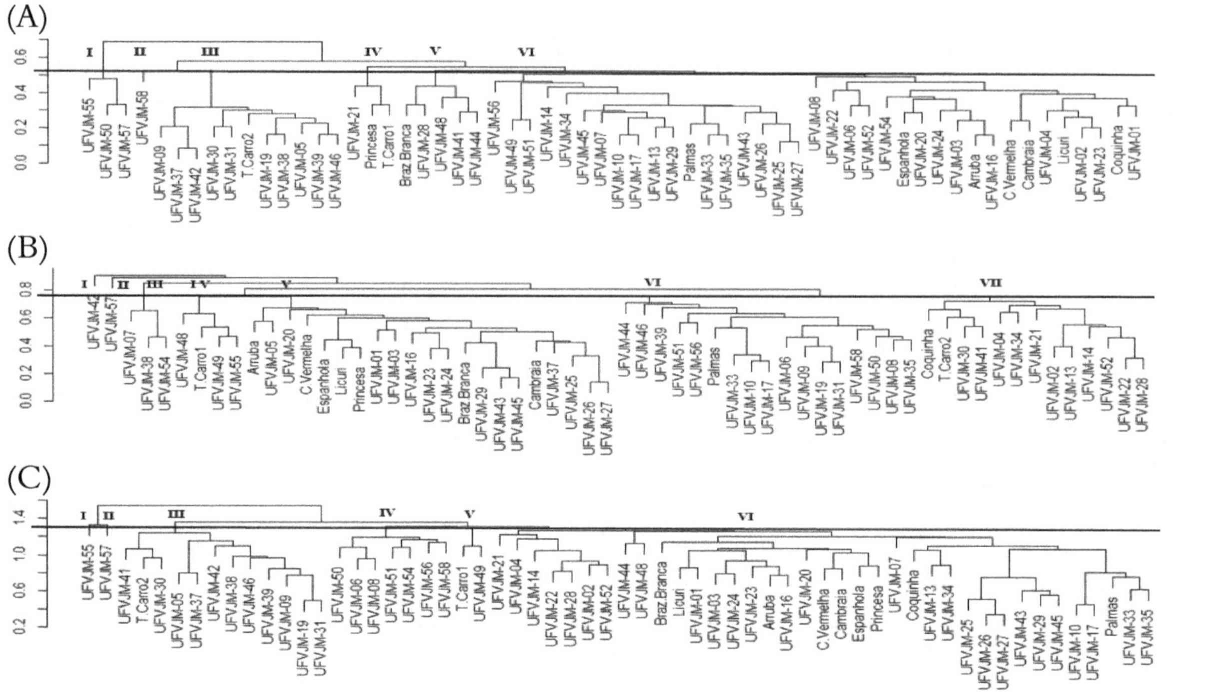

Figure 1. Dendrograms obtained by UPGMA method from dissimilarity matrices based on: multicategoric (A), molecular (B) and combined data (multicategoric + SSR molecular) (C), among 60 sweet potato genotypes (*Ipomoea batatas*). UFVJM, Diamantina, Minas Gerais State, 2016.

Cophenetic correlation estimates larger than 0.70 were found for all dissimilarity matrices considering the UPGMA dendrogram (Table 4). This indicated the viability of using dendrograms to summarize the information of dissimilarity matrices (Cruz et al., 2012). For multicategoric data, a cutoff equal to an average dissimilarity of 0.52 was applied, which allowed the formation of six groups (Figure 1). Group I (UFVJM-55, UFVJM-50, UFVJM-57) is characterized by having a semi-elliptical shape of the central lobe, petiole completely or almost purple, root bark predominantly purple-red with an intense dark colour. Group II consisted only of a UFVJM-58 accession. This differed from other groups by being the only one with pigmentation of green and purple spots throughout the petiole and predominantly orange brown pulp. Group III met eleven hits that generally had triangular-shaped mature leaves, thin diameter of the internode, predominantly cream-coloured root bark, and strong purple with absence of colour distribution in pulp. Group IV (UFVJM-21, Princesa, Tomba Carro 1) was formed by accessions characterized by having five lobes dispersed in the sheet length of the stem and no defects on the surface of roots. Group V was composed of five hits (Brazlândia Branca, UFVJM-28, UFVJM-48, UFVJM-41, UFVJM-44) that had completely purple veins, median mature leaf size, and cream-coloured bark and pulp. Group VI gathered the largest number of accessions, 37, which was 62% of the total. This group was characterized by the absence of colour distribution in pulp and secondary bark and pulp colour.

For molecular data, the grouping of accessions, which are represented by dendrogram (Figure 1), resulted in seven groups based on a cutoff equal to the average dissimilarity between all genotype pairs (0.76). Groups I and II were formed by only one accession each. Accession UFVJM-42 (group I) showed 19 specific polymorphic bands (9% of total). Accession UFVJM-57 (group II) amplified only 10 specific polymorphic bands, which corresponded to 4.8% of total. Group III gathered three accessions: UFVJM-38, UFVJM-54, UFVJM-07, which shared 8 common polymorphic bands. Group IV was formed by four accessions: UFVJM-48, Tomba Carro 1, UFVJM-49, and UFVJM-55, which showed 20 polymorphic bands for all primers used. Group V was formed by the largest number of accessions, totalling 21, which was 35% of studied accessions. Thirty-four polymorphic bands were observed for most accessions, which was more than 16% of the total polymorphic bands. Group VI was formed by 17 accessions, and most accessions present in this group showed 29 similar polymorphic fragments. Group VII gathered 13 accessions (Coquinho, Tomba Carro 2, UFVJM-30, UFVJM-41,

UFVJM-04, UFVJM-34, UFVJM-21, UFVJM-02, UFVJM-13, UFVJM-14, UFVJM-52, UFVJM-22, and UFVJM-28). In this group, there were 17 similar polymorphic fragments.

Based on molecular data, most accessions had a dissimilarity distinct from that represented by morphological characteristics. This dissimilarity demonstrates the importance of complementation of multicategoric data with molecular data. Genetic dissimilarity estimated by combined analysis (multicategoric + SSR molecular) varied from 0.35 to 1.8 with an average of 1.28 (Table 3). Accessions UFVJM-55 and UFVJM-21 had the largest dissimilarity. Accessions UFVJM-55 and UFVJM-21 had distances of 1.80, while accessions UFVJM-26 and UFVJM-27 were the most similar among each other with a distance of 0.35 (Figure 1).

The correlation between the distance matrix that was obtained with multicategoric data and the matrix of SSR marker was 0.133, which indicates a low association between these two methodologies (Table 4). Similarly, Maric, Bolarić, Martincic, Pejić, and Kozumplik (2004) reported a low correlation (r = 0.12) using RAPD markers and 12 morphologic characters in wheat. The correlation coefficient among combined and multicategoric matrices (r= 0.764) was similar to that of combined and molecular matrices (r = 0.74). Both correlations were significant, p = 0.01, by the Mantel test (Table 4).

Table 4. Descriptive statistics, correlations among dissimilarity matrices estimated and correlation cophenetic from multicategoric, molecular and combined data (multicategoric + SSR molecular), in genetic divergence study on 60 sweet potato genotypes (*Ipomoea batatas*). UFVJM, Diamantina, Minas Gerais State, 2016.

Matrices	Descriptive analysis		
	Minimum	Maximum	Average
$D_{multicategoric}$	0.08	0.92	0.52
$D_{molecular}$	0.14	1.00	0.76
$D_{(mult.+molec.)}$	0.35	1.80	1.28
Matrices	Correlation coefficient		
	$D_{multicategoric}$	$D_{molecular}$	$D_{(mult.+molec.)}$
$D_{multicategoric}$	-	0.13	0.76
$D_{molecular}$	**	-	0.74
$D_{(mult.+molec.)}$	**	**	-
Matrices	Correlation cophenetic		
$D_{multicategoric}$	0.77**		
$D_{molecular}$	0.87**		
$D_{(mult.+molec.)}$	0.72**		

**Significant correlation at 1% probability by Mantel test with 1.000 permutations.

For combined data (multicategoric + molecular), a cutoff equal to the average dissimilarity (1.28) was applied, which allowed the formation of seven groups (Figure 1). The use of multivariate techniques that simultaneously analyse morphologic and molecular data has been increasing in studies focused on genetic divergence among accessions of germplasm banks (Machado, Jesus, & Ledo, 2015). Groups I and II were formed by a single accession each. Access UFVJM-55 (group I) was the only one that had predominantly purple foliage. Fifty-three specific bands were observed, and they were polymorphic for all oligonucleotide primers. Accession UFVJM-57 (group II) was the only one with a secondary colour of foliage with reddish nodes, and the only one that only amplified 10 specific polymorphic bands, which corresponds to 4.8% of the total. Group III was formed by 12 accessions with an absence of secondary colour of foliage, absence of pubescence point in the apex, triangular mature leaf with average size, predominantly cream-coloured root bark and absence of secondary colour of bark. In this group, seven polymorphic bands were observed for most accessions.

Group IV was formed by seven accessions (Figure 1) and reunited genotypes with absence of pubescence points in foliage apex, mature leaves with average size and fine diameter of internode. This group was not polymorphic to primers Ib-255 and Ib316, and eight polymorphic fragments were observed. Group V was formed by only two accessions: Tomba Carro 1 and UFVJM-49. These genotypes were grouped by characteristics such as five leaf lobes, green-coloured mature leaf, petiole and internode with short length, predominantly white pulp, and absence of a secondary colour in bark and pulp. In this group, 31 polymorphic bands were observed. Group VI gathered the largest number of accessions, totalling 37, which was 61.7% of studied accessions. Cultivars from Embrapa, Brazilândia Branca, Princesa Coquinho, and Palmas were in this group.

The importance of using molecular markers is evident by the large differences that SSR markers have revealed, within groups that were initially formed based only on multicategoric descriptors. The possibility to join multicategoric and molecular data provides reproducibility in analysis, demonstrating that the entire information set can be used.

Conclusion

There is large amount of genetic variability among sweet potato genotypes of the germplasm bank at UFVJM; this variability is detected by both morphological and molecular characterizations. There was no strong association between morphological and molecular analyses of sweet potato accessions. There was no duplicate suspicion in dissimilarity studies. Access pairs UFVJM-55 and Tomba Carro 1, UFVJM-55, and UFVJM-21 of multicategoric data and accession pairs UFVJM-42 and Espanhola, UFVJM-07, and UFVJM-44 for molecular data are the most dissimilar. These pairs can be used as genitors to compose breeding programs and can provide progenies with high genetic variability.

Acknowledgements

The authors thank CAPES, CNPq, and FAPEMIG for the financial support and for conceding scholarships to the project development.

References

Alves, A. A., Bhering, L. L., Rosado, T. B., Laviola, B. G., Formighieri, E. F., & Cruz, C. D. (2013). Joint analysis of phenotypic and molecular diversity provides new insights on the genetic variability of the Brazilian physic nut germplasm bank. *Genetics and Molecular Biology, 36*(3), 371-381.

Andrade Júnior, V. C., Viana, D. J. S., Fernandes, J. S. C., Figueiredo, J. A., Nunes, U. R., & Neiva, I. P. (2009). Selection of sweet potato clones for the region Alto Vale do Jequitinhonha. *Horticultura Brasileira, 27*(3), 389-393.

Azevedo, A. M., Andrade Júnior, V. C., Fernandes, J. S. C., Pedrosa, C. E., & Oliveira, C. M. (2015). Desempenho agronômico e parâmetros genéticos em genótipos de batata-doce. *Horticultura Brasileira, 33*(1), 84-90.

Azevedo, A. M., Andrade Júnior, V. C., Viana, D. J. S., Elsayed, A. Y. A. M., Pedrosa, C. E., Neiva, I. P., & Figueiredo, J. A. (2014). Influence of harvest time and cultivation sites on the productivity and quality of sweet potato. *Horticultura Brasileira, 32*(1), 21-27.

Buteler, M. I., Jarret, R. L., & La Bonte, D. R. (1999). Sequence characterization of microsatellites in diploid and polyploid *Ipomoea*. *Theoretical and Applied Genetics, 99*(1), 123-132.

Chávez, R., Gallo, P., Rossel, G., Reynoso, D., Leva, H., & Vera, N. (2006). Caracterizaión morfológica y molecular de genótipos mejorados de camote (*Ipomoea batatas* L.) para ecosistemas Árido-Salino-Bórico. *Revista Ciencia y Desarrollo, 8*(1), 84-115.

Cruz, C. D., Regazzi, A. J., & Carneiro, P. C. S. (2012). *Modelos biométricos aplicados ao melhoramento genético.* Viçosa, MG: UFV.

Dray, S., & Dufour, A. B. (2007). The ade4 package: implementing the duality diagram for ecologists. *Journal of Statistical Software, 22*(4), 1-20.

Filgueira, F. A. R. (2008). *Novo manual de horticultura: agrotecnologia moderna na produção e comercialização de produtos hortícolas.* Viçosa, MG: UFV.

Huamán, Z. (1991). *Descriptors for sweet potato.* Rome, IT: CIP, AVRDC, IBPGR.

Jaccard, P. (1901). Étude comparative de la distribuition florale dans une portion des Alpes et des Jura. *Bulletin de la Societé Voudoise des Sciencies Natureller, 37*(142), 547-579.

Jarret, R. L., & Bowen, N. (1994). Simple Sequence Repeats (SSRs) for sweetpotato germplasm characterization. *Plant Genetic Resources Newsletter, 100*(1), 9-11.

Koussao, S., Gracen, V., Asante, I., Danquah, E. Y., Ouedraogo, J. T., Baptiste, T. J., ... Vianney, T. M. (2014). Diversity analysis of sweetpotato (*Ipomoea batatas* [L] Lam) germplasm from Burkina Faso using morphological and simple sequence repeats markers. *African Journal of Biotechnology, 13*(6), 729-742.

Machado, C. F., Jesus, F. N., & Ledo, C. A. S. (2015). Divergência genética de acessos de maracujá utilizando descritores quantitativos e qualitativos. *Revista Brasileira de Fruticultura, 37*(2), 442-449.

Maric, S., Bolaríc, S., Martincic, J., Pejić, I., & Kozumplik, V. (2004). Genetic diversity of hexaploid wheat cultivars estimated by RAPD markers, morphological traits and coefficients of parentage. *Plant Breeding, 123*(4), 366-369.

Monteiro, A. B. (2007). Silagens de cultivares e clones de batata-doce para alimentação animal visando sustentabilidade da produção agrícola familiar. *Revista Brasileira de Agroecologia, 2*(2), 978-981.

Moulin, M. M., Rodrigues, R., Gonçalves, L. S. A., Sudré, C. P., Santos, M. H., & Silva, J. R. P. (2012). Collection and morphological characterization of sweet potato landraces in north of Rio de Janeiro state. *Horticultura Brasileira, 30*(2), 286-292.

Oliveira, E. J., Pádua, J. G., Zucchi, M. I., Vencovsky, R., & Vieira, M. L. C. (2006). Origin, evolution and genome distribution of microsatellites. *Genetics and Molecular Biology, 29*(2), 294-307.

Oliveira, M. S. P., & Silva, K. J. D. (2008). Diferenciação genética entre procedências de açaizeiro por marcadores moleculares RAPD e SSR. *Revista Brasileira de Fruticultura, 30*(2), 438-443.

R Core Team (2014). *R: A language and environment for statistical computing.* Vienna, AU: R Foundation for Statistical Computing.

Roesler, P. V. S., Gomes, S. D., Moro, E., Kummer, A. C. B., & Cereda, M. P. (2008). Produção e qualidade de raiz tuberosa de cultivares de batata batata-doce no oeste do Paraná. *Acta Scientiarum. Agronomy, 30*(1), 117-122.

Souza, D. C. L. (2015). Técnicas moleculares para caracterização e conservação de plantas medicinais e aromáticas: uma revisão. *Revista Brasileira de Plantas Medicinais, 17*(3), 495-503.

Souza, G. A., Carvalho, M. R. O., Martins, E. R., Guedes, R. N. C., & Oliveira, L. O. (2008). Diversidade genética estimada com marcadores ISSR em populações brasileiras de *Zabrotes subfasciatus. Pesquisa Agropecuária Brasileira*, *43*(7), 843-849.

Veasey, E. A., Borges, A., Rosa, M. S., Queiroz-Silva, J. R., Bressan, E. A., & Peroni, N. (2008). Genetic diversity in Brazilian sweet potato (*Ipomoea batatas* (L.) Lam. Solanales, Convolvulaceae) landraces assessed with microsatellite markers. *Genetics and Molecular Biology, 31*(3), 725-733.

Veasey, E. A., Silva, J. R. Q., Rosa, M. S., Borges, A., Bressan, E. A., & Peroni, N. (2007). Phenology and morphological diversity of sweet potato (*Ipomoea batatas)* landraces of the Vale do Ribeira. *Scientia Agricola, 64*(4), 416-427.

20

Genetic divergence among maize hybrids and correlations with heterosis and combining ability

Rodrigo Oliboni, Marcos Ventura Faria*, Mikael Neumann, Guilherme Mendes Battistelli, Rafael Gallo Tegoni and Juliano Tadeu Vilela de Resende

*Departamento de Agronomia, Universidade Estadual do Centro Oeste, R. Simeão Camargo Varella de Sá, 3, Cx. Postal 3010, 85040-080, Guarapuava, Paraná, Brazil. *Author for correspondence. E-mail: mfaria@unicentro.com.br*

ABSTRACT. Twelve corn hybrids recommended for cropping in the south-central region of the Paraná State and 66 crossings obtained among these hybrids were evaluated for agronomic and morphological traits. These genotypes were evaluated in three experiments performed in Laranjeiras do Sul, Guarapuava and Cantagalo, Paraná State, Brazil. Heterosis and specific combining ability (SCA) were estimated for yield of husked ears. Data from measuring 22 morphological/agronomical traits of the parents were subjected to a multivariate joint analysis of variance and cluster analysis to group the parental hybrids via the neighbor method, using the generalized Mahalanobis distance. Pearson correlation coefficients between heterosis, SCA, D^2 and yield of husked ears were obtained. The cross P30F44 x Sprint displayed a high mean and a high heterosis for yield of husked ears, but a moderate estimate of genetic divergence. Estimates of genetic divergence were not effective at predicting the most heterotic crossings, as Pearson correlation coefficients between D^2 and heterosis and D^2 and CEC were not significant. Positive significant correlations were observed between yield means and CEC and heterosis.

Keywords: *Zea mays*, specific combining ability, Mahalanobis distance, plant breeding.

Introduction

The choice of germplasm is an essential and crucial step in any plant breeding program, whether for the development of varieties or to produce hybrids, and can determine the success or failure of the selection process (ARAÚJO; NASS, 2002).

The presence of genetic divergence among accessions of germplasm is essential; however, satisfactory results are obtained only if the germplasm employed in the cross also present high values for the traits of interest (CARVALHO et al., 2003).

In maize breeding programs, the choice of base populations with high potentials for grain yield is fundamental to the process of obtaining superior inbred lines. Accordingly, populations generated from commercial hybrids represent an interesting potential base population, as they have been previously tested in various environments. Thus, these hybrids generally present high yield and desirable agronomic characteristics, with large proportions of favorable loci previously selected (AMORIM; SOUZA, 2005). The use of populations with different heterotic patterns is a strategy that allows exploration and capitalization on heterosis to produce new corn hybrids (PINTO et al., 2001; VIEIRA et al., 2009).

Heterosis, expressed in crosses between individuals from different populations, depends on the presence of genes with non-additive effects in controlling desirable characteristics and the genetic divergence between them. Thus, in the choice of parents for these populations, divergent and previously adapted individuals are preferentially chosen (FUZATTO et al., 2002; GORGULHO; MIRANDA FILHO, 2001). In the case of any degree of dominance greater than zero, heterosis is due to differences in allele frequency between the parents (FALCONER; MACKAY, 1996).

The study of genetic divergence can assist in the choice of genotypes to be used in breeding programs for the development of new populations (CRUZ; REGAZZI, 1997). Genetic divergence is related to the degree of distance between populations in the set of genetic characters that differ between the populations. However, in most cases genetic distance is positively correlated with heterosis (PATERNIANI et al., 2008). Thus, the magnitude of heterosis is generally proportional to the genetic distance between the parents.

According Cargnelutti Filho et al. (2008), genetic divergence (evaluated based on the genetic distance between individuals) is a predictive feature that allows for the identification of crosses with a higher probability of success. Specifically, the evaluation of divergence can identify those crosses that will optimize heterosis while avoiding undesirable features.

In the prediction of genetic divergence between genotypes, multivariate methods, such as principal component analysis, canonical variables and agglomerative methods, can be applied. Agglomerative methods differ from the commonly utilized approaches, primarily because they rely on previously estimated dissimilarity measures, such as the Mahalanobis distance (CRUZ; REGAZZI, 1997).

The goal of this study was to evaluate the genetic divergence among 12 commercial corn hybrids, as well as the heterosis and specific combining the ability of the 66 crosses between these hybrids. Further, we aimed to evaluate the efficiency of the estimates of genetic divergence in predicting the most heterotic crosses between the parental hybrids.

Material and methods

Twelve commercial corn hybrids (P30F53, P30F44, AG8021, GNZ2005, GNZ2004, Penta, Premium Flex, Sprint, AS1575, 2B587, 2B688 and DKB234) were intercrossed with each other. The 66 crosses obtained, together with the 12 parental hybrids and three checks (P30R50, AS1560 and DKB214) were evaluated by a randomized block design with three replicates in three experiments conducted at different locations in the central-south region of Paraná State during the 2007/2008 harvest.

The first experiment was carried out in Laranjeiras do Sul, located at latitude 25°24'15"S and longitude 52°28'22"O, with an altitude of 790 meters and soil classified as Typic Eutrophic (EMBRAPA, 2006). The second experiment was conducted in Guarapuava, located at latitude 25°23'02"S and longitude 51°29'43"O, an altitude of 1026 meters, with soil classified as Distroferric Oxisoil (EMBRAPA, 2006). The third experiment was conducted in Cantagalo, located at latitude 25°22'28"S and longitude 52°07'35"O, at an altitude of 790 meters with soil classified as Typic Eutrophic (EMBRAPA, 2006).

Each plot consisted of two rows of 5 meters in length, spaced at 0.80 meters, with a total area of 8 m^2 with 5 plants per linear meter and a population equivalent to 62,500 plants ha^{-1}.

To obtain the estimates of heterosis of the 66 crossings, five agronomic traits were assessed at each location: male flowering (MF) - number of days, counted from sowing to the issuance of 50% mature tassels in the plot; prolificacy (PR) - the ratio between the number of ears per plot and the final stand; plant height (PH) and ear insertion height (EH) in meters, sampled from five competitive plants from each plot; yield of husked ears (YHE) t ha^{-1}, with 13% moisture and stand corrected by covariance (RAMALHO et al., 2000), considering the ideal stand as 50 plants per plot. Heterosis values were calculated based on average data from three locations: $H = F_1 - (MP_i + MP_j) / 2$, where H is heterosis, F_1 is the mean of the F_1 generation, and MP is the mean of parental P_i or P_j. The heterosis estimates were also calculated as a percentage and the significance was determined with a *t* test (FUZATTO et al., 2002).

For the assessment of genetic diversity among the 12 parental hybrids (in addition to the five characteristics mentioned) 20 other morphological characteristics were evaluated from five competitive plants per plot in experiments at Laranjeiras do Sul and Guarapuava: total number of internodes (NI); length of the internode below the ear (IL); diameter of the stalk at the internode below the ear (SD); leaf length at ear height (LL); leaf width at ear height (LW); number of leaves below the ear (NLB); number of leaves above the ear (NLA); angle between the leaf (at ear height) and the stalk (AL) - scores (1 = 20°, 2 = 30°, 3 = 40°, 4 = 50° and 5 ≥ 60°); length of the main stem of the tassel (LMS); length of the secondary stem of the first lateral branch of the tassel (LSS); number of secondary stems of the tassel (NSS); angle between the main and secondary stem of the tassel (ASS) - scores (1 = 20°, 2 = 30°, 3 = 40°, 4 = 50° e 5 ≥ 60°); ear placement (EP) - score of the angle between the ear and the stalk (1 = ear parallel to stalk , 2 = 30°, 3 = 45°, 4 = 60° and 5 = ear fully decumbent in relation to stalk); husk coverage of the ear (HC) - scores from 1 to 5 (1 = ear with excellent husk coverage; 5 = ear with very bad husk coverage); number of rows of kernels on the ear (NRK); number of kernels per row (NKR); length of the ear (EL); diameter of the ear (ED); diameter of the cob (CD); color of the cob (CC) - scores (1 = purple, 2 = dark red, 3 = clear red, 4 = beige and 5 = white).

Data were subjected to analysis of variance for each individual location, and after confirmation of homoscedasticity by the Hartley test (RAMALHO et al., 2000), joint analysis of variance was performed.

Data on morphological and agronomic characteristics obtained from the 12 parental hybrids were subjected to a multivariate joint analysis of variance to obtain the matrices of the sum of squares and sum of products of the error. The dissimilarity methodology was used to form the groups, by the method of single linkage cluster of the nearest neighbor (JOHNSON; WICKERN, 1988) from the dissimilarity matrix, based on the Mahalanobis distance (D^2).

To establish the cut-off point in the dendrogram for the identification of groups, the procedure used by Marchioro et al. (2003) was employed. However, as the number of genotypes (12) was smaller than the number of traits (22), the SAS routine (SAS, 1988) was modified, replacing the matrix of variance and covariance for the parametric matrix. The D^2 (cutoff line) was obtained with 95% of probability using 5,000 simulations.

Estimates of specific combining ability SCA ($\hat{s}_{ij}$) were obtained according to Method 2 of Griffing's Model I. Pearson correlation coefficients (STEEL; TORRIE, 1980) were obtained between D^2, the estimates of heterosis, SCA ($\hat{s}_{ij}$) and the yield of husked ears (YHE) to check whether the genetic diversity, estimated by multivariate distances, could be used as a predictive factor to identify the most promising hybrid combinations. Analyses were performed using the statistical software GENES (CRUZ, 2006).

Results and discussion

The values of heterosis for male flowering (MF) were positive for all crosses, indicating the absence of bidirectional dominance deviations. Among the 58 crosses that showed significant heterosis by *t* test, estimates ranged from 1.5 days (1.9%) to 7.5 days (10.1%) for the cross Premium Flex x Sprint (data not shown).

In the case of prolificacy (PR), only 14 crossings showed statistically significant heterosis values, ranging from -0.08 (-6.9%) to 0.15 (15.5%) (data not shown). These data indicate that expression of bidirectional dominance deviations. The positive estimates of the greatest values were presented by the crosses GNZ2005 x Sprint and GNZ2005 x Penta, which demonstrated a desirable increase in the average number of ears per plant.

Reduction in plant height and ear insertion height are desirable traits, and nine estimates of heterosis presented significant negative values for both for plant height (PH) and ear insertion height (EH) (data not shown). The cross P30F44 x 2B688 displayed the most negative heterosis values for PH (-0.15 m or -6.3%) and EH (-0.17 m or -12.3%).

Estimates of heterosis for yield of husked ears (YHE) are presented in Table 1. Only one significant negative estimate of heterosis of YHE was identified, the cross Premium Flex x Sprint (-2.22 t ha^{-1} or -21.2%).

All estimates of heterosis from crosses involving both parents from the same breeding program (P30F53 x P30F44, GNZ2005 x GNZ2004, Penta x Premium Flex, Premium Flex x Sprint, Penta x Sprint and 2B587 x 2B688) were negative or equivalent to zero (Table 1). This is likely the result of low genetic complementarity of loci with non-additive effects, possibly because these crosses displayed some degree of parental relationship. Pfann et al. (2009) identified negative estimates of specific combining ability (SCA) for grain yield for the cross Penta x Premium Flex in diallelic evaluation of two locations in the central-south region of Paraná State (Guarapuava and Goioxim). These data agree with the findings of this study, as SCA and heterosis are related parameters.

Significant positive estimates of heterosis for YHE were identified for 38 crosses, ranging from 1.18 (9.3%) for the cross AS1575 x 2B688 to 3.76 t ha^{-1} (35.5%) for the combination P30F44 x Sprint (Table 1). Hallauer and Miranda Filho (1981) and Falconer and Mackay (1996) argue that in addition to the existence of genes with some degree of dominance controlling the character, the expression of heterosis also depends on the divergence between genotypes, as differences in allele frequencies are required at loci involved in the expression of desirable characteristics.

The mean values for the traits obtained from the joint analysis of variance involving the two locations (Laranjeiras do Sul and Guarapuava) were used to evaluate genetic divergence between the 12 parental hybrids, since no significant 'genotypes x locations' interaction were identified for most traits (Table 2).

Table 1. Mean estimates of heterosis of yield of husked ears (YHE) of the 66 crosses obtained from 12 corn hybrids evaluated in three locations in the South-Central region of the Paraná State, Brazil.

Cross	Heterosis YHE		Cross	Heterosis YHE	
	t ha^{-1}	%		t ha^{-1}	%
P30F53 x P30F44	-0.39^{ns}	-3.2	GNZ2005 x Sprint	2.60*	23.3
P30F53 x AG8021	0.26^{ns}	2.2	GNZ2005 x AS1575	1.40*	12.5
P30F53 x GNZ2005	1.28*	11.4	GNZ2005 x 2B587	1.28*	11.6
P30F53 x GNZ2004	1.49*	12.8	GNZ 2005 x 2B688	0.21^{ns}	2.1
P30F53 x Penta	1.95*	16.6	GNZ2005 x DKB234	1.79*	16.7
P30F53 x Premium Flex	0.44^{ns}	3.7	GNZ2004 x Penta	2.44*	22.5
P30F53 x Sprint	2.37*	21.1	GNZ2004 x Premium Flex	2.39*	23.4
P30F53 x AS1575	0.73^{ns}	5.8	GNZ2004 x Sprint	1.68*	14.5
P30F53 x 2B587	-0.24^{ns}	-1.9	GNZ2004 x AS1575	1.28*	11.0
P30F53 x 2B688	-0.01^{ns}	-0.1	GNZ2004 x 2B587	1.56*	13.3
P30F53 x DKB234	0.41^{ns}	3.6	GNZ2004 x 2B688	1.43*	13.8
P30F44 x AG8021	1.28*	11.7	GNZ2004 x DKB234	1.32*	12.0
P30F44 x GNZ2005	1.79*	16.9	Penta x Premium Flex	0.29^{ns}	2.6
P30F44 x GNZ2004	1.70*	15.5	Penta x Sprint	-0.59^{ns}	-5.7
P30F44 x Penta	0.89^{ns}	8.0	Penta x A 1575	0.40^{ns}	3.4
P30F44 x Premium Flex	1.96*	17.4	Penta x 2B587	0.89^{ns}	7.6
P30F44 x Sprint	3.76*	35.5	Penta x 2B688	1.58*	13.4
P30F44 x AS1575	1.30*	10.9	Penta x DKB234	1.36*	13.0
P30F44 x 2B587	0.67^{ns}	5.6	Premium Flex x Sprint	-2.22*	-21.2
P30F44 x 2B688	1.59*	13.1	Premium Flex x AS1575	0.33^{ns}	2.8
P30F44 x DKB234	1.93*	18.0	Premium Flex x 2B587	1.58*	13.3
AG8021 x GNZ2005	2.60*	25.6	Premium Flex x 2B688	0.99^{ns}	8.3
AG8021 x GNZ2004	1.57*	14.9	Premium Flex x DKB234	0.68^{ns}	6.4
AG8021 x Penta	1.51*	14.2	Sprint x 1575	1.09^{ns}	9.8
AG8021 x Premium Flex	1.98*	18.3	Sprint x 2B587	1.85*	16.5
AG8021 x Sprint	2.15*	21.1	Sprint x 2B688	1.04^{ns}	9.2
AG8021 x AS1575	0.88^{ns}	7.6	Sprint x DKB234	1.85*	18.6
AG8021 x 2B587	0.69^{ns}	5.9	AS1575 x 2B587	0.80^{ns}	6.4
AG8021 x 2B688	0.10^{ns}	0.8	AS1575 x 2B688	1.18*	9.3
AG 8021 x DKB234	1.54*	14.9	AS1575 x DKB234	0.79^{ns}	7.0
GNZ2005 x GNZ2004	0.06^{ns}	0.6	2B587 x 2B688	-0.87^{ns}	-6.9
GNZ2005 x Penta	2.24*	21.5	2B587 x DKB234	0.90^{ns}	7.9
GNZ2005 x Premium Flex	1.95*	19.9	2B688 x DKB234	1.90*	16.6

*, ^{ns}: significant and not significant at 5% de probability by *t* test.

Table 2. Summary of joint analysis of variance concerning the morphological/agronomical traits evaluated in Laranjeiras do Sul and Guarapuava to study the genetic divergence among the 12 parental corn hybrids.

Trait	QM				
	Genotype	Genotypes x locations	Error	Mean	CV (%)
MF	21.6515*	16.1212^{ns}	8.0492	75.41	3.76
PR	0.0343*	0.0061^{ns}	0.0054	1.05	6.97
YHE	9.9931*	2.8629 *	2.8629	11.70	14.45
PH	0.0558^{ns}	0.0162^{ns}	0.0140	2.48	4.77
EH	0.0728*	0.0104^{ns}	0.0081	1.49	6.00
NI	3.3000*	0.3731^{ns}	0.2270	15.31	3.11
IL	9.9797*	0.7057^{ns}	0.6717	16.51	4.96
SD	0.0843*	0.0232^{ns}	0.0343	1.29	14.27
LL	0.0527^{ns}	0.0328^{ns}	0.0683	0.95	27.45
LW	1.9227^{ns}	0.8167^{ns}	1.2828	10.20	11.10
NLB	1.4634*	0.0840^{ns}	0.0817	6.08	4.69
NLA	2.1513*	0.3436^{ns}	0.1824	9.22	4.62
LA	0.5751*	0.2320^{ns}	0.3420	2.93	19.91
LMS	62.4149*	7.6839^{ns}	5.4761	42.60	5.49
LSS	25.0825*	8.4534^{ns}	5.7666	21.32	11.26
NSS	1.2357*	0.2284^{ns}	0.1191	1.41	24.32
ASS	28.5503*	3.0303^{ns}	1.6389	9.50	13.47
EP	0.8001*	0.4138^{ns}	0.1305	2.50	14.40
HC	1.7301*	0.6866^{ns}	0.2886	1.95	27.47
NRK	19.9589*	0.4304*	0.6504	15.85	5.08
NKR	23.4744*	7.0256*	4.2183	35.94	5.71
EL	4.7888*	2.2873^{ns}	1.1500	17.47	6.13
ED	0.8935*	0.0398^{ns}	0.0230	4.50	3.36
CD	0.4627*	0.0168*	0.0070	2.33	3.58
CC	0.7254*	0.0454^{ns}	0.0985	4.51	6.94

MF = male flowering, PR = prolificacy, YHE = yield of husked ears, PH = plant height, EH = ear insertion height, NI = number of internodes, IL= internode length, SD = diameter of the stalk, LL = Leaf length, LW = leaf width, NLB = number of leaves below the ear, NLA = number of leaves above the ear, LA = angle of the leaf, LMS = length of main stem of the tassel, LSS = length of secondary stem of the tassel, NSS = number of secondary stems of the tassel, ASS = angle of the secondary stem of the tassel, EP = ear placement, HC = husk coverage of the ear = NRK = number of rows of kernels on the ear, NKR = number of kernels per row, EL = length of the ear, ED = diameter of the ear, CD = diameter of the cob, CC = color of the cob. *, ^{ns}: significant and not significant at 5% probability by F test.

The analysis of variance revealed no significant differences for plant height (PH), leaf length (LL) or width of leaves (WL), indicating that there are no genetic differences between the genotypes for these traits (Table 2). Thus, these traits were disregarded in the assessment of genetic diversity.

The amplitude of the generalized Mahalanobis distances (D^2) ranged from 406.7 for the most divergent pair of hybrids (DKB234 and 2B688) to 26.8 between the least divergent hybrids (Penta and Premium Flex) (Table 3). These data are supported by the findings of by Pfann et al. (2009). Larger magnitudes of divergence, in descending order, were found between the pairs AG8021 and Sprint, GNZ2005 and DKB234, Sprint and DKB234 and the pair GNZ2004 and 2B688. The lowest estimates of divergence were observed for the pairs P30F44 and Penta, Premium Flex and 2B587, followed by P30F44 and Premium Flex.

Miranda et al. (2003) reported that estimates of the generalized Mahalanobis distances (D^2) clearly indicate that the pairs of popcorn cultivars are more divergent and more similar genetically. Fonseca and Silva (1999) reported the effectiveness of the cluster technique, with measures of genetic diversity represented by the Mahalanobis distance, for the identification of duplicate accessions of beans. This method has also been applied to several studies with corn (FERREIRA et al., 1995; FUZATTO et al., 2002; MELO et al., 2001).

The cross between the most divergent hybrids (2B688 x DKB234) (Table 3) showed an increase of 1.90 t ha^{-1} over the average of these two parental hybrids, representing a heterosis of 16.6% (Table 1). Moreover, the cross between hybrids P30F44 and Sprint, which presented a low estimate of genetic divergence (Table 3), displayed a YHE of 14.34 t ha^{-1}, which represents an increase of 3.76 t ha^{-1} over the parental average, representing for heterosis of 35.5% (Table 1). These data indicate that estimates of genetic divergence may not be enough to reveal which combinations of genotypes can be successfully used for breeding. Our data suggest that the means of yield and heterosis should be considered jointly. According to Hallauer and Miranda Filho (1981), it is preferable to cross genotypes with high yield that display intermediate genetic divergence than to cross genotypes with intermediate yield and wide divergence.

We performed a cluster analysis based on multivariate distances from the dissimilarity using the nearest neighbor method. The dendrogram from this analysis revealed the formation of five groups (Figure 1) when adopting a cutoff of 67.73 in the distance scale with 95% probability, according to Marchioro et al. (2003).

According to the magnitude of divergence, the five groups discriminated: group I, formed by the hybrid DKB234, group II formed by the hybrid AG8021, group III by 2B688, group IV by Spirnt and group V by Penta, Premium Flex, P30F44, 2B587, P30F53, GNZ005, GNZ2004 and AS1575, a group with an internal divergence lower than the cutoff level.

The interaction between SCA and location was not significant for YHE (data not shown). Thus, estimates of the SCA ($\hat{s}_{ij}$) are presented as the average of three locations and the respective means of the YHE of the crosses (Table 4). The Pearson correlation coefficient demonstrated a significant ($p < 0.05$) relationship between $\hat{s}_{ij}$ and heterosis (0.885), YHE and $\hat{s}_{ij}$ (0.818) and between heterosis and YHE (0.646) (Table 5).

Moreover, the estimates of correlations between D^2 and all other parameters were not significant ($p > 0.05$), indicating poor predictability between multivariate parameters and YHE behavior of the crosses. Ferreira et al. (1995) also obtained low estimates for correlations between D^2 and other parameters in maize. Fuzatto et al. (2002) found low correlations between the estimates of D^2 and $\hat{s}_{ij}$ (0.051), D^2 and heterosis (0.398) and D^2 and grain yield (0.082).

Table 3. Estimates of genetic dissimilarity between pairs of the 12 parental hybrids based on 22 morphological characteristics based on Mahalanobis distance (D^2).

Hybrid	P30F44	AG8021	G2005	G2004	Penta	Premiun Flex	Sprint	AS1575	2B587	2B688	DKB234
P30F53	65.5	141.2	136.4	96.8	84.2	92.9	107.4	141.9	100.1	130.7	162.9
P30F44		148.6	106.5	141.6	53.0	58.3	90.3	168.5	71.9	165.8	201.1
AG8021			222.8	132.8	180.0	157.4	321.8	191.9	117.9	245.3	129.4
G2005				160.2	64.7	80.7	98.7	128.2	99.1	151.7	321.6
G2004					174.5	169.1	218.8	177.3	141.7	256.3	158.7
Penta						26.8	92.3	75.7	87.9	127.8	237.0
P. Flex							127.2	105.0	54.2	112.2	253.4
Sprint								193.7	139.7	183.0	299.2
AS1575									134.9	217.5	254.4
2B587										150.0	192.8
2B688											406.7

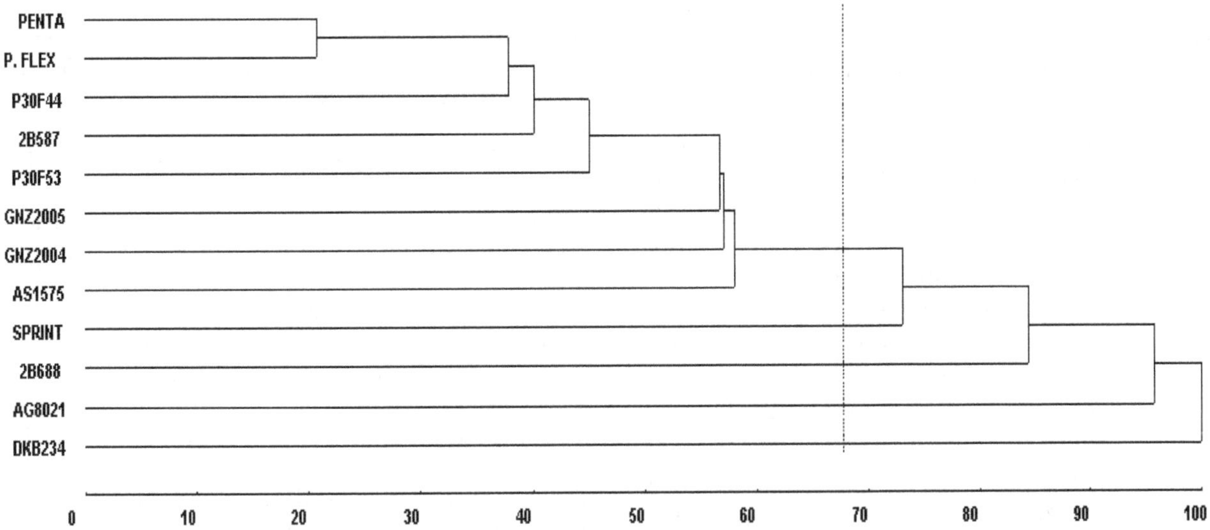

Figure 1. Dendrogram illustrating the pattern of dissimilarity, established through a single linkage obtained from the cluster analysis via the nearest neighbor method, based on the Mahalanobis distance (D^2) of 22 morphological traits among the 12 parental hybrids.

Table 4. Mean estimates of specific combining ability ($\hat{s}_{ij}$) (up diagonal) and yield of husked ears (t ha^{-1}) (down diagonal) of the 66 crosses obtained from the 12 parental corn hybrids evaluated in three locations in the South-Central region of the Paraná State, Brazil.

	P30F53	P30F44	AG8021	GNZ 2005	GNZ 2004	Penta	Premium Flex	Sprint	AS175	2B587	2B688	DKB 234
P30F53	**-0,352**[1]	-1.389	-0.540	0.368	0.632	1.200	0.069	1.498	0.248	-0.575	-0.426	-0.382
P30F44	11.59 a	**-1,650**	-0.165	0.231	0.198	0.379	0.934	2.235	0.167	-0.317	0.530	0.496
AG 8021	11.81 a	12.20 a	**-1,248**	1.240	0.264	0.309	1.155	0.829	-0.051	-0.087	-0.762	0.302
GNZ2005	12.47 a	12.35 a	12.73 a	**-1,479**	-1.362	0.920	0.998	1.166	0.347	-0.622	-0.760	0.434
GNZ2004	13.07 a	12.65 a	12.06 a	10.22b	**-1,356**	1.291	-0.469	0.304	0.297	0.716	0.517	0.023
Penta	13.64 a	12.84 a	12.14 a	12.51a	13.21a	**-1,162**	-0.491	-1.870	-0.487	-0.146	0.761	0.165
P. Flex	12.29 a	13.18 a	12.77 a	12.37a	11.53b	11.22b	**-0,407**	-3.112	-0.192	1.208	0.553	-0.138
Sprint	13.58 a	14.34 a	12.30 a	12.39a	11.86a	9.70b	8.23b	**-1,395**	0.090	0.992	0.113	0.543
AS1575	13.28 a	13.22 a	12.37 a	12.52a	12.80a	12.03a	12.11a	12.24a	**-0,625**	0.328	0.640	-0.137
2B587	12.35 a	12.63 a	12.22 a	11.44b	13.12a	12.56a	13.40a	13.04a	13.32 a	**-0,324**	-1.265	0.124
2B688	12.68 a	13.65 a	11.73 a	11.48b	13.09a	13.35a	12.92a	12.33a	13.81 a	11.80a	**-0,471**	1.039
DKB234	11.75 a	12.64 a	11.82 a	11.70a	11.63a	11.79a	11.25b	11.79a	12.06 a	12.21a	31.31a	**-1,235**

[1]Diagonal = ($\hat{s}_{ii}$). Means followed by same letters do not differ by Scott Knott test at 5% of probability.

Table 5. Coefficients of linear Pearson correlations obtained from estimates of means of yield of husked ears (YHE), specific combining ability (SCA, $\hat{s}_{ij}$), heterosis and generalized Mahalanobis distance (D^2).

	CEC ($\hat{s}_{ij}$)	Heterosis	(D^2)
YHE	0.818*	0.646*	-0.279
CEC ($\hat{s}_{ij}$)		0.885*	0.105
Heterosis			0.161

*Significant at 5% of probability by *t* test.

The literature asserts that heterosis and combining ability depend directly on genetic divergence between the parents and that there are more chances of generating promising combinations when more divergent genotypes are crossed (FERREIRA et al., 1995; PATERNIANI et al., 2008). However, in this study, the correlation between genetic divergence (D^2) and heterosis and between D^2 and $\hat{s}_{ij}$ were not significant (Table 5), contradicting the idea that to express high heterosis, it is necessary for the parental hybrids involved in the cross to be divergent. In fact, divergence of the parental is necessary for high levels of heterosis; however, divergence does not ensure high levels of heterosis, because heterosis depends not only on differences in allele frequencies but also on dominance and epistatic interactions that were not considered in this work.

The lack of correlation between $\hat{s}_{ij}$ and D^2 was confirmed by the crosses GNZ 2005 x Penta, Premium Flex x 2B587 and P30F53 x Penta, which displayed high positive estimates of $\hat{s}_{ij}$ for YHE (1.291, 1.208 and 1.200 t ha^{-1}, respectively) (Table 4) while belonging to the same group (Figure 1). On the other hand, crosses between divergent parents, classified into distinct groups (Figure 1), did not always present satisfactory $\hat{s}_{ij}$. For example, the crosses P30F53 x AG 8021, P30F53 x DKB 234, P30F53 x 2B688, AG 8021 x 2B688, GNZ 2005 x GNZ 2004, GNZ 2005 x 2B688, GNZ 2004 x Premium Flex and 2B587 x 2B688 displayed negative estimates of $\hat{s}_{ij}$ for YHE (Table 4). These results are in agreement with those obtained by Melo et al. (2001) and Paterniani et al. (2008), which indicated that crosses between some genotypes, even those with high estimates of divergence, do not guarantee high heterosis, resulting in high values of CEC.

Guimarães et al. (2007) reported non-significant correlation between mean yield and genetic distance of hybrids obtained by diallele crosses. Paterniani et al. (2008) reported that there was no correlation between heterosis, specific combining ability or yield of hybrids with genetic distance calculated by AFLP and SSR markers, indicating that it was not possible to infer the behavior of corn hybrids from the genetic divergence between the parental lines.

The results of our study show that genetic divergence is not sufficient to secure favorable heterosis and $\hat{s}_{ij}$. One hypothesis to explain this finding is the fact that when dealing with a quantitative characteristic, such as grain yield, one can encounter cases of bidirectional dominance, i.e. some loci are dominant in one direction and other loci are dominant in the opposite direction (FALCONER; MACKAY, 1996). Another hypothesis is that some characteristics used to estimate divergence did not significantly influence grain yield.

Conclusion

Estimates of genetic divergence are not always sufficient to reveal which combinations should be used in breeding crosses. Instead, data related to the mean yield and heterosis should be assessed together in evaluating potential crosses. Genetic divergence is necessary for heterosis, however it is not sufficient to ensure heterosis at a high level.

Acknowledgements

The authors wish to thank CNPq and Fundação Araucária for financial support and Capes for providing scholarships.

References

AMORIM, E. P.; SOUZA, J. C. Híbridos de milho inter e intrapopulacionais obtidos a partir de populações S0 de híbridos simples comerciais. **Bragantia**, v. 64, n. 3, p. 561-567, 2005.

ARAÚJO, P. M.; NASS, L. L. Caracterização e avaliação de populações de milho crioulo. **Scientia Agricola**, v. 59, n. 3, p. 589-593, 2002.

CARGNELUTTI FILHO, A.; RIBEIRO, N. D.; REIS, R. C. P.; SOUZA, J. R.; JOST, E. Comparação de métodos de agrupamento para o estudo da divergência genética em cultivares de feijão. **Ciência Rural**, v. 38, n. 8, p. 2138-2145, 2008.

CARVALHO, L. P.; LANZA, M. A.; FALLIERI, J.; SANTOS, J. W. Análise da diversidade genética entre acessos de banco ativo de germoplasma de algodão. **Pesquisa Agropecuária Brasileira**, v. 38, n. 10, p. 1149-1155, 2003.

CRUZ, C. D. **Programa genes**: biometria. Viçosa: UFV, 2006.

CRUZ, C. D.; REGAZZI, A. J. **Modelos biométricos aplicados ao melhoramento genético**. 2nd ed. Viçosa: UFV, 1997.

EMBRAPA-Empresa Brasileira de Pesquisa Agropecuária. **Sistema brasileiro de classificação de solos**. 2nd ed. Rio de Janeiro: Embrapa Solos, 2006.

FALCONER, D. S.; MACKAY, T. F. C. **Introduction to quantitative genetics**. Edinburgh: Addison Wesley Longman, 1996.

FERREIRA, D. F.; OLIVEIRA, A. C.; SANTOS, M. X.; RAMALHO, M. A. P. Métodos de avaliação da divergência genética em milho e suas relações com os cruzamentos dialélicos. **Pesquisa Agropecuária Brasileira**, v. 30, n. 9, p. 1189-1194, 1995.

FONSECA, J. R.; SILVA, H. T. Identificação de duplicidades de acessos de feijão por meio de técnicas multivariadas. **Pesquisa Agropecuária Brasileira**, v. 34, n. 3, p. 409-414, 1999.

FUZATTO, S. R.; FERREIRA, D. F.; RAMALHO, M. A. P.; RIBEIRO, P. H. E. Divergência genética e sua relação com os cruzamentos dialélicos na cultura do milho. **Ciência e Agrotecnologia**, v. 26, n. 1, p. 22-32, 2002.

GORGULHO, E. P.; MIRANDA FILHO, J. B. Estudo da capacidade combinatória de variedades de milho no esquema de cruzamento dialélico parcial. **Bragantia**, v. 60, n. 1, p. 1-8, 2001.

GUIMARÃES, P. S.; PATERNIANI, M. E. A. G. Z.; LÜDERS, R. R.; SOUZA, N. A. P.; LABORDA, P. R.; OLIVEIRA, K. M. Correlação da heterose de híbridos de milho com divergência genética entre linhagens. **Pesquisa Agropecuária Brasileira**, v. 42, n. 6, p. 811-816, 2007.

HALLAUER, A. R.; MIRANDA FILHO, J. B. **Quantitative genetics in maize breeding**. Ames: Iowa State University Press, 1981.

JOHNSON, R. A.; WICKERN, D. W. **Applied multivariate statistical analysis.** 2nd ed. New York: Englewood Cliffs; Prentice Hall, 1988.

MARCHIORO, V. S.; CARVALHO, F. I. F.; OLIVEIRA, A. C.; CRUZ, P. J.; LORENCETTI, C.; BENIN, G.; SILVA, J. A. G.; SCHMIDT, D. A. M.; Dissimilaridade genética entre genótipos de aveia. **Ciência e Agrotecnologia**, v. 27, n. 2, p. 285-294, 2003.

MELO, W. M. C.; VON PINHO, R. G.; FERREIRA, D. F. Capacidade de combinatória e divergência genética em híbridos comerciais de milho. **Ciência e Agrotecnologia**, v. 25, n. 4, p. 821-830, 2001.

MIRANDA, G. V.; COIMBRA, R. R.; GODOY, C. L.; SOUZA, L. V.; GUIMARÃES, L. J. M.; MELO, A. V. Potencial de melhoramento e divergência genética de cultivares de milho-pipoca. **Pesquisa Agropecuária Brasileira**, v. 38, n. 6, p. 681-688, 2003.

PATERNIANI. M. E. A. G. Z.; GUIMARÃES, P. S.; LÜDERS, R. R.; GALLO, P. B.; SOUZA, A. P.; LABORDA, P. R.; OLIVEIRA, K. M. Capacidade combinatória, divergência genética entre linhagens de milho e correlação com heterose. **Bragantia**, v. 67, n. 3, p. 639-648, 2008.

PFANN, A. Z.; FARIA, M. V.; ANDRADE, A. A.; NASCIMENTO, I. R.; FARIA, C. M. D. R.; BRINGHENTTI, R. M. Capacidade combinatória entre híbridos simples de milho em dialelo circulante. **Ciência Rural**, v. 39, n. 3, p. 128-134, 2009.

PINTO, R. M. C.; GARCIA, A. A. F.; SOUZA JÚNIOR, C. L. Alocação de linhagens de milho derivadas das populações BR-105 e BR-106 em grupos heteróticos. **Scientia Agricola**, v. 58, n. 3, p. 541-548, 2001.

RAMALHO, M. A. P.; FERREIRA, D. F.; OLIVEIRA, A. C. **Experimentação em genética e melhoramento de plantas**. Lavras: UFLA, 2000.

SAS-Statistical Analisys System. **SAS/STAT guide for personal computers**: statistics. 6th ed. Cary: Statistical Analysis System Institute, 1988.

STEEL, R. G. D.; TORRIE, J. H. **Principles and procedures of statistics**: a biometrical approach. 2nd ed. New York: McGraw-Hill, 1980.

VIEIRA, R. A.; SOUZA NETO, I. L.; BIGNOTTO, L. S.; CRUZ, C. D.; AMARAL JÚNIOR, A. T.; SCAPIM, C. A. Heterotic parametrization for economically important traits in popcorn. **Acta Scientiarum. Agronomy**, v. 31, n. 3, p. 411-419, 2009.

21

Prediction of genetic gains by selection indexes and REML/BLUP methodology in a population of sour passion fruit under recurrent selection

Fernando Higino de Lima e Silva*, Alexandre Pio Viana, Eileen Azevedo Santos, Jôsie Cloviane de Oliveira Freitas, Daniele Lima Rodrigues and Antonio Teixeira do Amaral Júnior

*Laboratório de Melhoramento Genético Vegetal, Universidade Estadual do Norte Fluminense Darcy Ribeiro, Av. Alberto Lamego, 2000, 28013-602, Parque Califórnia, Campos dos Goytacazes, Rio de Janeiro, Brazil. *Author for correspondence. E-mail: fernandohiginolima@gmail.com*

ABSTRACT. Breeding programmes must be improved to accelerate the development of new cultivars due to the commercial importance of passion fruit. This study compared four selection indexes and the REML/BLUP methodology in an assessment of predicted genetic gains in the traits of interest. A total of 81 full-sib progenies derived from the third cycle of recurrent selection were assessed for one harvest in one environment. The experiment was arranged in a randomized complete block design with five plants per plot. The following traits were assessed: number of fruits, total yield, fruit mass, fruit longitudinal diameter, fruit transverse diameter, fruit pulp percentage, shell thickness and content of soluble solids. The Mulamba & Mock index produced the best results for the selection of progenies. The REML/BLUP method was the most efficient and selected progenies with predicted genetic gains better than the selection indexes tested.

Keywords: mixed models, *Passiflora edulis* Sims, predicted genotypic values, simultaneous selection.

Introduction

Passion fruit belongs to the Passifloraceae family, and *Passiflora edulis* Sims is the main cultivated species (sour passion fruit) (Bernacci, Soares-Scott, Junqueira, Passos, & Meletti, 2008), accounting for approximately 90% of Brazilian orchards, mainly due to its quality, productivity, vigour and juice yield.

According to the (Instituto Brasileiro de Geografia e Estatística [IBGE], 2014) records, Brazil is the largest producer of sour passion fruit, with an estimated production of 780 thousand tons. Due to the importance of this culture, breeding programmes must be improved to accelerate the development of new cultivars (Silva et al., 2014).

The literature offers little information on *Passiflora* breeding programmes that use recurrent selection. Therefore, this effort represents a pioneering research proposal that allows the generation of a large amount of knowledge on the genetic characteristics of this fruit as well as agreement on breeding methods that can be applied to passion fruit (Silva et al., 2014).

Use of the recurrent selection breeding method is expected to increase the frequency of favourable alleles, thus improving the phenotypic expression of the trait under selection and preserving genetic variability to allow genetic gains in the following cycles (Hallauer, Miranda Filho, & Carena, 2010).

The use of selection indexes is an additional character established by the optimal combination of several analysed characteristics that allows efficient and simultaneous selection for many characters and thus the correct classification of genotypes (Cruz, Regazzi, & Carneiro, 2004). The adoption of selection indexes has often been applied in sour passion fruit breeding with the aim of progeny selection (Gonçalves, Viana, Bezerra Neto, Pereira, & Pereira, 2007; Neves, Bruckner, Cruz, Viana, & Barelli, 2011; Freitas, Oliveira, Jesus, Cruz Neto, & Santos, 2012).

Genetic assessment based on REML/BLUP mixed models (maximum non-biased restricted likelihood/best linear unbiased prediction) has received special attention from researchers because of the potential for outstanding genetic progress and great accuracy in the selection process. The mixed model allows simultaneous modelling of fixed and random effects. Currently, BLUP is most widely used in the latter. The REML/BLUP methodology has been frequently applied in fruit breeding aimed at selection of progeny and/or individuals in pequi trees (Giordani, Fernandes, Titon, & Santana, 2012), peach trees (Bruna , Moreto, & Dalbo, 2012), açaí trees (Farias Neto, Resende, & Oliveira, 2011) and coconut trees (Farias Neto, Lins, Resende, & Muller, 2009).

The objective of the current study is to compare four selection indexes and the REML/BLUP methodology in an evaluation of predicted genetic gains for the traits of interest in the UENF passion fruit breeding programme.

Material and methods

Eighty-one full-sib progenies derived from the third cycle (C_{03}) of recurrent selection of passion fruit genetic breeding were evaluated in Itaocara-Ilha Barra do Pomba (21°40' S, 42°04' W, and altitude of 76 m) in Northwestern Rio de Janeiro State, Brazil. The progenies were planted in November 2011, and the harvest season occurred from August 2012 to April 2013. During this period, the average annual temperature was 22.5°C, and the average precipitation was 1,041 mm.

The field experiment was arranged as a randomized complete block design with two replicates and five plants per plot. A vertical trellis was chosen as the conduction system and consisted of treated eucalyptus fence posts with 2.5 m height, 4 m spacing and one number 12 wire located 1.80 m above ground, with spacing of 3.5 m between the planting lines.

All plants received the treatments recommended for crop cultivation, such as plant conduction and control of weeds, pests and diseases. Daily drip irrigation was used during the dry season. Seedlings were fertilized monthly with 22.5 g urea and 45 g potassium chloride per plant. In addition, 100 g of limestone and 200 g of simple superphosphate were applied to each plant at planting.

The following traits were assessed: number of fruits (NF) (obtained from a count of the total number of fruits in the plot produced by the first flowering season), yield (obtained from weekly harvest. At the end of the evaluation period, the masses of the fruits per plot were summed and converted into tons per hectare), fruit mass in grams (MF) (given as the arithmetic mean mass of ten fruits sampled per plot with the aid of a digital scale), longitudinal fruit diameter in mm (DLF) (obtained from longitudinal measurements of the fruits with a digital calliper and sampling of ten fruits per plot), transverse diameter of the fruit in mm (DTF) (obtained by measuring the transverse dimension of the fruits with a digital calliper and sampling of ten fruits per plot), percentage of fruit pulp (PP) (obtained by the ratio between the pulp mass and the total fruit mass and sampling of ten fruits per plot), shell thickness in mm (EC) (determined by the arithmetic mean of the measurements of four points of the outer shell in the middle portion of the fruits, cross-sectioned in the direction of the wider diameter, with a digital calliper using samples of ten fruits per plot), soluble solids content in °Brix (SST) (obtained from refractometry using a portable ATAGO N1 digital refractometer with readings in the range of 0 to 32° Brix. The readings were performed on aliquots of juice from the pulp in samples of ten fruits).

The Smith (1936), Hazel (1943), Williams (1962) Pesek and Baker (1969) and Mulamba & Mock (1978) selection indexes were tested for selection of the best 25 progenies. In the computational analysis, economic weights were assigned by trials for each analysed characteristic according to relevance. The selection of superior progenies using selection indexes was performed with the aid of the Genes software system (Cruz, 2013).

The mixed models were assessed by the Selegen-REML/BLUP software system (Resende, 2007a) for the model that considered the randomized block design in one location and one harvest, according to the model below:

Y = Xr + Zg + Wp + Ɛ,

where Y is the vector of phenotypic values, r is the vector of repetition effects (assumed to be fixed) added to the overall average, g is the vector of individual genotypic values (assumed to be random), p is the vector of plot effects (assumed to be random), and Ɛ is the error vector or residue-vector (random). The capital letters refer to the incidence matrices for these purposes. The distributions and structures of the variance associated with the terms of the model are described by the following:

$$y|b,V \sim N(Xb, V)$$
$$g\ |I\sigma_g^2 \sim N(0, I\sigma_g^2)$$
$$p|\sigma_p^2 \sim N(0, I\sigma_p^2)$$
$$\varepsilon|\sigma_\varepsilon^2 \sim N(0, I\sigma_\varepsilon^2)$$
$$V = Z\ \sigma_g^2 Z' + WI\sigma_p^2 W' + I\sigma_e^2$$

The breeding values of each progeny were obtained by adding each genotypic effect (g) to the overall mean of the experiment (μ). The genetic gain is equivalent to the average of the vectors of the predicted genetic effects for the progenies selected. The overall mean added to the genetic gain results in an improved population average.

The following components of variance (individual REML) were estimated:

σ^2_g: Genotypic variance among full-sib progenies, equivalent to half the additive genetic variance plus one quarter of the genetic variance of dominance and disregarding epistasis;

σ^2_f: Phenotypic variance;

h^2_{mp}: Heritability of average progeny, assuming complete survival;

Acprog: Accuracy of progeny selection, assuming complete survival;

The index of coincidence was obtained using the ratio between twice the number of progenies in which both selection indexes coincide and the sum of the total number of progenies containing the selection index A plus the total number of progenies containing the selection index B (Pedrozo, Benites, Barbosa, Resende, & Silva, 2009).

Results and discussion

The Pesek and Baker (1969) index resulted in undesirable gains for the traits NF, MF and DLF, with values of -0.26, -1.89 and -1.17%, respectively (Table 1).

Table 1. Estimates of the percentage gains based on the selection differential for simultaneous selection in eight traits in the third cycle of intrapopulation recurrent selection in full-sib progenies of passion fruit.

Trait[(1)]	Selection indexes				REML/ BLUP
	Pesek and Baker	Smith and Hazel	Mulamba and Mock	Williams	
NF	- 0.26	28.00	22.35	28.02	77.98
Yield	1.72	10.54	10.15	13.02	19.26
MF	- 1.89	1.69	0.69	1.94	25.22
DLF	- 1.17	- 0.48	0.18	0.27	23.85
DTF	0.79	1.44	0.54	1.43	20.65
PP	0.69	- 0.47	3.11	- 0.86	19.40
EC	- 1.80	0.55	- 4.08	0.29	0.00
SST	0.52	- 0.24	0.84	0.43	19.71

[(1)]NF, (1) NF, number of fruits; yield, in t ha^{-1}; MF, fruit mass in grams; DLF, longitudinal diameter of the fruit in mm; DTF, transverse diameter of the fruit in mm; PP, percentage of pulp; EC, shell thickness in mm; SST total soluble solids in °Brix.

Gonçalves et al. (2007) assessed six productive traits of a population of yellow passion fruit structured in Design I in two locations (Viçosa, Minas Gersis State and Miracema, Rio de Janeiro State) and observed that the Pesek and Baker (1969) index showed the highest gain for the trait of number of fruits per plant, unlike the results observed in this study.

The objective of passion fruit breeding programmes is development of plants with a greater number of fruits, which should also be larger and have greater masses, to meet the demands of the market. However, the work of Pesek and Baker (1969) stands out based on the desired gain in the trait EC because fruits with a thinner peel tend to have a larger amount of pulp.

The Smith (1936) and Hazel (1943) indexes delivered genetic gains for the traits NF and yield, with values of 28.00 and 10.54%, respectively (Table 1). A positive gain was noted for MF, but the value was low (1.69%). However, undesirable gains were observed for the traits DLF (-0.48%), PP (-0.47%), EC (0.55%) and SST (-0.24%). Thus, this method is not indicated for selection of progenies in that population because undesirable gains were observed for most traits of economic interest. Gonçalves et al. (2007) observed results similar to those obtained in this study. The authors found that the Smith (1936) and Hazel (1943) indexes showed the lowest predicted gain in several alternatives, which resulted in unsatisfactory gains compared with the other indexes assessed.

The genetic gains predicted by the Mulamba and Mock (1978) index were satisfactory for all traits (Table 1) assessed. Genetic gains of 22.35, 10.15, 3.11 and -4.08% were observed for NF, yield, PP and EC, respectively. Gonçalves et al. (2007) analysed various selection indexes in a sour passion fruit population structured in Design I and indicated the Mulamba and Mock (1978) selection index because it produced greater predicted genotypic gain and is considered promising for selection of superior progenies. In a cycle of recurrent selection that preceded the population evaluated in this study, Silva, Viana, Gonçalves, Amaral Junior, and Pereira (2009) observed that when using the Mulamba and Mock (1978) index, the predicted genetic gains were 1.03% for number of fruits, 3.18% for fruit mass, 0.47% for fruit length, 1.36% for fruit width, and 0.65% for shell thickness. In the cycles of recurrent selection that preceded this study, the results obtained by the authors mentioned above led to the conclusion that an increase in gains occurred with the advance of recurrent selection for the main traits of sour passion fruit, which indicates that the chosen selection method is satisfactory. Freitas et al. (2012) used the Mulamba and Mock index (1978) in sour passion fruit and observed satisfactory simultaneous gains in productivity, fruit mass, pulp yield and content of total soluble solids as well as reduced disease severity.

The Williams (1962) index allowed satisfactory genetic gains for the traits NF, yield and MF, with values of 28.02, 13.02 and 1.94%, respectively (Table 1). These gains were greater than those obtained by the Mulamba and Mock (1978) index. Positive but low gains were observed for DLF (0.27%), DTF (1.43%) and SST (0.43%). Simultaneous selection was not satisfactory for PP and EC because they presented gains of -0.86 and 0.29%, respectively. These results demonstrate that the Williams (1962) index is suitable for selection of progenies in the current population, but undesirable gains were observed for the traits of interest PP and EC. Neves et al. (2011) analysed the feasibility of using the Williams index (1962) in sour passion fruit and concluded that the index created the selection of 35 families with gains in productivity and pulp weight.

In mixed models, estimators and predictors are obtained using iterative processes. To begin the iterative process, the initial value adopted for h^2 was equal to 0.10, as recommended by Resende (2007a). The process produced results for the components of genotypic (σ^2_g) and individual phenotypic (σ^2_f) variance, heritability (h_{mp}^2), and accuracy of selection of progenies (Acprog), as included in Table 2.

Table 2. Estimates of genetic and phenotypic parameters of full-sib progenies of sour passion fruit for the traits of number of fruits (NF), yield and fruit mass (MF) in Itaocara, Rio de Janeiro State.

Estimate	NF	Yield	MF
σ^2_g	5253.48	17.21	200.82
σ^2_f	38792.86	187.65	949.82
h^2_{mp}	0.23	0.18	0.34
A_{cprog}	0.48	0.36	0.59
Average	305.13	15.9	167.46

[(1)]NF, (1) NF, number of fruits; Yield, in t ha^{-1}; MF, fruit mass in grams.

The main evaluated traits of NF, yield and MF presented estimated values for the genotypic variance of 5253.48, 17.21, and 200.82, respectively (Table 2). These values indicate that the population studied presents high genetic variability for these characters, which demonstrates that it is possible to select progenies with increased yield and increased fruit mass. Knowledge of the genotypic variance is highly important for breeding programmes because it indicates the range of genetic variation of a character that breeders intend to improve.

Estimates of phenotypic variance were 38792.86, 187.65 and 949.82 for NF, yield and MF, respectively (Table 2). In a cycle of recurrent selection preceding that of the population assessed in the current work, Silva,Viana, Amaral Junior, Gonçalves, and Reis (2012) observed estimates of phenotypic variance of 1789.00, 14.96, and 328.28 in the progenies of full-sibs for the traits NF, yield and MF, respectively. The higher values of phenotypic variance observed in this study indicate a stronger influence of the environment compared with the previous cycle. This result is probable because Itaocara was more heavily affected by the environment than the municipality of Campos dos Goytacazes, where the previous cycles were conducted.

The h^2_{mp} estimates were 0.23, 0.18, and 0.34 for NF, yield and MF, respectively (Table 2), which classifies their magnitude as moderate ($0.15 < h^2 < 0.50$). However, it cannot be concluded that genetic gains obtained by selection will be lower because estimates of heritability values of high magnitude might occur for characters of small genetic variance, provided that the environmental effect on the trait is of small magnitude. Thus, heritability is not an immutable characteristic and is a property not only of the trait but also of the population and environmental conditions to which the population submits.

The accuracy values of 0.48, 0.36, and 0.59 found for NF, yield and MF, respectively, are classified according to Resende and Duarte (2007) as low (Acgen ≤ 0.50) and moderate (0.50 ≤ Acgen ≤ 0.70), respectively. The accuracy value indicates the precision in the access to the actual genetic variation based on the phenotypic variation observed in each trait. Low accuracy indicates that the data for these traits are less reliable, possibly because the trait is highly affected by the environment.

The mean values of the traits NF, yield and MF were 305.13, 15.9 t ha^{-1} and 167.46 grams, respectively. In a cycle preceding that of this study, the mean values reported by Silva et al. (2009) were lower except for yield, with mean values of 141.88 for NF, 16.3 t ha^{-1} for yield and 144.53 g for MF. This result highlights the potential of this population to achieve genetic gains from the UENF sour passion fruit intrapopulation breeding programme.

Compared with the selection indexes studied, the REML/BLUP methodology allowed higher predicted gains for all traits, except for EC (Table 1). The highest predicted gains were observed for the traits NF (77.98%) and MF (25.22%), which is desirable for the breeding programme. The difference between the REML/BLUP estimates of gain compared with the selection indexes investigated might be explained by the fact that the REML/BLUP methodology uses the predicted genotypic effects, the coefficient matrices of kinship, and the selection gains for each family as a solution vector. The correction of the values for environmental effects predicts the genotypic values accurately and without bias and leads to increased genetic gain with selection (Rodrigues, Vieira, Barbosa, Souza filho, & Candido, 2013). The result observed in this study indicates the REML/BLUP methodology for recombination of superior progenies in the subsequent cycles.

Resende (2007b) reported that simultaneous analysis of various traits could be efficiently performed using the multivariate REML/BLUP approach. Selection indexes commonly use economic weights, which become arbitrary genetic gains. In passion fruit, this result is even more evident due to the negative correlation between the characteristics of economic importance, namely, shell thickness and pulp percentage. Because these traits are negatively correlated, direct selection via genotypic values of the characteristics becomes compromised. Using the REML/BLUP approach for the same group of progenies and traits evaluated in this work, Silva, Muñoz, Vicent, and Viana (2016) found the highest genetic correlation between shell thickness and pulp percentage (-0.98). This high genetic correlation was expected because fruits with more pulp tend to have thinner skin, which is an objective of the UENF passion fruit breeding programme.

Santos et al. (2015) evaluated hybrids of *Passiflora* for the main traits using the REML/BLUP method and identified superior genotypes derived from interspecific cross between *P. edulis* and *P. setacea* for generation advance in the sour passion fruit breeding programme aimed at resistance to *Cowpea aphid-borne mosaic virus* (CABMV). Oliveira, Fraife Filho, Freitas, Dantas, and Resende (2012) used the REML/BLUP methodology to rate F2 segregating populations of papaya and found high values of individual heritability for length and diameter of fruit and low values for SST.

The coefficients of coincidence of 25 full-sib progenies chosen by the selection indexes and the REML/BLUP methodology are presented in Table 3. The coefficients are generally low except for the coefficients found between the indexes of Smith (1936) and Hazel (1943) and Williams (1962), with a value of 0.92, i.e., 23 coincident progenies, and between the selection index of Mulamba and Mock (1978) and the REML/BLUP methodology, which presented a value of 0.84, i.e., 21 coincident progenies (Table 3).

Table 3. Coefficients of coincidence of 25 progenies selected using a selection index and REML/BLUP methodology for the eight traits evaluated in the fruits of sour passion fruit from Itaocara, Northwestern State of Rio de Janeiro.

Selection index	Pesek and Baker	Smith and Hazel	Mulamba and Mock	Williams	REML/ BLUP
Pesek and Baker	-				
Smith and Hazel	0.48	-			
Mulamba and Mock	0.48	0.64	-		
Williams	0.48	0.92	0.64	-	
REML/BLUP	0.44	0.60	0.84	0.56	-

According to Pedrozo et al. (2009) the higher the coefficient of coincidence between two selection indexes, the greater the agreement between them will be with respect to the results of selection.

With the aim of selecting progenies of sour passion fruit, Silva and Viana (2012) evaluated the indexes based on the sum of ranks of Mulamba and Mock (1978) and the distance between the genotype and the ideotype (Cruz, 2013) and observed a coefficient of coincidence of 0.97. This result allowed simultaneous selection of superior progenies for the traits of number of fruits, total productivity, average fruit weight, fruit diameter, fruit length, fruit width and weight of fruit pulp.

Out of the 81 progenies assessed, the best 25 for the variables NF, yield and MF were selected through the REML/BLUP methodology and account for 30.86% of the progenies. Genetic gains were predicted, and the new estimated averages were higher than the overall average for all variables (Table 4).

Table 4. Estimates and ranking of the best 25 full-sib progenies, predicted genotypic values (g), genotypic values (μ + g), and new predicted average (BLUP) for the traits number of fruits, yield and fruit mass.

	Number of fruits					Yield (t ha^{-1})					Fruit mass (g)				
Ranking	Progeny	g	μ+g	Gain	New mean	Progeny	g	μ+g	Gain	New mean	Progeny	g	μ+g	Gain	New mean
1	26	125.82	430.95	125.82	430.96	26	0.73	16.65	0.73	16.65	69	18.70	186.17	18.70	186.17
2	56	110.96	416.09	118.40	423.53	24	0.46	16.38	0.59	16.52	5	16.08	183.55	17.39	184.86
3	24	83.86	389.00	106.89	412.02	70	0.44	16.37	0.54	16.47	3	15.65	183.12	16.81	184.28
4	70	67.45	372.58	97.03	402.16	69	0.38	16.30	0.50	16.43	8	15.44	182.91	16.47	183.94
5	1	61.19	366.33	89.86	395.00	68	0.33	16.26	0.47	16.39	68	15.27	182.74	16.23	183.70
6	68	59.11	364.25	84.74	389.87	1	0.32	16.25	0.44	16.37	42	14.14	181.61	15.88	183.35
7	19	50.38	355.52	79.83	384.97	40	0.27	16.20	0.42	16.34	47	13.91	181.38	15.60	183.07
8	69	46.47	351.61	75.66	380.80	50	0.23	16.15	0.39	16.32	72	13.78	181.25	15.37	182.84
9	47	46.09	351.22	72.38	377.51	56	0.22	16.15	0.38	16.30	75	12.75	180.22	15.08	182.55
10	34	42.18	347.31	69.36	374.49	45	0.21	16.14	0.36	16.29	51	11.32	178.79	14.71	182.17
11	45	39.31	344.44	66.62	371.76	22	0.20	16.13	0.34	16.27	78	10.45	177.92	14.32	181.79
12	38	33.84	338.97	63.89	369.03	19	0.20	16.13	0.33	16.26	2	9.72	177.18	13.94	181.40
13	16	32.80	337.93	61.50	366.64	16	0.20	16.13	0.32	16.25	41	9.57	177.04	13.60	181.07
14	10	31.89	337.02	59.39	364.52	34	0.19	16.11	0.31	16.24	6	9.11	176.58	13.28	180.75
15	22	31.89	337.02	57.55	362.69	38	0.19	16.11	0.30	16.23	33	9.05	176.52	13.00	180.47
16	23	30.58	335.71	55.87	361.00	55	0.18	16.10	0.30	16.22	7	8.06	175.53	12.69	180.16
17	40	30.32	335.45	54.36	359.50	60	0.17	16.10	0.29	16.22	53	7.69	175.16	12.39	179.86
18	55	29.28	334.41	52.97	358.11	47	0.16	16.09	0.28	16.21	73	6.40	173.87	12.06	179.53
19	50	29.02	334.15	51.71	356.85	10	0.16	16.08	0.28	16.20	1	6.33	173.80	11.76	179.23
20	60	25.89	331.02	50.42	355.55	5	0.15	16.08	0.27	16.20	24	6.05	173.52	11.47	178.94
21	17	25.37	330.50	49.23	354.36	23	0.14	16.07	0.26	16.19	52	5.19	172.66	11.18	178.64
22	11	21.98	327.11	47.99	353.12	79	0.14	16.06	0.26	16.18	44	5.17	172.64	10.90	178.37
23	54	20.81	325.94	46.81	351.94	17	0.13	16.05	0.25	16.18	29	4.82	172.29	10.64	178.11
24	5	19.38	324.51	45.66	350.80	62	0.13	16.05	0.25	16.17	59	4.81	172.28	10.40	177.86
25	9	15.08	320.21	44.44	349.58	7	0.11	16.03	0.24	16.17	48	4.26	171.73	10.15	177.62

Progeny 26 was best ranked in the identification of promising progenies for number of fruits and yield, i.e., more productive progenies. For NF, the gain predicted by the selection of this progeny was 125.82, with 431 fruits on average, and progeny 9 was ranked in the 25th position, with a gain of 44.44%, thus producing an average increase of 29.26 (Table 4). A total of 22 progenies selected for NF were also selected for yield. According to Silva et al. (2016), the high correlation observed between NF and yield (0.97) for this same group of progenies implies that it is not necessary to evaluate both characteristics in subsequent cycles, and thus the easiest or less costly trait should be measured. Considering the importance of selecting progenies with greater fruit mass (MF), the highest-ranking position was obtained by progeny 69, the expected gains of which exceeded 18.70% of the overall average. Progeny 48 was ranked at the lowest position, with an estimated gain of 4.26%.

The genotypic values predicted by BLUP refer to those observed without the environmental effect. Therefore, contrary to observations in species of vegetative propagation, in which all genotypic values are utilized, in allogamous or outcrossing species, which are subjected to progeny tests, only the additive effects are transmitted to the descendants and should be considered during the selection of genotypes that will be used as parents in the following generations (Resende, 2002).

The REML/BLUP method was more efficient than the selection indexes and allowed selection of progenies with promising predicted genetic gains for all traits assessed, primarily NF, yield and MF, which are the main traits required in the sour passion fruit breeding programme. Based on the results found in this work, the REML/BLUP methodology was indicated for the recombination of superior progenies in the subsequent selection cycles.

Conclusion

Among the four selection indexes tested, Mulamba and Mock and Williams are most suitable for selection of full-sib progenies in the sour passion fruit population under study.

The REML/BLUP method offers the best predicted genetic gains compared with those obtained by the selection indexes and is efficient in selecting progenies of sour passion fruit.

Acknowledgements

The authors thank FAPERJ and CNPq for financial support of experiments and CAPES for granting a doctoral scholarship for the first author.

References

Bernacci, L. C., Soares-Scott, M. D., Junqueira, N. T. V., Passos, I. R. S., & Meletti, L. M. M. (2008). *Passiflora edulis* Sims: the correct taxonomic way to cite the yellow passion fruit (and of others colors). *Revista Brasileira de Fruticultura, 30*(2), 566-576.

Bruna, E. D., Moreto, A. L., & Dalbo, M. A. (2012). Uso do coeficiente de repetibilidade na seleção de clones de pessegueiro para o litoral sul de Santa Catarina. *Revista Brasileira de Fruticultura, 34*(1), 206-215.

Cruz, C. D. (2013). Genes: a software package for analysis in experimental statistics and quantitative genetics. *Acta Scientiarum. Agronomy, 35*(3), 271-276.

Cruz, C. D., Regazzi, A. J., & Carneiro, P. C. S. (2004). *Modelos biométricos aplicados ao melhoramento genético.* Viçosa, MG: UFV.

Farias Neto, J. T., Lins, P. M. P., Resende, M. D. V., & Muller, A. A. (2009). Seleção genética em progênies híbridas de coqueiro. *Revista Brasileira de Fruticultura, 31*(1), 190-196.

Farias Neto, J. T., Resende, M. D. V., & Oliveira, M. S. P. (2011). Seleção simultânea em progênies de açaizeiro irrigado para produção e peso de fruto. *Revista Brasileira de Fruticultura, 33*(2), 532-539.

Freitas, J. P. X., Oliveira, E. J., Jesus, O. N., Cruz Neto, A. J., & Santos, L. R. (2012). Formação de população base para seleção recorrente em maracujazeiro-azedo com uso de índices de seleção. *Pesquisa Agropecuária Brasileira, 47*(3), 393-401.

Giordani, S. C. O., Fernandes, J. S. C., Titon, M., & Santana, R. C. (2012). Parâmetros genéticos para caracteres de crescimento em pequizeiro em estádio precoce. *Revista Ciência Agronômica, 43*(1), 146-153.

Gonçalves, M. G., Viana, A. P., Bezerra Neto, F. V., Pereira, M. G., & Pereira, T. N. S. (2007). Seleção e herdabilidade na predição de ganhos genéticos em maracujá-amarelo. *Pesquisa Agropecuária Brasileira, 42*(2), 193-198.

Hallauer A. R., Miranda Filho, J. B., & Carena, M. J. (2010). *Quantitative Genetics in Maize Breeding.* New York, US: Springer.

Hazel, L. N. (1943). The genetic basis for constructing selection indexes. *Genetics, 28*(6), 476-490.

Instituto Brasileiro de Geografia e Estatística [IBGE]. (2014). *Sistema IBGE de recuperação automática – SIDRA, 2014.* Produção agrícola municipal: produção de maracujá (production of passion fruit). Retrieved on June 27, 2014 from http://www.sidra.ibge.gov.br/

Mulamba, N. N., & Mock, J. J. (1978). Improvement of yield potential of the Eto Blanco maize (*Zea mays* L.) population by breeding for plant traits. *Egyptian Journal of Genetics and Cytology,* 7(1), 40-51.

Neves, L. G., Bruckner, C. H., Cruz, C. D., Viana, A. P., & Barelli, M. A. A. (2011). Predição de ganhos, com diferentes indices de seleção, para características de frutos do maracujazeiro-amarelo. *Revista Brasileira de Fruticultura, 33*(4), 1322-1330.

Oliveira, E. J., Fraife Filho, G. A., Freitas, J. P. X., Dantas, J. L. L., & Resende, M. D.V. (2012). Plant selection in F2 segregating populations of papaya from commercial hybrids. *Crop Breeding and Applied Biotechnology, 12*(3), 191-198.

Pesek, J., & Baker, R. J. (1969). Desired improvement in relation to selection indexes. *Canadian Journal of Plant Science, 49*(6), 803-804.

Pedrozo, C. A., Benites, F. R. G., Barbosa, M. H. P., Resende, M. D. V., & Silva, F. L. (2009). Eficiência de índices de seleção utilizando a metodologia REML/BLUP no melhoramento da cana-de-açúcar. *Scientia Agraria, 10*(1), 31-36.

Resende, M. D. V. (2002). *Genética biométrica e estatística no melhoramento de plantas perenes. Embrapa Informação Tecnológica.* Colombo, PR: Embrapa Florestas.

Resende, M. D. V. (2007a). *Selegen-REML/BLUP: statistical system and computerized genetic selection throught linear mixed models.* Colombo, PR: Embrapa Florestas.

Resende, M. D. V. (2007b). *Matemática e Estatística na Análise de Experimentos e no Melhoramento Genético.* Colombo, PR: Embrapa Florestas.

Resende, M. D. V., & Duarte, J. B. (2007). Precisão e controle de qualidade em experimentos de avaliação de cultivares. *Pesquisa Agropecuária Brasileira, 37*(3), 182-194.

Rodrigues, W. P., Vieira, H. D., Barbosa, D. H., Souza filho, G. R., & Candido, L. S. (2013). Adaptability and genotypic stability of *Coffea arabica* genotypes based on REML/BLUP analysis in Rio de Janeiro State, Brazil. *Genetics and Molecular Research, 12*(3), 2391-2399.

Santos, E. A., Viana, A. P., Freitas, J. C. O., Rodrigues, D. L., Tavares, R. F., Paiva, C. L., & Souza, M. M. (2015). Genotype selection by REML/BLUP methodology in a segregating population from na interspecific *Passiflora* spp. crossing. *Euphytica, 204*(1), 1-11.

Silva, F. H. L., Muñoz, P. R., Vicent, C., & Viana, A. P. (2016). Generating relevant information for breeding *Passiflora edulis*: genetic parameters and population structure. *Euphytica, 208*(3), 609-619.

Silva, F. H. L., Viana, A. P., Ferreira, R. T., Freitas, J. C. O., Santos, J. O., & Rodrigues, D. L. (2014). Measurement of genetic diversity in progenies of sour passion fruit by Ward-MLM methodology: a strategy for heterotic group formation. *Ciência e Agrotecnologia, 38*(3), 1234-1239.

Silva, M. G. M., & Viana, A. P. (2012). Alternativas de seleção em população de maracujazeiro-azedo sob seleção recorrente intrapopulacional. *Revista Brasileira de Fruticultura, 34*(2), 525-531.

Silva, M. G. M., Viana, A. P., Amaral Junior, A. T., Gonçalves, L. S. A., & Reis, R. V. (2012). Biometria aplicada ao melhoramento intrapopulacional do maracujazeiro amarelo. *Revista Ciência Agronômica, 43*(3), 493-499.

Silva, M. G. M., Viana, A. P., Gonçalves, G. M., Amaral Junior, A. T., & Pereira, M. G. (2009). Seleção recorrente intrapopulacional no maracujazeiro amarelo: alternativa de capitalização de ganhos genéticos, *Ciência e Agrotecnologia*, *33*(1), 170-176.

Smith, H. F. (1936). A discriminant function for plant selection. *Annals of Eugenics*, 7(3), 240-250.

Williams, J. S. (1962). The evaluation of a selection index. *Biometrics*, *18*(3), 375-393.

22

Genetics of the concentration of copper in common bean seeds

Nerison Luís Poersch[1], Nerinéia Dalfollo Ribeiro[1*], Daniele Piano Rosa[1], Sandra Maria Maziero[1] and Evandro Jost[2]

*[1]Centro de Ciências Rurais, Departamento de Fitotecnia, Universidade Federal de Santa Maria, Av. Roraima, 1000, 97105-900, Santa Maria, Rio Grande do Sul, Brazil. [2]Instituto Federal de Educação, Ciência e Tecnologia Farroupilha, São Vicente do Sul, Rio Grande do Sul, Brazil. *Author for correspondence. E-mail: nerineia@hotmail.com*

ABSTRACT. The common bean (*Phaseolus vulgaris* L.) is an important source of copper in human nutrition. The objective of this research was to investigate the possible maternal effects on copper concentrations in the seeds of the common bean, to estimate the heritability and the selection gains in early hybrid generations for the characteristic of copper concentration and to evaluate the potential for genetic improvement. The controlled crossings were performed among IAPAR 44 x IAPAR 31 and Diamante Negro x TPS Bonito cultivars. The reciprocal F_1, and F_2 generations, as well as the backcross populations, were produced by each hybrid combination. The range of variation in the copper concentration among the tested progenies was from 1.07 to 7.75 mg kg^{-1} dry matter, and no significant maternal effect was discovered. A narrow-sense heritability was of an intermediate (51.98%) to high (61.99%) value. A 46.78% increase in the copper concentration was obtained in the seeds using only the four parents tested in this study.

Keywords: *Phaseolus vulgaris*, embryo, maternal effect, heritability, early generation.

Introduction

The common bean (*Phaseolus vulgaris* L.) represents 50% of the grain legumes consumed worldwide (TALUKDER et al., 2010). The common bean grains contain a high concentration of copper in its seeds, and a genetic variability has been observed in the copper concentration character. In the germplasm of the common bean cultivated in Brazil, copper values ranged from 11.37 to 17.73 mg kg^{-1} of dry matter (DM) (MESQUITA et al., 2007).

Genetic breeding is a strategy often used to increase the mineral concentration in common bean seeds. Controlled crossing and selection among segregated progenies resulted in concentration increases of 19.1% in phosphorus (RIBEIRO et al., 2011), 33.6% in calcium (JOST et al., 2009a), 94.0% in iron (JOST et al., 2009b) and 37.3% in zinc (ROSA et al., 2010) of the common bean seeds of early hybrid generations. There is no documentation concerning the use of genetic breeding to increase the copper concentration in common bean seeds. Increasing the common bean seeds' copper concentration is important for human health because this micronutrient is essential for the satisfactory operation of the immune system, carrying iron, metabolizing glucose and cholesterol, myocardial contractility and brain development and is also a component of various enzymes (CUNHA; CUNHA, 1998).

In common bean seeds, the embryo (cotyledons and embryonic axis) represents approximately 90.1% of the entire DM, while the seed coat consists of 9.9% of the entire DM (MORAGHAN et al., 2006).

These tissues accumulate minerals in different ways. The embryo contains 93% of the zinc, 76% of the iron and 16% of the entire calcium content of the whole seed (MORAGHAN; GRAFTON, 2002). The embryo originates from fertilization, and the seed coat represents the maternal tissue (RAMALHO et al., 2008). That these tissues are present in different generations of the same seed this will have implications for the selection of segregating populations in breeding programs.

No significant maternal effect was discovered for the zinc concentration in the common bean seeds (ROSA et al., 2010). Accordingly, most zinc is in the embryo and the F_1 generation seeds (embryo in the F_1 generation) exhibited the products of crossing between the parents. However, because most iron is in the seed coat, a significant maternal effect was observed for the iron concentration in the common bean seeds (JOST et al., 2009b). The iron concentration in the seed coat varies from 47.8% to 83.2% of the total DM of seeds of the common bean cultivars grown in Brazil (RIBEIRO et al., 2012). In this case, the appearance of the genotype is delayed for one generation. As the iron concentration was dependent on the seed coat, the selection for high iron in the common bean seeds begins with the F_3 seeds (embryo in F_2 generation) (JOST et al., 2009b).

A maternal effect on the copper concentration in the common bean seeds has not yet been investigated, and there is doubt as to whether the copper concentration displays a qualitative or quantitative inheritance. The objective of this work was to investigate the possible maternal effects on the copper concentration of the common bean seeds, as well as to estimate the heritability and selection gains in the early hybrid generations and to evaluate the potential for genetic improvements in high copper common bean germplasm.

Material and methods

The copper concentration inheritance pattern was studied in hybrids obtained from crosses between four cultivars from the Mesoamerican gene pool: IAPAR 44 x IAPAR 31 and Diamante Negro x TPS Bonito. These crosses were produced in a greenhouse at the Plant Science Department of the Federal University of Santa Maria (UFSM) in Santa Maria, Rio Grande do Sul State, Brazil. The seed mass of the cultivars varied from 22 to 23 g per 100 seeds. The cultivars were selected from common bean accessions of the UFSM Bean Germplasm Bank based on their different seed copper concentration and their agronomic value. IAPAR 44 and Diamante Negro cultivars represented the black commercial group and have an indeterminate growth habit with a small to medium guide development (type II). Also, IAPAR 31 and TPS Bonito cultivars belonged to the Carioca type seeds (beige with brown streaks) and have an indeterminate growth habit with long guides (type III).

An interlacing method with emasculation of the flower bud was used to obtain 50 F_1 (♀ P_1 x ♂ P_2) and 50 F_1 reciprocal (♀ P_2 x ♂ P_1) seeds for each hybrid combination in the fall/winter of 2008. In the spring/summer of 2008, natural self-fertilization of the F_1 and F_1 reciprocal plants produced the 48 F_2 and 48 F_2 reciprocal generations, while backcrossing of the F_1 (♀) to the P_1 (BC_1) and F_1 (♀) to P_2 (BC_2) produced the 36 backcrossed generations. This process was repeated in the fall/winter of 2009, meaning that the seeds of the F_1, F_1 reciprocal, F_2, F_2 reciprocal and backcrossed (BC_1 and BC_2) generations could be evaluated under the same cultivation conditions.

The seeds were sown in plastic pots with a 5 L capacity with two plants per pot. Pots were filled with soil, carbonized rice husks and Plantmax® commercial substrate that was mixed at a 2:1:1 ratio according to volume. The soil was typical alitic Argisol, Hapludalf with the following chemical composition: pH (H_20): 6.5; organic matter: 2%; phosphorus: 14.4 mg dm^{-3}; potassium: 40 mg dm^{-3}; sulfur: 17.6 mg dm^{-3}; calcium: 5.2 $cmol_c$ dm^{-3}; magnesium: 2.1 $cmol_c$ dm^{-3}; copper: 0.2 mg dm^{-3}; zinc: 0.6 mg dm^{-3}.

Soil fertility was corrected, and nitrogen was applied to the surface based on the chemical soil analysis. Plants were watered daily to maintain the soil water content close to that of field capacity. Plant disease and insect infestation were controlled as necessary to avoid compromising the normal development of common bean plants and to maintain the integrity of the flower buds.

At the end of the growth cycle, seeds from each generation were collected and dried in a greenhouse (65 to 70°C) until the average moisture content was 13%. The samples (13% moisture) were ground in a micro-mill to obtain particles less than 1 mm in diameter. Samples (0.5 g) of the bean flour from each generation were digested in 5 mL of a nitric-perchloric solution (HNO_3 + $HClO_4$ at a 5:1 ratio by volume). After 12 hours of cold digestion (no heat applied), the temperature was gradually elevated by 30°C every 30 minutes until it reached 180°C. At the end of the digestion, when 1 mL of the transparent solution remained, the sample was diluted by adding distilled water to a total volume of

50 mL. Approximately 20 mL of this dilution was removed, and the copper concentration was measured by atomic absorption spectrophotometer (Perken Elmer®, model Analyst 200®) at wavelength settings of 324.8 nm.

The copper genetic study was performed in a randomized complete block design with five replications for parents and F_1 reciprocal generations and 24 replications for the F_2 reciprocal generations. To test the maternal effect hypothesis, a comparison was made between the P_1 vs. P_2, the F_1 vs. F_1 reciprocal and the F_2 vs. F_2 reciprocal mean values by Student's t-test at 5% probability.

The genetic parameters were estimated from the variances of the parents' (P_1 and P_2) and the F_1, F_2, BC_1 and BC_2 generations based on embryo generation for each hybrid combination. The broad-sense heritability and the narrow-sense heritability were estimated with the backcross method (WARNER, 1952), by the following expressions:

$h_a^2 = \sigma_G^2/\sigma_P^2$ and $h_r^2 = \sigma_A^2/\sigma_P^2$ respectively,

where:

σ_G^2 refers to the genotypic variance, estimated by: $\sigma_G^2 = \sigma_P^2 - \sigma_E^2$, where the σ_E^2 represents the environmental variance in F_2 ($\sigma_E^2 = (\sigma_{F_1}^2 + \sigma_{P_1}^2 + \sigma_{P2}^2)/3$); σ_P^2 is the phenotypic variance ($\sigma_P^2 = \sigma_{F_2}^2$); and σ_A^2 is the additive variance, estimated by: $\sigma_A^2 = 2\sigma_{F_2}^2 - (\sigma_{BC_1}^2 + \sigma_{BC_2}^2)$.

Heterosis in the F_1 generation was quantified as a percentage of both the heterosis related to the mean of the parents:

$$H\% = [(F_1 - P)/P]x100$$

and of the heterobeltiosis:

$$HT\% = [(F_1 - MP)/MP] x100$$

where:

P is the average of parents ($P = (P_1 + P_2)/2$), and MP is the best male parent.

To predict selection gains, we took a selection of 25% of plants with F_2 seeds (embryos in the F_2 generation) with the highest copper concentration. The gain expected when accounting for the selection and recombination of superior plants with F_2 seeds (embryos in the F_2 generation) was estimated by the following equation:

$$\Delta G(\%) = (\Delta Gx100)/\overline{F}_2$$

where:

ΔG refers to the selection gain, obtained by $\Delta G = DSxh_r^2$, in which DS is the selection differential, expressed by $\overline{X}_s - \overline{X}_0$, where $\overline{X}_s$ is the average of selected plants with seeds in F_2 and $\overline{X}_0$ is the average of plants with seeds in F_2. Analyses were performed with the aid of Microsoft Office Excel spreadsheets and Genes software (GENES, 1998).

Results and discussion

In the reciprocal hybrids IAPAR 44 (5.28 mg copper kg^{-1} DM) x IAPAR 31 (3.49 mg copper kg^{-1} DM) and Diamante Negro (4.36 mg copper kg^{-1} DM) x TPS Bonito (1.60 mg copper kg^{-1} DM), the contrast of the copper concentration between P_1 vs. P_2 was significant (Figure 1), demonstrating a genetic difference between the parents in their copper concentration.

No significant difference was observed between the mean copper concentration values of the reciprocal F_1 hybrids (Figure 1), suggesting that there was no significant maternal effect on the copper concentration phenotype in the common bean seeds. This result was observed because the embryo from the common bean seeds contains 40.4% to 72.0% of the entire copper content (RIBEIRO et al., 2012). Thus, the copper concentration of common bean seeds is dependent on the embryo and the F_1 generation seeds represent the fertilized product of the parents. Similarly, the common bean seeds' zinc concentration exhibited no significant maternal effect (ROSA et al., 2010). Zinc, similar to copper, was accumulated predominantly in the seeds of the common bean's embryos (MORAGHAN; GRAFTON, 2002; MORAGHAN et al., 2002).

In the IAPAR 44 x IAPAR 31 hybrid, no significant difference was observed for the F_2 vs. F_2 reciprocal contrast (Figure 1), reinforcing that the phenotypes of these seeds were similar in terms of copper concentration and represented the expression of the genotype of the F_1 generation. Therefore, F_2 seeds showed embryos in the F_2 generation. For this reason, selecting for a high copper concentration in the common bean seeds should begin with the F_2 seeds, a generation where ample genetic variability is observed. A similar result was observed for the zinc concentration in common bean seeds (ROSA et al., 2010). As no significant maternal effect for the zinc concentration was observed, the selection of the F_2 seeds harvested from the one plant was efficient.

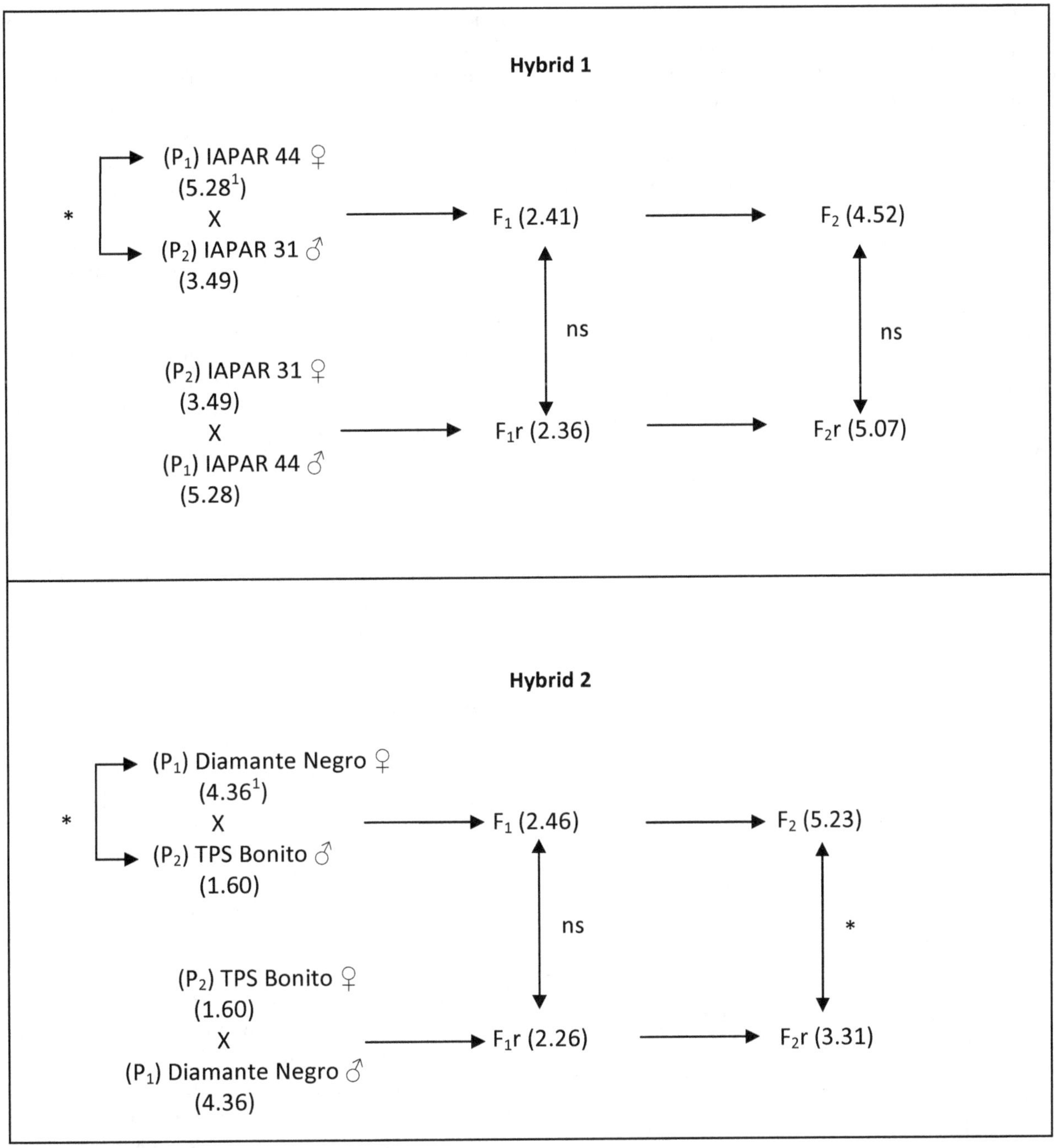

Figure 1. Breeding and selection process schematics for copper concentration of common bean seeds of the IAPAR 44 x IAPAR 31 and Diamante Negro x TPS Bonito hybrids. The mean copper concentration (mg kg^{-1} of DM) is in parenthesis. P_1 and P_2 are the parental seeds, F_1r is the F_1 and F_1 reciprocal seeds, and F_2r is the F_2 and F_2 reciprocal seeds; the significance level for differences in the mean values was evaluated for P_1 vs. P_2, F_1 vs. F_1 reciprocal and F_2 vs. F_2 reciprocal from the IAPAR 44 x IAPAR 31 and Diamante Negro x TPS Bonito hybrids. [1]Copper concentration (mg kg^{-1} of DM); [ns]non-significant; *Significant by Student's test at 5% probability.

However, a significant maternal effect was observed for the iron concentration in common bean seeds (JOST et al., 2009b). Because the iron concentration was dependent on the seed coat, selection began only on the F_3 seeds (embryos in the F_2 generation) when segregation was verified. Ribeiro et al. (2012) found that 47.8 to 83.2% of the whole iron content accumulated in the seed coat of the common bean seeds. The migration of the iron was lower for the embryos of the common bean seeds. Thus, a breeding program should investigate of the existence of a maternal effect on mineral concentrations in germplasm subjected to selection. The embryo generation should be considered for the selection process and the progression of segregating populations in genetic breeding programs.

A breakdown of the phenotypic variance showed the predominance of genetic effects in relation to the environmental effects for each hybrid combination (Table 1). As result, a high broad-sense heritability was observed in the IAPAR 44 x IAPAR 31 hybrid (h^2a: 71.90%) and Diamante Negro x TPS Bonito hybrid (h^2a: 79.65%). This finding was observed because the genetic variance was responsible for most of the phenotypic variance.

Table 1. Genetic parameters and predicted selection gains for copper concentration in the common bean seeds of the IAPAR 44 x IAPAR 31 and Diamante Negro x TPS Bonito hybrids.

Estimate	Copper concentration (mg kg^{-1} of DM)	
	IAPAR 44 x IAPAR 31	Diamante Negro x TPS Bonito
Mean	4.32	3.78
Environmental coefficient of variation (%)	26.39	24.96
Phenotypic variance (σ^2_P)	1.86	1.99
Parent 1 variance ($\sigma^2 P_1$)	0.53	0.20
Parent 2 variance ($\sigma^2 P_2$)	0.35	0.36
F_1 variance ($\sigma^2 F_1$)	0.69	0.66
Environmental variance ($\sigma^2_E F_2$)	0.52	0.40
Genetic variance (σ^2_G)	1.34	1.59
Additive variance (σ^2_A)	0.97	1.23
Broad-sense heritability (h^2_a)	71.90	79.65
Narrow-sense heritability (h^2_r)	51.98	61.99
Heterosis (H%)	-44.97	-17.48
Heterobeltiosis P_1 (HT%)	-54.28	-43.63
Maximum value in parents	6.27	5.00
Minimum value in parents	2.60	1.07
Maximum value in F_2	7.75	7.53
Minimum value in F_2	2.47	1.98
Selection gain (ΔG %)	18.55	29.03

A narrow-sense heritability of an intermediate (h^2r: 51.98%) to high magnitude (h^2r: 61.99%) was observed in the IAPAR 44 x IAPAR 31 and Diamante Negro x TPS Bonito hybrids, respectively (Table 1). Previous studies have shown that iron and zinc concentrations in seeds of the common bean have a narrow-sense heritability ranging from an intermediate to high value (JOST et al., 2009b; ROSA et al., 2010). As the additive variance represented the higher fraction of the genetic variance, the selection for high copper, iron and zinc in the seeds of the common bean is possible. The fixation of these characters could be observed in advanced generations.

Given that genetic factors exert a large effect on the phenotype and an intermediate to high narrow-sense heritability was observed, the copper concentration appears to exhibit a qualitative inheritance. Thus, we expected to easily select for the measured character. Consequently, it is likely that it will be easy to develop common bean cultivars with higher nutritional quality for the use in foods based on the identification of seeds with a high copper concentration. A more in-depth study of the genetic effects and the number of genes that control the copper concentration was not carried out in this study, as the efficiency of hybridization was low, and little replications for the F_2 generation and backcross (BC_1 and BC_2) populations were obtained.

For conventional heterosis, negative values in two of the hybrids were observed (Table 1). The mean value for the F_1 generation was lower compared to those of the parents (Figure 1). A similar response was observed for heterobeltiosis with negative estimates for the IAPAR 44 x IAPAR 31 (-54.28%) and Diamante Negro x TPS Bonito (-43.63%) hybrids. Negative heterosis estimates were also obtained for plant height (-9.54%), number of pods per plant (-27.87%), and seeds per pod (-18.57%) in the common bean (GONÇALVES-VIDIGAL et al., 2008). Therefore, the epistasis effect on the phenotype of the F_1 generation must be considered and included in the model for the partition of the genetic components of variance (BARONA et al., 2009).

Copper concentrations ranged from 1.07 mg kg^{-1} (Diamante Negro) to 7.75 mg kg^{-1} (F_2 plant from IAPAR 44 x IAPAR 31 hybrid) (Table 1). The observed values were approximately half those verified in the genotypes cultivated in Brazil (MESQUITA et al., 2007). Despite this difference, the copper concentration values verified in the present study are sufficient to supply approximately 20% of the daily requirement for an individual (ANDRADE et al., 2004). Thus, populations of high nutritional value for the human diet were obtained.

Transgressive segregation was observed for the common bean seed's copper concentration. Among the F_2 generation seeds examined, individuals with 7.75 mg copper kg^{-1} DM were identified, which represented an increase of 46.78% in the copper concentration relative to the IAPAR 44 parent (5.28 mg copper kg^{-1} MS). This was the first report of an increased copper concentration in the common bean seeds by a genetic improvement. There have been reports of increases of 94.0% in iron (JOST et al., 2009b) and of 37.3% in zinc (ROSA et al., 2010) in the common bean seed by use of controlled crossing and selection among segregation progenies in early hybrid generations.

In relation to selection gains, assuming a retention of 25% of plants with F_2 seeds (embryo in F_2 generation) with the highest copper concentration, gains of 18.55 and 29.03% could be expected for the IAPAR 44 x IAPAR 31 and Diamante Negro x TPS Bonito hybrids, respectively (Table 1). Thus, an early selection of the common bean germplasm with a high copper concentration can be a successful strategy, given that significant gains in the first cycle after selection were obtained.

Given that it was possible to increase the copper concentration in common bean seeds by genetic breeding, this strategy could result in health benefits because copper is an essential mineral for humans (CUNHA; CUNHA, 1998). Additionally, increasing the seed copper concentration could contribute to agricultural sustainability, thereby reducing the amount of copper-based fertilizer required and the accumulation of heavy metals in the soil.

Conclusion

The early generation narrow-sense heritability of the copper concentration of the common bean seed falls between intermediate (51.98%) and high (61.99%) values, and there is no maternal effect on the expression of this characteristic. It is possible to increase the copper concentration in common bean seeds by 46.8% using genetic breeding.

Acknowledgements

We thank the National Council of Technological and Scientific Development (CNPq) and the Coordination for the Improvement of Higher Level (CAPES) for supporting this research with awarded grants.

References

ANDRADE, E. C. B.; BARROS, A. M.; MAGALHÃES, A. C. P.; CASTRO, L. L. C.; TAKASEI, I. Comparação dos teores de cobre e zinco em leguminosas cruas e após processadas termicamente em meio salino e aquoso. **Ciência e Tecnologia dos Alimentos**, v. 24, n. 3, p. 316-318, 2004.

BARONA, M. A. A.; COLOMBARI FILHO, J. M.; GERALDI, I. O. Epistasia para a produção de grãos em soja. **Bragantia**, v. 68, n. 2, p. 313-318, 2009.

CUNHA, D. F.; CUNHA, S. F. C. Microminerais. In: DUTRA-DE-OLIVEIRA, J. E.; MARCHINI, J. S. (Ed.). **Ciências nutricionais**. São Paulo: Sarvier, 1998. p. 141-165.

GENES. Aplicativo Computacional em Estatística Aplicada à Genética (GENES - Software for Experimental Statistics in Genetics). **Genetics and Molecular Biology**, v. 21, n. 1, p. 135-138, 1998.

GONÇALVES-VIDIGAL, M. C.; SILVÉRIO, L.; ELIAS, H. T.; VIDIGAL FILHO, P. S.; KVITSCHAL, M. V.; RETUCI, V. S. Combining ability and heterosis in common bean cultivars. **Pesquisa Agropecuária Brasileira**, v. 43, n. 9, p. 1143-1150, 2008.

JOST, E.; RIBEIRO, N. D.; MAZIEIRO, S. M.; CERUTTI, T.; ROSA, D. P. Efeitos gênicos do teor de cálcio em grãos de feijão. **Ciência Rural**, v. 39, n. 1, p. 31-37, 2009a.

JOST, E.; RIBEIRO, N. D.; CERUTTI, T.; POERSCH, N. L.; MAZIEIRO, S. M. Potencial de aumento do teor de ferro em grãos de feijão por melhoramento genético. **Bragantia**, v. 68, n. 1, p. 35-42, 2009b.

MESQUITA, F. R.; CORRÊA, A. D.; ABREU, C. M. P.; LIMA, R. A. Z.; ABREU, A. F. B. Linhagens de feijão (*Phaseolus vulgaris* L.): composição química e digestibilidade protéica. **Ciência e Agrotecnologia**, v. 31, n. 4, p. 1114-1121, 2007.

MORAGHAN, J. T.; GRAFTON, K. Distribution of selected elements between the seed coat and embryo of two black bean cultivars. **Journal of Plant Nutrition**, v. 25, n. 1, p. 169-176, 2002.

MORAGHAN, J. T.; ETCHEVERS, J. D.; PADILHA, J. Contrasting accumulations of calcium and magnesium in seed coats and embryos of common bean and soybean. **Food Chemistry**, v. 95, n. 4, p. 554-561, 2006.

MORAGHAN, J. T.; PADILHA, J.; ETCHEVERS, J. D.; GRAFTON, K.; ACOSTA-GALLEGOS, J. A. Iron accumulation in seed of common bean. **Plant and Soil**, v. 246, n. 2, p. 175-183, 2002.

RAMALHO, M. A. P.; SANTOS, J. B.; PINTO, C. A. B. P. **Genética na agropecuária**. Lavras: UFLA, 2008.

RIBEIRO, N. D.; ROSA, S. S.; JOST, E.; ROSA, D. P.; POERSCH, N. L.; MAZIERO, S. M. Genetics of phosphorus content in common bean seeds. **Pesquisa Agopecuária Brasileira**, v. 11, n. 3, p. 250-256, 2011.

RIBEIRO, N. D.; MAZIERO, S. M.; PRIGOL, M.; NOGUEIRA, C. W.; ROSA, D. P.; POSSOBOM, M. T. D. F. Mineral concentrations in the embryo and seed coat of common bean cultivars. **Journal of Food Composition and Analysis**, v. 26, n. 1-2, p. 89-95, 2012.

ROSA, S. S.; RIBEIRO, N. D.; JOST, E.; REINIGER, L. R. S.; ROSA, D. P.; CERUTTI, T. Potential for increasing the zinc content in common bean using genetic improvement. **Euphytica**, v. 175, n. 2, p. 207-213, 2010.

TALUKDER, Z. I.; ANDERSON, E.; MIKLAS, P. N.; BLAIR, M. W.; OSORNO, J.; DILAWARI, M.; HOSSAIN, K. G. Genetic diversity and selection of genotypes to enhance Zn and Fe content in common bean. **Canadian Journal of Plant Science**, v. 90, n. 1, p. 49-60, 2010.

WARNER, J. N. A method for estimating heritability. **Agronomy Journal**, v. 44, n. 8, p. 427-430, 1952.

23

Increased transient genetic transformation in immature embryos of Brazilian BR 451 maize co-cultivated with *Agrobacterium tumefaciens*

Dielli Aparecida Didoné, Marilia Rodrigues de Silva, Cássia Canzi Ceccon, Tiago Teixeira, Marilei Suzin and Magali Ferrari Grando*

*Laboratório de Biotecnologia de Plantas, Universidade de Passo Fundo, BR-285, 99052-900, Passo Fundo, Rio Grande do Sul, Brazil. *Author for correspondence. E-mail: magali@upf.com*

ABSTRACT. Genetic engineering has amplified the possibilities of crop breeding and supported sustainability ideals. *Agrobacterium tumefaciens* is the preferred transformation system, since it produces transgenic plants with more stable transgene expression and inheritance. The agrobacteria and plant tissue must be co-cultivated in conditions that allow gene transfer. This study aimed to evaluate how co-cultivation time and temperature affect the transformation of immature maize embryos of the Hi-II hybrid (model genotype) and the Brazilian BR 451 variety with *A. tumefaciens* EHA101:pTF102. The pTF102 plasmid carries a *uidA* reporter gene that enables transient transformation to be quickly verified by GUS histochemical assays. Increasing the co-cultivation period from three to five days at 20°C resulted in a higher number of GUS positive embryos and blue spots per embryo in the BR 451 Brazilian variety, indicating better bacterial T-DNA transfer into the target explant cells. This condition raised the BR 451 response level to match the response level of the Hi-II control genotype, indicating that this Brazilian variety is suitable for genetic transformation.

Keywords: agrobacteria; GUS histochemical assay; genetic engineering; *Zea mays*.

Introduction

Maize (*Zea mays* L.) is a highly important crop for food production, occupying 13% of the agricultural area in the world (USDA, 2014) and being extensively used for human and animal nutrition. Maize has been a target species for biotechnological innovation, being the first genetically modified cereal released in the world market. In Brazil, 88.4% of maize is genetically modified (ISAAA, 2016), but there is no Brazilian transgenic maize commercially released in the current market.

Despite the unmatched commercial success of maize transformation and significant progress made in the molecular techniques for this species, the effective rate of genetic transformation of maize is still insufficient (Yadava et al., 2016). The inefficiency of genetic transformation is attributable to various limitations of the available maize tissue culture and transformation protocols.

Genetic transformation by *A. tumefaciens* allows large DNA segments to be transferred and inserted into the plant genome with a low gene copy number (Zhao et al., 2001). The first transgenic maize plants obtained by agrobacteria were reported by Ishida et al. (1996) followed by other groups (Zhao et al., 2001; Frame et al., 2002; Ombori, Muoma, & Machuka, 2014; Lee & Zhang, 2016; Souza et al., 2017). To succeed in maize transformation, it is necessary to use target tissues with high rates of cell division, such as immature zygotic embryos. Maize transformation also requires use of highly *in vitro* responsive maize genotypes and specific *A. tumefaciens* strains, since this bacterium does not naturally infect monocotyledonous plants.

While several maize genotypes have been effectively transformed (Ji, Xu, & Wang, 2013, Ombori et al., 2014; Souza et al., 2017), Hi-II hybrid maize has been widely used for genetic transformation (Zhao et al., 2001; Frame et al., 2002; Frame, Main, Schick, & Wang, 2011; Lee & Zhang, 2016; Nahampun, Lopez-Arredondo, Xu, Herrera-Estrella, & Wang, 2016) and is considered the model for maize plant transformation. However, the poor agronomic traits of the Hi-II hybrid (Ishida, Hiei, & Komari, 2007; Que et al., 2014) motivate the search for other maize genotypes that respond well to transformation. Since the Brazilian BR 451 maize variety presents high protein quality, good agronomical performance and adaptation to different regions of Brazil (Guimarães, Parentoni, & Pacheco, 2004), as well as a suitable *in vitro* response (Carvalho, Bohorova, & Bordallo, 1997; Petrillo et al., 2008), it offers a Brazilian alternative to the American Hi-II hybrid.

Co-cultivation is a critical step in the maize genetic transformation protocol (Ishida et al., 2007), and a three-day co-cultivation period is most often used (Frame et al., 2002; 2011; Ishida et al., 2007; Nahampun et al., 2016). In addition, the co-cultivation temperature also influences transformation efficiency (Huang & Wey, 2005).

This study aimed to test whether modifying co-cultivation period and temperature can improve *Agrobacterium*-mediated maize genetic transformation of Hi-II hybrid and BR 451 maize and verify if the Brazilian BR 451 variety is suitable for genetic transformation.

Material and method

Experiments were conducted in the Plant Biotechnology Laboratory at the University of Passo Fundo (CQB 272/08). The BR 451 variety (Guimarães et al., 2004), courtesy of Embrapa Maize and Sorghum, and the Hi-II American hybrid (A188 x B73) were established at a greenhouse for zygotic embryo production.

The *A. tumefaciens* disarmed strain EHA 101 containing the pTF102 binary plasmid was used. This plasmid harbors the *uidA* reporter gene under control of CaMV35S constitutive promoter and contains an intron in the coding region (Paz et al., 2004). Bacteria from a glycerol stock were grown in YEP solid medium (5 g L^{-1} yeast extract, 10 g L^{-1} peptone, 5 g L^{-1} NaCl, and 15 g L^{-1} Bacto-agar, pH 6.8) supplemented with 100 mg L^{-1} spectinomycin and 50 mg L^{-1} kanamycin for two days at 28°C. Isolated colonies from this main board were cultured in the same medium supplemented with 100 μM acetosyringone for three days at 19°C. Two loops of this culture were suspended in 5 mL infection medium (pH 5.2) (Frame et al., 2006) containing N6 salts (Chu et al., 1975), vitamins (0.5 mg L^{-1} thiamine HCl, 2.0 g L^{-1} glycine, 0.5 mg L^{-1} pyridoxine HCl, 0.5 mg L^{-1} nicotinic acid), 100 mg L^{-1} hydrolyzed casein, 2.88 g L^{-1} L-proline, 2.0 mg L^{-1} 2,4-D, 68.5 g L^{-1} sucrose, 36 g L^{-1} glucose, supplemented with 100 μM acetosyringone. Cell density was adjusted to OD_{550nm} = 0.3 - 0.4, and bacteria were incubated for two hours at room temperature in a horizontal shaker at 100 rpm.

Approximately 11 days after pollination, 1.2 to 1.8 mm immature zygotic embryos of Hi-II and BR 451 were collected and immersed in 1 mL infection medium containing *A. tumefaciens* for five min in the dark. After infection, explants were transferred to solid co-cultivation medium containing the same infection medium formulation with the following modifications: without glucose; sucrose reduced to 30 g L^{-1}, with 0.5 g L^{-1} MES; 10 μM silver nitrate; 8.0 g L^{-1} agar; 5.8 pH (Frame et al., 2006). Embryos were placed in Petri dishes with scutella in contact with the culture medium. Petri dishes were sealed with porous tape and kept in the dark. Four treatments were conducted, with co-cultivation time and temperature variations, as follows: (T1) three days at 20°C, (T2) three days at 28°C, (T3) five days at 20°C, and (T4) five days at 28°C. After the co-cultivation, all the agrobacteria co-cultured explants were transferred to the resting medium, containing the same co-cultivation medium formulation, with the following modifications: without acetosyringone, with 100 mg L^{-1} cefotaxime and with 100 mg L^{-1} vancomycin. Petri dishes were sealed with porous tape and kept in the dark at 28 °C for seven days. This step aimed to eliminate the bacteria and allow embryogenic callus development.

Transient transformation was verified by *uidA* reporter gene expression in immature embryos by the β-glucuronidase (GUS) histochemical assay, performed in accordance to Jefferson, Kavanagh, and Bevan (1987) five days after infection in all treatments. Embryos were incubated in a solution containing X-glucuronidase substrate for 16 hours at 37°C. After incubation, the reaction was removed and embryos were stored in 70% ethanol until analysis time. The *uidA* gene encodes the GUS enzyme that produces an indigo blue color dimer in the presence of the substrate, enabling transformation events to be visualized and counted.

The frequency of embryos with blue spots (GUS positive embryos), and the number of blue spots per embryo were accessed. Embryos not subjected to infection were co-cultivated in each treatment and used as control.

The genotypes (Hi-II and BR 451) were evaluated in independent experiments carried out in the same manner, as follows. The experimental design was completely randomized, with five replicates per treatment (co-cultivation time and temperature). The experimental unit was a Petri dish containing 30 explants. For transient transformation analysis, a random sample of five explants per repetition was taken, totaling 25 embryos per treatment evaluated by GUS histochemical assay. Means ± 1 standard deviation was used for treatment comparison. The experiment was repeated three times.

Result and discussion

Differences were not observed in the number of GUS positive embryos or number of blue sports per embryo in Hi-II maize when the co-cultivation was performed in different conditions. On average, 72% of embryos were transformed, and 14.6 blue spots per embryo were observed (Figure 1A and B). The blue spots represent the location where the reporter gene was inserted in the embryo and the number of blue spots per embryo represents the intensity of the transient genetic transformation. The most commonly used co-culture conditions for Hi-II maize genetic transformation protocols are three-day co-cultivations at 20°C (T1) (Frame et al., 2002; 2006; 2011; Omer, Matheka, Ali, & Machuka, 2013; Ombori et al., 2014; Nahampun et al., 2016). In our analysis, a longer period of co-cultivation (five days) or higher temperature did not increase the T-DNA transfer from bacteria to Hi-II embryos.

The overall transformation frequency observed in the Brazilian BR 451 variety was lower (43%) compared to Hi-II (72%). Likewise, the number of transformation sites (blue spots per embryo) was also lower for the BR 451 (11.7) compared to Hi-II (14.6) (Figure 1). Differences among maize genotypes in the capacity to be infected have been widely reported in the literature (Huang & Wei, 2005; Frame et al., 2006; Omer et al., 2013; Ombori et al., 2014; Souza et al., 2017), with genotype dependence being the main limitation in maize genetic transformation by this method (Hiei, Ishida, & Komari, 2014). Genotypic differences can be explained by compatibility variation between maize genotype and the bacterial strain employed in the transformation or the T-DNA transference machinery encoded by the bacteria and plant cell genomes (Ombori et al., 2014; Lowe et al., 2016). The molecular and chemical signaling that allow agrobacteria to connect and insert T-DNA into the host plant cell is regulated by turning genes on and off, which is strongly influenced by the plant genotype (Tzfira, Li, Lacroix, & Citovsky, 2004).

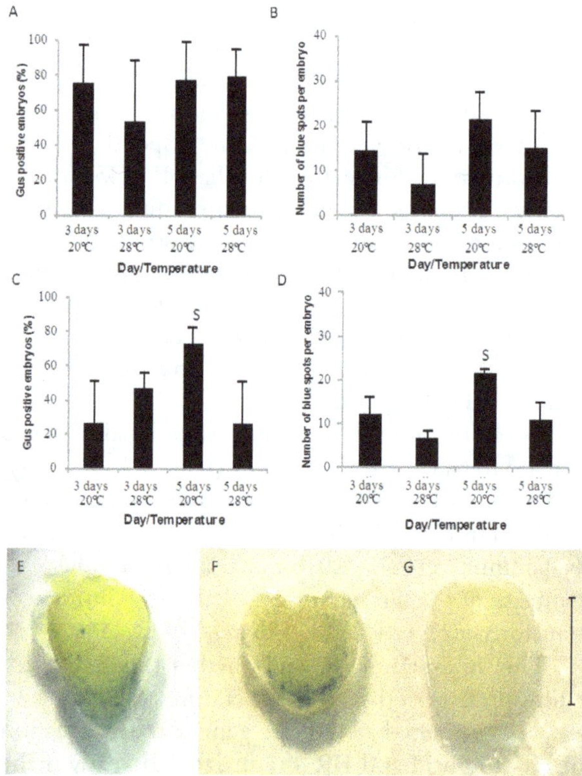

Figure 1. Transient expression of GUS observed in immature zygotic maize embryos co-cultivated with *A. tumefaciens* at different conditions: (A) Frequency of GUS positive Hi-II embryos, (B) mean number of blue spots per Hi-II embryo, (C) Frequency of GUS positive BR 451 embryos, (D) mean number of blue spots per BR 451 embryo, (E) Gene *uidA* expression in Hi-II immature embryos co-cultivated for five days at 20°C, (F) Gene *uidA* expression in BR 451 embryos co-cultivated for five days at 20°C, and (G) negative control not infected with agrobacteria. Bar = 1 mm, S = (superior) Means +1 SD.

The co-cultivation conditions affected transient transformation of BR 451 embryos. The T3 treatment (five days at 20°C) promoted superior GUS positive embryos (73%) (Figure 1C), as well as the superior number of blue spots observed per embryo (23.1) compared to other treatments (Figure 1D). Compared with the standard co-cultivation conditions used for maize (three days at 20°C), two more days in co-cultivation doubled the number of embryos transformed and the number of blue spots per embryo. Increased transformation suggests that prolonged co-cultivation increases the chance of agrobacteria binding to the embryo surface, which increases the chance of T-DNA transference from agrobacteria to the plant cell nuclei. In contrast, Huang and Wey (2005) found no influence of co-cultivation period (from three to six days) on bacterial T-DNA transient transformation in immature embryos from three elite maize strains.

In accordance with Alimohammadi and Gherieh-Najjar (2009), co-cultivation temperature should be investigated, as there are different responses among monocots, genotypes, explants and agrobacteria strains. Low temperatures (19°C) favor agrobacteria T-pili formation, and higher temperatures (28°C) inhibit T-pili formation (Fullner, Lara, & Nestert, 1996). The T-pilus is an appendix or filamentous channel attached to the surface of agrobacteria, composed of Ti plasmid-encoded proteins. It is thought that this secretion system is the primary means of T-DNA transport from the bacterial cell to the plant cell (Mccullen & Binns, 2006).

According to Hiei et al. (2014), high temperature during co-cultivation may also result in agrobacteria overgrowth in the co-cultivation medium. The present study showed that high temperature (28°C) was unfavorable for transient transformation but did not promoted bacterial overgrowth even during a prolonged co-culture period. Bacterial overgrowth is undesirable because it can cause plant tissue death.

In our experiment, five day co-cultivation at 20°C increased the BR 451 response level to the level of Hi-II, since in these same conditions, the Hi-II genotype produced 71% GUS positive embryos with approximately 21 blue spots per embryo. Therefore, by modifying the BR 451 co-cultivation condition, it was possible to match the transformation performance of the Hi-II model genotype. The *uidA* gene expression in Hi-II and BR 451 immature embryos co-cultivated under these conditions is shown in Figure 1E and F.

The success of obtaining genetically transformed maize plants depends on the plant genotype, which must respond well to both *in vitro* culture and genetic transformation. The Hi-II maize hybrid and its source lines (A188 and B73) have these characteristics and have been used in transgenic plants production worldwide (Frame et al., 2002; 2011; Zhao et al., 2001). However, the Hi-II hybrid presents poor agronomic performance (Ishida et al., 2007; Que et al., 2014). The use of a genotype showing good *in vitro* cultivation response and good agronomic characteristics, such as the BR 451 (Grando et al., 2013; Petrillo et al., 2008), accelerates commercial transgenic plant production, since the gene inserted in the transgenic event can be transferred to elite lines without the transfer of additional unwanted traits. Therefore, improvement in the genetic transformation of BR 451 opens the possibility of implementing its use in genetic engineering.

Genetic engineering is an invaluable tool that enables the introduction of new traits, endogenous gene overexpression and silencing of undesirable features in important crop plants. Hence, the "New Breeding Techniques" era, including the newer genome editing technologies, such as the CRISPR/Cas9 system, is opening up a new dimension in transformation protocols and workflow (Yadava et al., 2016). The utilization of these newly arising molecular techniques to explore and modify the genome will require optimization of DNA delivery, plant *in vitro* regeneration and a more efficient transformation platform (Char et al., 2017).

This paper reports an increase in agrobacteria gene transfer for the BR 451 Brazilian genotype, enabling its use for genetically modified maize production. This aspect is particularly important for Brazil, where there are few research groups applying this technology for maize.

Conclusion

By prolonging the period of co-cultivation of immature embryos with agrobacteria, it is possible to increase the transient genetic transformation of the Brazilian genotype BR 451, enabling its use in genetic transformation studies.

Acknowledgements

The authors would like to thank the Plant Transformation Facility of Iowa State University for technical support and Embrapa Milho e Sorgo for providing the BR 451 seeds. The authors are also grateful to the Fundação de Amparo à Pesquisa do Estado do Rio Grande do Sul (Fapergs) and Coordenação de Aperfeiçoamento de Pessoal de Nível Superior (Capes) for student scholarships and

Polo de Inovação Tecnologia, Governo do Estado do Rio Grande do Sul for financial support.

References

Alimohammadi, M., & Gherieh-Najjar, M. B. (2009). Agrobacterium-mediated transformation of plants: sic principles and influencing factors. *African Journal of Biotechnology*, *8*(20), 5142-5148.

Carvalho, C. H. S., Bohorova, N., & Bordallo, P. (1997). Type II callus production and plant regeneration in tropical maize genotypes. *Plant Cell Reports*, 7(1), 73-76, 1997. doi 10.1007/s002990050355.

Char, S. N., Neelakandan, A. K., Nahampun, H., Frame, B., Main, M., Spalding, M. H., ... Yang, B. (2017). An Agrobacterium-delivered CRISPR/Cas9 system for high-frequency targeted mutagenesis in maize. *Plant Biotechnology Journal*, *15*(2), 257–268. doi 10.1111/pbi.12611.

Chu, C. C., Wang, C. C., Sun, C. S., Hsu, C., Yin, K. C., & Chu, C. Y. (1975). Establishment of an efficient medium for another culture of rice through comparative experiments on the nitrogen source. *Scientia Sinica*, *18*(5), 659-668.

Frame, B. R., Shou, H., Chikwam, R. K., Zhang, Z., Xiang, C., Fonger, T. M., Pegg, ... Wang, K. (2002). *Agrobacterium tumefaciens*-mediated transformation of maize embryos using a standard binary vector system breakthrough technologies. *Plant Physiology*, *129*(1), 13-22. doi 10.1104/pp.000653

Frame, B. R., Mcmurray, J. M., Fonger, T. M., Main, M. L., Taylor, K. W., Paz, M. M., ... Wang, K. (2006). Improved *Agrobacterium*-mediated transformation of three maize inbred lines using MS salts. *Plant Cell Reports*, *25*(10), 1024-1034. doi 10.1007/s00299-006-0145-2.

Frame, B., Main, M., Schick, R., & Wang, K. (2011). Genetic transformation using maize immature zygotic embryos. In: Thorpe, T., Yeung, E. *Methods in Molecular Biology (Methods and Protocols)*, *710*, 327-341. Humana press. doi 10.1107/978-1-61737-988-8_22.

Fullner, J. K., Lara, C. J., & Nestert, E. W. (1996). Pilus assembly by agrobacterium T-DNA transfer genes. *Science*, *273*(5278), 1107-1109.

Grando, M. F., Varnier, M. L., Silva, M. R., Emydio, B. M., Pereira, L. R., & Suzin, M. (2013). Immature tassels as alternative explants in somatic embryogenesis and plant regeneration in South Brazilian maize genotypes. *Acta Scientiarum. Agronomy*, *35*(1), 39-47. doi 10.4025.actasciagrob.v35i1.15545.

Guimarães, P. E. O., Parentoni, S. N., & Pacheco, C. A. P. (2004). BR 451 – Milho de alta qualidade proteica. Sete Lagoas, MG: Embrapa Milho e Sorgo. (Comunicado Técnico, 111).

Hiei, Y., Ishida, Y., & Komari, T. (2014). Progress of cereal transformation technology mediated by *Agrobacterium tumefaciens*. *Frontiers in Plant Science*, 5(628), 1-11. doi 10.3389/fpls.2014.00628.

Huang, X. Q., & Wei, Z. M. (2005). Successful *Agrocterium*-mediated genetic transformation of maize elite inbred lines. *Plant Cell, Tissue and Organ Culture*, *83*(2), 187-200. doi 10.1007/s11240-005-5772-8.

ISAAA. (2016). *Global Status of Commercialized Biotech/GM Crops*: 2016. ISAAA Brief No. 52. Ithaca, NY: ISAAA.

Ishida, Y., Saito, H., Ohta, S., Hiei, Y., Komari, T., & Kumashiro, T. (1996). High efficiency transformation of maize (*Zea mays* L.) mediated by *Agrobacterium tumefaciens*. *Nature Biotechnology*, *14*(6), 745–750. doi 10.1038/nbt0696-745.

Ishida, Y., Hiei, Y., & Komari, T. (2007). *Agrobacterium* mediated transformation of maize. *Nature Protocols*, *2*(7), 1614-1621. doi 10.1038/nprot.2007.241.

Jefferson, R. A., Kavanagh, T. A., & Bevan, M. W. (1987). GUS fusions: Beta-glucuronidase as a sensitive and versatile gene fusion marker in high plants. *Embo Journal*, *6*(13), 3901-3907.

Ji, Q., Xu, X., & Wang, K. (2013). Genetic transformation of major cereal crops. *International Journal of Developmental Biology*, *57*(6-8), 495–508. doi 10.1387/ijdb.130244kw.

Lee, H., & Zhang, Z. J. (2016). Maize (*Zea mays*) Hi-II transformation via Agrobacterium-mediated T-DNA transfer. *Current Protocols in Plant Biology*, *1*(1), 121-137. doi 10.1002/cppb.20016.

Lowe, K., Wu, E., Wang, N., Hoerster, G., Hastings, C., Cho, M. J., ... Wang, L. (2016). Morphogenic regulators baby boom and wuschel improve monocot transformation. *The Plant Cell*, *28*(9), 1998-2015. doi 10.1105/tpc.16.00124.

Mccullen, A. C., & Binns, A. N. (2006). *Agrobacterium tumefaciens* and plant cell interactions and activities required for interkingdom macromolecular transfer. *Annual Review of Cell Development Biology, 22(1)*, 101-127. doi 10.1146/annurev.cellbio.22.011105.102022.

Nahampun, H. N., López-Arredondo, D., Xu, X., Herrera-Estrella, L., & Wang, K. (2016). Assessment of ptxD gene as an alternative selectable marker for Agrobacterium-mediated maize transformation. *Plant Cell Reports*, *35*(5), 1121-1132. doi 10.1007/s00299-016-1942-x.

Ombori, O., Muoma, J., & Machuka, J. (2014). *Agrobacterium*-mediated genetic transformation of selected tropical inbred and hybrid maize (*Zea mays* L.) lines. *Plant Cell Tissue Organ Culture*, *113*(1), 11–23. doi 10.1007/s11240-012-0247-1.

Omer, R. A., Matheka, J. M., Ali, A. M., & Machuka, L. (2013). Transformation of tropical maize with the NPK1 Gene for drought tolerance. *International Journal of Genetic Engineering*, *3*(2), 7-14. doi j.ijge.20130302.01.

Paz, M. M., Shou, H., Guo, Z., Zhang, Z., Banerjee, A. K., & Wang, K. (2004). Assessment of conditions affecting *Agrobacterium*-mediated soybean transformation using the cotyledonary node explant. *Euphytica*, *136*(2), 167-179. doi 10.1023/B:EUPH.0000030669.75809.dc

Petrillo, C. P., Carneiro, N. P., Purcino, A. A. C., Carvalho, C. H. S., Alves, J. D., & Carneiro, A. A. (2008). Optimization of particle bombardment parameters for the genetic transformation of Brazilian maize inbred lines. *Pesquisa Agropecuária Brasileira, 43*(3), 371-378. doi 10.1590/S01oo-204X2008000300012.

Que, Q., Elumalai, S., Li, X., Zhong, H., Nalapalli, S., Schweiner, M., ... Chilton, M. D. M. (2014). Maize transformation technology development for commercial event generation. *Frontiers in Plant Science, 5*(379), 1-19. doi 10.3389/fpls.2014.00379.

Souza, R. A. V. D., Alves, M. D. C., Carneiro, N. P., Barros, B. D. A., Borém, A., & Carneiro, A. A. (2017). *Agrobacterium*-mediated genetic transformation of a tropical elite maize line. *Crop Breeding and Applied Biotechnology, 17*(2), 133-140. doi 10.1590/1984-70332017v17n2a20.

Tzfira, T., Li, J., Lacroix, B., & Citovsky, V. (2004). Agrobacterium T-DNA integration: molecules and models. *Trends in Genetics, 20*(8), 375-83. doi 10.1016/j.tig.2004.06.004.

United States Department of Agriculture [USDA]. (2014). Retrieved on Jan. 15, 2015 from http://apps.fas.usda.gov/psdonline/psdhome.aspx.

Yadava, P., Abhishek, A., Singh, R., Singh, I., Kaul, T., Pattanayak, A., & Agrawal, P. K. (2016). Advances in maize transformation technologies and development of transgenic maize. *Frontiers in Plant Science,* 7(1949), 10-12. doi 10.3389.fpls.2016.01949.

Zhao, Z., Gu, W., Cai, T., Tagliani, L., Hondred, D., Bond, D., ... Pierce, D. (2001). High throughput genetic transformation mediated by *Agrobacterium tumefaciens* in maize. *Molecular Breeding, 8*(4), 323-333. doi 10.1023/A:1015243600325.

24

Genetic control of partial resistance to Asian soybean rust

Juliana Araújo Santos Martins[1*] and Fernando Cézar Juliatti[2]

[1]*Instituto Federal de Educação, Ciência e Tecnologia do Triângulo Mineiro, Fazenda Sobradinho, s/n, Cx. Postal 592, 38400-974, Uberlândia, Minas Gerais, Brazil.* [2]*Instituto de Ciências Agrárias, Universidade Federal de Uberlândia, Uberlândia, Minas Gerais, Brazil. *Author for correspondence. E-mail: julianaaraujo@iftm.edu.br*

ABSTRACT. The genetic control of rust resistance was studied using the Caiapônia x IAC-100 and Luziânia x Potenza crosses. The F_2 and F_3 generations were evaluated. Rust severity was quantified through visual assessment of the middle third of three leaflets per plant and performed by three different evaluators; the average score was calculated foreach individual plant. From these data, we estimated the mean and variance of the genetic components by employing the weighted least squares method. The estimated number of genes and broad- and narrow-sense heritabilities were also obtained. It was concluded that rust resistance is a characteristic controlled by 2 to 23 genes that are predominantly dominant. The estimate of narrow-sense heritability was greater than 70% for the Caiapônia x IAC-100 cross, and the wide-sense heritability was greater than 60% for the Luziânia x Potenza cross; thus, it is possible to successfully select resistant individuals in early generations.

Keywords: *Glycine max*, *Phakopsora pachyrhizi*, dominance variance, heritability, race non-specific disease resistance.

Introduction

The Asian soybean rust (ASR) caused by the fungal pathogen *Phakopsora pachyrhizi* was identified in Brazil in 2001 and quickly infected soybeans in several countries in South America. ASR is responsible for severe economic losses because of its high damage potential and its constant presence in almost all soybean-producing areas recently (HARTMAN et al., 2005).

Brazilian breeding programs have not yet developed resistant cultivars, and therefore, chemical control is still used to reduce the losses caused by ASR. Currently, the primary form of control is based on the use of fungicides of different classes and action modes (MILES et al., 2007). This is partially because the fungus has several races with multiple virulence alleles (SINCLAIR; HARTMAN, 1999), which suggests that the ASR pathogen has high genetic variability, mainly in the Brazilian savanna (YORINORI et al., 2005). Thus, the identification of resistant cultivars is necessary.

Genetic studies confirm the existence of additive genetic effects for ASR (PIEROZZI et al., 2008; RIBEIRO et al., 2007). Quantitative studies of the genes that express resistance to ASR genotypes are needed to support the development of effective longer-lasting resistance in new cultivars. However, in Brazil, because of the recent re-emergence of the pathogen, there are few studies concerning the inheritance of resistance to soybean rust (AKAMATSU et al., 2004; PIEROZZI et al., 2008; RIBEIRO et al., 2007).

Genetic models were fitted to these data to obtain estimates of the additive, dominance, epistatic, and environmental components. The fits enabled predictions of the cross potential to generate inbred lines with greater resistance to ASR

(CRUZ, 2005). Previous studies by Azevedo et al., 2007, Santos et al., 2007 and Silva et al., 2007, demonstrated partial resistance for IAC-100 and Potenza.

This study investigates on genetic control of resistance to Asian soybean rust in lines derived from IAC-100 and Potenza.

Material and methods

The experiment was carried out in a greenhouse at the Agricultural Sciences Institute, which belongs to the Federal University of Uberlândia, Uberlândia, Brazil, and also at CapimBranco farm, which belongs to the Federal University of Uberlândia, Uberlândia, Minas Gerais State (located at 55°18'23"S and 48°17'19"W). The populations were assessed at the Agroteste Experimental Station, Uberlândia, Minas Gerais State (located at 21°12'58"S and 45°03'18"W). In July 2007, (Caiapônia x IAC-100) and (Luziânia x Potenza) were crossed in greenhouse conditions to obtain the F_1 population. The breeding of the F_1 and F_2 generations and the multiplication of the parents was also performed in a greenhouse in the 2008 to 2009 period.

For the 2009-2010 crop, the parents and the F_2 and F_3 generations were assessed under field conditions. The trials were arranged in a randomised complete block design with four replications. The experimental conditions were those commonly used in soybean crop management. During the experiment, chemical control of pests and manual weed control were performed as necessary.

To reduce the risk of the evolving ASR epidemic (because sowing was delayed) and to better assess partial resistance, a fungicide of the chemical group strobilurin (azoxystrobin) was sprayed. 300 mL of commercial product per hectare were used for the first spraying. In the second spraying, a mixture of triazole and strobilurin (pyraclostrobin + epoxiconazole) was used; the dose was 500 mL per hectare. The spraying was performed at stages R_1 and R_5 (FEHR; CAVINESS, 1977). This action, therefore, was intended to improve the reliability of the assessment of partial resistance.

The assessment was performed at stage $R_{5.5}$ (FEHR; CAVINESS, 1977), 80 days after sowing, because this stage was the developmental stage in which there was the greatest expression of the severity of symptoms in susceptible strains. The disease severity was scored by three evaluators who inspected the middle third of three leaflets per plant and used the diagrammatic scale of disease severity scores, which considers the percentage of leaf tissue that is infected (the scale uses thresholds of 0.5, 5, 20, 40, 60, and 80%).

Statistical and genetic analysis

Using the scores of individual plants, the analysis of variance (ANOVA) was calculated for each cross using the parents and the F_2 and F_3 generations. The means and variances of the severity scores of the parents and the F_2 and F_3 generations were estimated. Using these data, the mean components were also estimated using the weighted least squares method (WLS) (MATHER; JINKS, 1984; RAMALHO et al., 1993; ROWE; ALEXANDER, 1980). The additive-dominance model without epistasis was considered. The genetic variance and environmental components were estimated using the WLS method iterative (MATHER; JINKS, 1984; RAMALHO et al., 1993). The components of the mean and variance were estimated using the computer software Mapgen. The software was also used to estimate the number of genes and the broad- and narrow-sense heritability coefficients (BERNARDO, 2002):

$$h^2_{rF_2} = \hat{\sigma}^2_A / \hat{\sigma}^2_{F_2},\ h^2_{rF3} = 1{,}5\,\hat{\sigma}^2_A / \hat{\sigma}^2_{F_3},\ h^2_{aF_2} = \hat{\sigma}^2_{F_2} - \hat{\sigma}^2_E / \hat{\sigma}^2_{F_2} \text{ and } h^2_{aF_3} = \hat{\sigma}^2_{F_3} - \hat{\sigma}^2_E / \hat{\sigma}^2_{F_3}.$$

To estimate the number of genes (k) involved in the trait control and to consider what fraction are dominant, the expression of Burton (1951) was used by Ramalho et al. (1993):

$$k = \frac{\frac{1}{4}(\frac{3}{4} - \mu + \mu^2)D^2}{\hat{\sigma}^2_{F_2} - \hat{\sigma}^2_E},$$

where $\mu = \frac{\overline{F}_1 - \overline{P}_1}{\overline{P}_2 - \overline{P}_1}$, D = $\overline{P}_2 - \overline{P}_1$, $\overline{P}_1$ is the mean trait for parent 1, $\overline{P}_2$ is the mean trait for parent 2, $\hat{\sigma}^2_{F_2}$ is the phenotypic variance of the F_2 family, $\hat{\sigma}^2_E$ is the environmental variance and $\overline{F}_1$ is the mean of the F_1 family.

Results and discussion

The results of the variance analysis shown in Table 1 indicate that there were significant differences ($p \leq 0.01$) among the parent and segregating generations for the two crosses used. The existence of variation is an essential condition for studying the genetic control of the disease resistance trait.

Table 1. Analysis of the variance of the severity scores of Asian rust for the parent and the cross-derived F_2 and F_3 generations of soybeans.

		MSQ		MSQ
Source of variation	Degrees of freedom	Caiapônia x IAC100	Degrees of freedom	Luziânia x Potenza
Blocks	3	8,36	3	11,12
Generations	5	513,17**	3	671,17**
Residual	15	30,10	9	13,0
Mean		27,01		24,0
Coefficient Of variation (%)		20,31		15,02

**: significant at the level of 1% of probability (p > 0.01) by F test.

The frequency distributions of the F_2 and F_3 severity scores of each cross are shown in Figures 1 and 2. The frequency distribution in all generations suggests that the trait is controlled polygenically; if the trait were monogenic, without the environmental effect, two or three phenotypic classes would be expected, but this was not the case. The absence of a distribution with well-defined phenotypic frequencies has already been noted by others (PIEROZZI et al., 2008; RIBEIRO et al., 2007).

For the Caiapônia x IAC-100 cross, the results shown in Figure 1 indicate the presence of dominance in the trait control. IAC-100 demonstrated strong resistance and had a score below 30%. The F_2 and F_3 generations demonstrated similar performance to that of the parents, but with some individuals had higher scores. The results presented by Garcia et al. (2008) and Pierozzi et al. (2008) for PI200526 reinforce the observation that some dominant genes may control the trait.

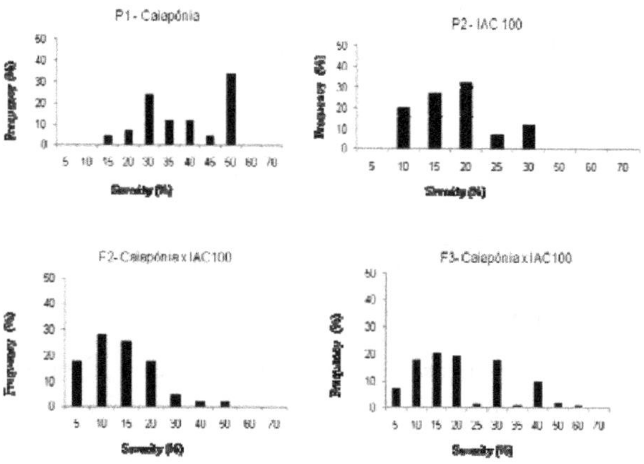

Figure 1. Frequency distribution of these verity of soybean rust in each generation for the Caiapônia x IAC-100 cross grown in Uberlândia, Minas Gerais State, 2011.

The Luziânia x Potenza cross generations showed a continuous distribution for the trait (see Figure 2). Although the parent Potenza is partially resistant to ASR (SILVA et al., 2007), in this experiment, the scores of disease severity for Potenza were significantly higher than for the parent Luziânia. This fact maybe related to environmental conditions and the variability of the pathogen (JULIATTI et al., 2003). The frequency distributions of the severity scores suggested that quantitative inheritance controls the resistance to ASR, which has been observed in other studies that used if different parents (GARCIA et al., 2008; RIBEIRO et al., 2007).

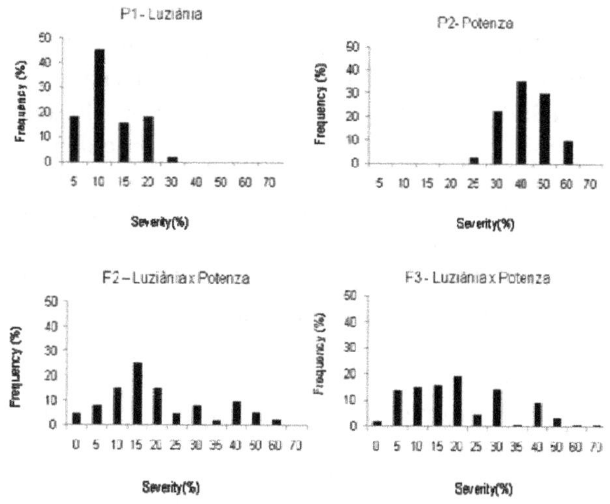

Figure 2. Frequency distribution oftheseverity scores of soybean rustin each generation for the LuziâniaxPotenza cross grown in Uberlândia, Minas Gerais State, 2011.

Components of the means and variances

Caiapônia x IAC-100 cross

The data inTable2 demonstrate that the means of the F_2 and F_3 generations' scores were different from the mean scores of the parents. In principle, this result demonstrates that dominant genes participate in the control of partial resistance to ASR in the IAC-100 parent.

This last observation may be confirmed by the estimates of the mean components. Observe that the estimate of *d*, the contribution of heterozygous loci in relation to the midpoint of the two homozygous parents, was non zero and negative, which indicates that the dominance is to reduce the expression of the trait (Table 2). It was also shown that the mean dominance degree (MDD) was greater than 1 (the value was 2.64).

Estimates of the variance components for the generations were also obtained. The values of the dominance variance were negative (Table 2). Scott and Jones (1989) also reported negative estimates of the dominance variance, which indicates a small or negligible dominance effect; they believe this is due to the small population size used.

Table 2. The number of individuals per population cross and the mean, the variance, and the components of the mean and the variance for the soybean rust severity scores of Caiapônia x IAC-100.

Population	Nº individuals	Mean (%)	Variance
Caiapônia	40	37.63	120.50
IAC 100	40	18.25	39.17
F_2	120	15.13	89.09
F_3	400	21.46	146.41
Mean Components[1]	Estimates	Variance components	Estimates
m	27.89 ± 0.06[2]	$\hat{\sigma}^2_A$	62.70 [55.94; 70.76][3]
a	9.66 ± 0.08	$\hat{\sigma}^2_D$	-[4] -
d	-25.59 ± 0.21	$\hat{\sigma}^2_E$	79.84 [59.70; 112.27]
MDD	2.64	-	-
χ^2	0.01[ns]		
R^2 (%)	99.9	-	96.6

[1]*m*: mean of the homozygous genotypes; *a*: deviation of the homozygous genotypes relative to the midpoint; d: deviation of the heterozygote relative to themidpoint. [2]Standard error. [3]Lower and upper limits of the reliable interval for the variances. [4]The estimates $\hat{\sigma}^2_D$ are negative.

Negative estimates can occur when the variance component is very small. Another cause may have been the failure of the assumptions of the Warner-Mather method, in which it is assumed that the sources of genetic variation are only dominant additive and that the environmental variance component for the F_2 generation is the mean of the environmental variance components for the backcross generations. Furthermore, there is the possibility of human error because the disease assessment used to assign the scores to the plants was made visually (LOBO et al., 2005).

Luziânia x Potenza cross

In this study, the heterozygous dominance effect (*d*) and the additive effect contribute to decreased susceptibility or increased resistance to ASR (Table 3).

The component of the additive effect (a) would never be negative, whereas the signal component of the effects of dominance deviations (d) depends on the predominant direction of dominance (MATHER; JINKS, 1984). The existence of negative additive components (a) is because that the parent P_1 does not always refer to the genotype with the highest trait expression, which in this case would be the genotype most susceptible to the disease. P_1, Luziânia, was more resistant and received the lowest scores and thus the lowest mean; therefore, it had less disease severity. The negative value for the additive effect (a) is a result of the low mean (LOBO et al., 2005).

It has been found that soybean lines' resistance to the pathogen may vary over time. In principle, this variance indicates that in some strains, the resistance mechanism depends on environmental factors or physiological effects. It has been observed that environmental conditions are the main factor that controls the disease severity and that more severe disease cases occur under conditions of prolonged leaf wetness (DEL PONTE et al., 2006). This fact was confirmed in the severity assessment because high variance in the disease severity was observed among the parents, which indicates greater instability because of environmental effects on the disease. The instability of the rust severity displayed by some soybean lines has been observed by others (PIEROZZI et al., 2008).

Table 3. The number of individuals per population cross and the mean, the variance, and the components of the mean and the variance for the soybean rust severity scores of Luziânia x Potenza.

Population	Nº individuals	Mean (%)	Variance
Luziânia	40	12.16	31.86
Potenza	40	42.43	96.76
F_2	226	20.77	194.62
F_3	400	20.29	162.54
Mean components[1]	Estimates	Variance components	Estimates
m	25.52 (3.28)[2]	$\hat{\sigma}^2_A$	32.4 [29.25; 36.07][3]
a	-14.13 (3.67)	$\hat{\sigma}^2_D$	96.55 [83.23; 113.36]
d	-13.45 (10.83)	$\hat{\sigma}^2_E$	65,68 [49.11; 92.36]
MDD	0.95	-	-
χ^2	17.44*		
R^2 (%)	99.20	-	-

[1]*m*: mean of the homozygous genotypes; *a*: deviation of the homozygous genotypes relative to the midpoint; *d*: deviation of the heterozygote relative to the midpoint. [2]Standard error. [3]Lower and upper limits of the reliable interval for the variances.

The estimate of *d* was lower than the estimate *a*, which demonstrates that the additive effects were more important than non-additive effects. The MDD calculated using the components of the mean was estimated to be 0.95, which indicates that dominance is present in the control of the resistance. However, when the variance components were considered, the non-additive component $\hat{\sigma}^2_D$ was greater than $\hat{\sigma}^2_A$. This is because the value *d* is calculated using the sum of the deviations such that values with opposite signs can cancel. The high value of $\hat{\sigma}^2_D$ associated with the low value (d) demonstrates that this trait is primarily controlled by dominant alleles. However, these alleles do not act in the same manner, i.e., there are dominant alleles that cause ASR resistance, but there are also dominant alleles that decrease the resistance.

Estimating the variance components enabled a determination of the true estimate of the genetic effects because, even when there are deviations of opposite signs, the individual effects are not clearly indicated because the individual effects of each locus were squared.

However, unlike the results found in this study, some previous studies have reported a predominance of additive effects in the genetic control of soybean rust resistance (PIEROZZI et al., 2008; RIBEIRO et al., 2007). According to Ribeiro et al. (2007) and Pierozzi et al. (2008), quantitative genetic resistance to rust is controlled predominantly by genes that act additively and which are dispersed in the parents.

Estimation of the heritability and the number of genes

Estimates of the heritability and the number of genes are shown in Table 4. It is noteworthy that the heritability is a property of the trait that depends on the population and the environmental conditions to which the individuals were submitted (RAMALHO et al., 1993).

Table 4. Estimates of heritability in broad and narrow sense and the likely number of genes.

Crosses	F_2 population (%)		F_3 population (%)		k
	h_a^2	h_r^2	h_a^2	h_r^2	
Luziânia x Potenza	66.25	16.65	60.0	30.0	2.0
Caiapônia x IAC 100	10.40	70.40	45.5	64.23	23.0

The estimates of the narrow-sense heritability (h_r^2) for the F_3 population were 30 and 64.23%; these values are similar to those observed by Pierozzi et al. (2008) for populations in the F_2 generation. The estimates of Ribeiro et al. (2007) ranged from 23.12 to 55.83%, which demonstrates that resistant individuals can be successfully selected in early generations. Because the values are not very high, the breeder should use a low selection intensity in early generations such that in later generations, the most superior progeny can be identified.

The differences between the narrow-sense heritability values (h_r^2) of two crosses for the two generations are noticeable; the values are higher in the F_2 generation for the Caiapônia x IAC-100 cross ($h_r^2 = 70.40\%$), and in the Luziânia x Potenza cross, the highest value was for the F_3 generation ($h_r^2 = 30\%$). The high values of h_r^2, which are higher than the estimate of h_a^2, for the Caiapônia x IAC-100 cross may be related to the negative values of the dominance variance.

In the Luziânia x Potenza cross, even the highest estimate of h_r^2 was not large. These results demonstrate the effect of the environment on the expression of resistance to soybean rust.

For the F_2 population, the broad-sense heritability estimates ranged from low (Caiapônia x IAC-100) to moderately high (Luziânia x Potenza) (Table 4). High or moderately high estimates for some traits indicate the prevalence of the genetic environmental effects (MARIGUELE et al., 2008). Estimates of h_r^2 were low for the Luziânia x Potenza cross. The low estimate indicates that although there was a predominance of genetic environmental effects, the efficiency of the selection of individual plants tends to be low in segregating generations, which are not significantly advanced because of low values of h_r^2 (h_r^2 is a function of additive genetic variance in addition to the portion of genetic variance that can be fixed with the selection).

When selection is used to reduce the symptoms or the disease severity in the presence of the pathogen (by artificial or natural inoculation), the effects of different genes and alleles in their allelic forms and the dominance variance derived from them, which directly influences the broad-sense heritability, are highly likely to be fixed by selection. Thus, one can successfully increase the ASR resistance by performing selection on the F_2 or F_3 populations. It is easier to understand this concept when the occurrence of disease in cultivated plants and the host-pathogen interaction is examined based on the key-lock phenomenon (the product of a gene or allele in the plant blocks recognition by the pathogen or delays the vital cycle, which reduces the disease progress). A failure of the pathogen to recognise the host or a delay in this process is caused by the effects of dominance (RIBEIRO et al., 2007). Thus, a protein or toxin that is produced is at least partially responsible for decreasing the severity of the symptoms. These effects were studied in the ASR pathosystem and were measured by the reduction in the disease progress rate (AZEVEDO et al., 2007; SANTOS et al., 2007; SILVA et al., 2007), the incubation and latent periods and the area under the disease progress curve, which directly relates to partial or horizontal resistance.

Estimates of the broad-sense heritability (h_a^2) of the F_3 populationwere moderately elevated (60.0 and 45.5% for Luziânia x Caiapônia and Potenza x IAC-100, respectively) and there was a predominance of genetic environmental effects.

The numbers of genes estimated to control the resistance, which were calculated assuming the absence of dominance, were two for the Luziânia x Potenza cross and 23 for the Caiapônia x IAC-100 cross (Table 4). According to Lobo et al. (2005), these values cannot be interpreted as the actual number but rather as an indication of the inheritance type (mono-, oligo- or polygenic). The estimated number of genes observed in this study confirms the quantitative inheritance of ASR resistance. The values estimated for the number of genes indicate that each intersection is particular and depends on the genotypes used in the crosses and on biotic and abiotic factors that are present.

The development of partial resistance to ASR is viable. However, cycles of recurrent selection for smaller groups must be used. One can use genes in different genotypes, with or without the use of molecular markers, or pyramid these genes to increase partial resistance to soybean rust (JULIATTI et al., 2010).

Conclusion

It was concluded that rust resistance is a characteristic controlled by 2 to 23 genes that are predominantly dominant. The estimate of narrow-sense heritability was greater than 70% for the Caiapônia x IAC-100 cross, and the wide-sense heritability was greater than 60% for the Luziânia x Potenza cross; thus, it is possible to successfully select resistant individuals in early generations.

Acknowledgements

The authors are indebted to the Fundação de Amparo a Pesquisa do Estado de Minas Gerais (FAPEMIG) and the Universidade Federal de Uberlândia (UFU) for funding and support.

References

AKAMATSU, M. A.; BARRETO, M.; AKAMATSU, R. Detecção e distinção de *Phakopsora pachyrhizie P. meibomiae* em amostras do herbário uredinológico do Instituto Biológico. **Fitopatologia Brasileira**, v. 29, Suplemento, p. 277-278, 2004.

AZEVEDO, L. A. S.; JULIATTI, F. C.; BARRETO, M. Resistência de genótipos de soja à *Phakopsora pachyrhizi*. **Summa Phytopathologica**, v. 33, n. 3, p. 364-371, 2007.

BERNARDO, R. **Breeding for quantitative traits in plants**. Minnesota: Dept. of Agronomy and Plant Genetics, 2002.

CRUZ, C. D. **Princípios de genética quantitativa**. Viçosa: Universidade Federal de Viçosa, 2005.

DEL PONTE, E. M.; GODOY, C. V.; LI, X.; YANG, X. B. Predicting severity of Asian soybean rust epidemics with empirical rain fall models. **Phytopathology**, v. 96, n. 7, p. 797-803, 2006.

FEHR, W. P.; CAVINESS, C. E. **Stages of soybean development**. Agriculture and home economics experiment station and cooperative extension service. Ames: Iowa State University, 1977. (Special Report 80).

GARCIA, A.; CALVO, E. S.; KIIHL, R. A. S.; HARADA, A.; HIROMOTO, D. M.; VIEIRA, L. G. E. Molecular mapping of soybean rust (*Phakopsora pachyrhizi*) resistance genes: discovery of a novel locus and alleles. **Theoretical and Applied Genetics**, v. 117, n. 4, p. 545-553, 2008.

HARTMAN, G. L.; MILES, M. R.; FREDERICK, R. D. Breeding for resistance to soybean rust. **Plant Disease**, v. 89, n. 6, p. 664-666, 2005.

JULIATTI, F. C.; BORGES, E. N.; PASSOS, R. R.; CALDEIRA JÚNIOR, J. C.; BRANDÃO, A. M. Doenças da soja. **Cultivar**, v. 47, p. 3-14, 2003.

JULIATTI, F. C.; BALARDIN, R. S.; MARTINS, J. A. S.; REZENDE, A. A.; LENTZ, G. Resistência genética da soja à *Phakopsora pachyrhizi* e uso de fungicidas no manejo sustentável da doença. **Revisão Anual de Patologia de Plantas**, v. 18, n. 1, p. 77-118, 2010.

LOBO, V. L. S.; GIORDANO, L. B.; LOPES, C. A. Herança da resistência à mancha-bacteriana em tomateiro. **Fitopatologia Brasileira**, v. 30, n. 4, p. 343-349, 2005.

MARIGUELE, K. H.; MALUF, W. R.; GOMES, L. A. A.; LOPES, M. J. C.; MELO, O. D. Controle genético da qualidade da vagem em cruzamento de feijão-vagem e feijão-comum. **Pesquisa Agropecuária Brasileira**, v. 43, n. 1, p. 47-52, 2008.

MATHER, K.; JINKS, L. L. **Introdução à genética biométrica**. Ribeirão Preto: Sociedade Brasileira de Genética, 1984.

MILES, M. R.; LEVY, C.; MOREL, W.; MUELLER, T.; STEINLAGE, T.; VAN RIJ, N.; FREDERICK, R. D.; HARTMAN, G. L. International fungicide efficacy trials for the management of soybean rust. **Plant Disease**, v. 91, n. 11, p. 1450-1458, 2007.

PIEROZZI, P. H. B.; RIBEIRO, A. S.; MOREIRA, J. U. V.; LAPERUTA, L. D. C.; RACHID, B. F.; LIMA, W. F.; ARIAS, C. A. A.; OLIVEIRA, M. F.; TOLEDO, J. F. F. New soybean (Glycine max Fabales, Fabaceae) sources os qualitative genetic resistance to asian soybean rust caused by *Phakopsora pachyrhizi* (Uredinales, Phakopsoraceae). **Genetics and Molecular Biology**, v. 31, n. 2, p. 505-511, 2008.

RAMALHO, M. A. P.; SANTOS, J. B.; ZIMMERMANN, M. J. O. **Genética quantitativa em plantas autógamas**: aplicações ao melhoramento do feijoeiro. Goiânia: UFG, 1993.

RIBEIRO, A. S.; MOREIRA, J. U. V.; PIEROZZI, P. H. B.; RACHID, B. F.; TOLEDO, J. F. F.; ARIAS, C. A. A.; SOARES, R. M.; GODOY, C. V. Genetic control of Asian rust in soybean. **Euphytica**, v. 157, n. 1-2, p. 15-25, 2007.

ROWE, K. E.; ALEXANDER, W. L. Computations for estimating the genetic parameters in joint-scaling test. **Crop Science**, v. 20, n. 1, p. 109-110, 1980.

SANTOS, J. A.; JULIATTI, F. C.; SANTOS, V. A.; POLIZEL, A. C.; JULIATTI, F. C.; HAMAWAKI, O. T.; Caracteres epidemiológicos e uso da análise de agrupamento para resistência parcial à ferrugem da soja. **Pesquisa Agropecuária Brasileira**, v. 42, n. 3, p. 443-447, 2007.

SCOTT, J. W.; JONES, J. B.; Inheritance of resistance to foliar bacterial spot of tomato incited by Xanthomonas campestris pv. vesicatoria. **Journal of the American**

Society for Horticultural Science, v. 114, n. 1, p. 111–114, 1989.

SILVA, V. A. S.; JULIATTI, F. C.; SILVA, L. A. S. Interação entre resistência genética parcial e fungicidas no controle da ferrugem asiática da soja. **Pesquisa Agropecuária Brasileira**, v. 42, n. 9, p. 1261-1268, 2007.

SINCLAIR, J. B.; HARTMAN, G. L. Soybean diseases. In: HARTMAN, G. L.; SINCLAIR, J. B.; RUPE, J. C. (Ed.). **Compendium of soybean diseases**. 4th ed. St. Paul: American Phytopathological Society, 1999. p. 3-4.

YORINORI, J. T.; PAIVA, W. M.; FREDERICK, R. D.; COSTAMILAN, L. M.; BERTAGNOLLI, P. F.; HARTMAN, G. E.; GODOY, C. V.; NUNES JUNIOR, J. Epidemics of soybean rust (*Phakopsora pachyrhizi*) in Brazil and Paraguay. **Plant Disease**, v. 89, n. 6, p. 675-677, 2005.

25

Resistance to *Enneothrips flavens* Moulton and genetic parameters estimation in interspecific genotypes of peanut

Melina Zacarelli Pirotta[1*], Fabiana Mota da Silva[1], Marcos Doniseti Michelotto[2], Alessandra Pereira Fávero[3], Ignácio José de Godoy[4] and Sandra Helena Unêda-Trevisoli[1]

*[1]Departamento de Produção Vegetal, Universidade Estadual Paulista, Via de Acesso Prof. Paulo Donato Castellane, s/n., 14884-900, Jaboticabal, São Paulo, Brazil. [2]Agência Paulista de Tecnologia dos Agronegócios, Polo Regional Centro Norte, Pindorama, São Paulo, Brazil. [3]Empresa Brasileira de Pesquisa Agropecuária, Embrapa Pecuária Sudeste, São Carlos, São Paulo, Brazil. [4]Instituto Agronômico de Campinas, Campinas, São Paulo, Brazil. *Author for correspondence. E-mail: melina_pirotta@hotmail.com*

ABSTRACT. Peanut is an oilseed crop of great importance for Brazilian agribusiness. A major factor affecting its production is pest incidence, mainly thrips. This study aimed to evaluate the potential for resistance to *Enneothrips flavens* in genotypes derived from the cross between IAC 503 and the amphidiploid (*A. magna* x *A. cardenasii*)^{4x} and to estimate the genetic and phenotype parameters in these genotypes, allowing for better targeting in the selection. The experiments were conducted in a Federer augmented block design with additional checks in two generations (F_3 and F_4). Resistance to thrips was evaluated by its natural infestation and the symptoms of attacks by the insect. They were also evaluated using agronomic trait indicators of interspecific segregating with cultivated species. The results indicated that the selected progeny exhibited high resistance to thrips compared to commercial genotypes, and they had the amphidiploid as the insect resistance source. Some progenies selected as resistant also had good production traits, but with the degree of suitability to the *A. hypogaea* L. genotypes still low, the use of a backcross as an alternative for the introgression of resistance genes and the consequent recovery of adapted genotypes of superior recurring parents is suggested.

Keywords: *Arachis hypogaea* L., thrips, resistance to insects, REML/BLUP.

Introduction

Peanut crop (*Arachis hypogaea* L.) has great importance for Brazilian agribusiness, particularly for export, generating dividends for the country. Thus, the expansion and sustainability of the crop depends on its technological development. Brazilian production is concentrated in the Southeast states and Central South regions, especially the state of Sao Paulo, which is the largest peanut producer of the first crop and accounts for approximately 90% of the country's supply (Companhia Nacional de Abastecimento [CONAB], 2015).

One of the limitations to peanut production is the high agricultural costs due to the need for numerous applications of pesticides for pest and disease control. Thus, the search for resistant cultivars is the most appropriate way to reduce costs while making production more profitable and more environmentally sustainable (Godoy, Moraes, Santos, & Zanotto, 2005).

Among the pests occurring in Brazil, thrips, *Enneothrips flavens*, Moulton, 1941 (Thysanoptera: Thripidae), is considered a key pest of the crop (Gallo et al., 2002). Some studies on current commercial cultivars have suggested that there is limited variability for resistance or tolerance to this insect (Boiça Júnior, Chagas Filho, Godoy, Lourenção, & Souza, 2012; Perozini, Silva, & Goussain, 2014), which can restrict the selection of superior genotypes.

Thus, it has been suggested that a more effective genetic resistance can be found in other species of the genus *Arachis* L. (Wynne & Halward, 1989). However, the use of wild species in breeding has been constrained by sterility barriers, mostly due to differences in the genome constitution and ploidy (Fávero, Godoy, & Suassuna, 2011).

The cultivated allotetraploid species has two genomes (A and B), while wild diploid species have, alternatively, genomes A or B. The cross between species of these two groups results in sterile diploid hybrid AB. To overcome this incompatibility, Simpson (1991) suggested the use of an amphidiploid, resulting from a cross between diploid species with genomes A and B, followed by chromosome doubling with the use of colchicine and then a cross between the tetraploidized amphidiploid with a cultivated peanut.

Pre-breeding research on crossings of *A. hypogaea* L. with wild germplasm has been conducted in Brazil with the aim of identifying sources of resistance to major pests and foliar diseases (Fávero, Simpson, Valls, & Vello, 2006; Fávero, Moraes, Garcia, Valls, & Vello, 2009; Janini, Boiça Júnior, Godoy, Michelotto, & Fávero, 2010; Michelotto et al., 2015; Santos et al., 2011; Santos, Godoy, Michelotto, & Fávero, 2013). One of the crosses made involved an elite cultivar from the Instituto Agronômico de Campinas (IAC 503) and the amphidiploid (*A. magna* x *A. cardenasii)*^{4x}. This amphidiploid has been studied as a source of resistance to thrips for having *A. cardenasii* as one of its wild progenitors, and this species is cited in the literature as a source of resistance to various biotic agents, representing an important source of germplasm to use in peanut breeding (Holbrook & Stalker, 2003).

There are few studies on the inheritance of resistance to pests in peanuts, and for thrips, no studies are known. Thus, segregating populations in the early generations of this cross can result in valuable material for the study of the genetic parameters related to pest resistance.

In studies of early generations, heritability estimates, variance components and genetic gains are important to defining the genetic or environmental origin of the observed variability and to contributing to the correct direction of the breeding program for the advancement of future generations and the selection of superior genotypes (Ramalho, Abreu, Santos, & Nunes, 2012).

This study aimed to (i) assess the early segregating populations of the cross involving the cultivar IAC 503 and the amphidiploid (*A. magna* x *A. cardenasii*)^{4x} as a source of resistance to thrips, (ii) estimate the genetic and phenotypic population parameters, and (iii) monitor the agronomic traits of interspecific segregating genotypes of the cultivated species.

Material and methods

The experiments were conducted in two growing seasons (2013/2014 and 2014/2015) in the experimental area of Agência Paulista de Tecnologia dos Agronegócios (Apta), Polo Regional Centro Norte, in Pindorama, São Paulo State, under a natural insect infestation.

We evaluated 92 progenies in F_3 (2013/2014) and F_4 (2014/2015) generations that were derived from the cross between the commercial cultivar IAC 503 (*A. hypogaea* L., female parent) with the synthetic amphidiploid (*A. magna x A. cardenasii*)^{4x} and advanced by the pedigree method.

In addition to segregating progenies, additional checks were used, namely, cultivar Runner IAC 886, IAC 503, line 335 of the IAC breeding program, amphidiploid An.8 (*A. magna* x *A. cardenasii*)^{4x} and progenitors of the amphidiploid (*A. magna* - V 13751 and *A. cardenasii* - GKP 10017), all seeded in an augmented block design by Federer (1956). Controls were chosen to cover a wide variability with respect to thrips resistance, according to prior knowledge of the behavior of such genotypes.

At first, seeds were left to germinate in plastic cups (200 mL) containing substrate Bioflora® and then placed in a greenhouse until the seedlings were approximately 15 cm high, after approximately 30 days, and then transplanted in the field.

Segregating populations and the genotypes of parents and controls were transplanted in 10 m-individual rows, spaced 1 m in the plant row and 1.80 m between the rows, to allow maximum individualization of each plant with no interference of one plant on the other. The experimental area was previously fertilized with NPK (8:28:16) at 250 kg ha^{-1}, according to the soil chemical analysis.

Evaluations of thrips infestation were performed on individual plants when they had a minimum of 5 terminal buds, which occurred 88, 98, 109, 123, 130, and 138 days after seeding (DAS) in the F_3 generation, and at 54, 63, 79, and 107 DAS in the F_4 generation. Analyses were conducted on the occurrence of thrips: (a) counting the number of insects (nymphs and adults) on 5 young leaflets, still closed, randomly sampled per plant; and (b) visual symptomology of a thrips attack: an application of visual symptoms scoring, ranging from 1 - leaf without symptoms to 5 - totally attacked leaves, in 5 newly opened leaves, also randomly sampled per plant (Moraes, Lourenção, Godoy, & Teixeira, 2005). The visual symptom scoring scale was applied 6 days after the infestation evaluation since, according to previous knowledge, this refers to the average opening time of leaflets and allows easy viewing of any symptoms.

Agronomic characterization was made after the manual harvest of each plant and pods were subjected to oven drying at 30°C for 72 hours, reducing moisture to the standard 13%, where the following traits were evaluated: (a) production of pods per plant (grams); (b) production of grains per plant (grams); (c) number of pods per plant; (d) number of grains per plant; and (e) percentage of pods with two locules per plant (%).

Individual and combined analyses of variance were performed. In the combined analysis, we used adjusted means for the progenies in each year, where the effects of block and generation were considered fixed and the progenies were considered random. Genetic gains in percentage for each trait in each generation were estimated by taking into account a 20% selection pressure, based on the average behavior of the controls.

Estimates of genetic parameters were based on the mixed models theory (Henderson, Kempthorne, Searle, & Krosigk, 1959), completing the estimation of the variance components with the restricted maximum likelihood method (REML) developed by Patterson and Thompson (1971). The BLUP method (best linear unbiased prediction) was used for the prediction of genetic values (Henderson, 1973).

Statistical analyses were performed using Statistical Analysis System software (SAS® 9.3, 2011). Raw data on the number of insects were transformed into $(x+0.5)^{1/2}$ to fit the data to a normal distribution. In turn, data from the production of pods, the production of grains, and the number of pods and number of grains were transformed according to the Box-Cox method (Box & Cox, 1964), whereas the percentage of pods with two seeds was subjected to arcsine transformation $(\text{percentage}/100)^{1/2}$.

For a better interpretation of the genetic gain obtained for each trait, a confidence interval was constructed through the expression of Resende, Araújo, Sampaio, and Wiecheteck (1995). In this study, the accuracy of selection was calculated by the square root of the heritability. The construction of confidence intervals were made with the aid of R software (R Development Core Team, 2013).

Results and discussion

According to the results on the infestation of thrips, there were some intervals of lower and higher occurrences because population dynamics can vary according to the region and can be influenced by biotic and abiotic factors (Chagas Filho, Boiça Júnior, Godoy, Lourenção, & Ribeiro, 2008).

In the F_3 generation (Figure 1), the largest population of *E. flavens* occurred at 130 days after seeding (DAS) in the genotypes of *A. hypogaea* L. (Runner IAC 886, IAC 503, and L. 335) and in the segregating progenies. At 123 DAS, the highest occurrence was found in the amphidiploid (*A. magna* x *A. cardenasii*)^{4x} and its wild progenitors (*A. magna* e *A. cardenasii*). The highest scores of visual symptoms were observed at 98 DAS in the segregating progenies, at 130 DAS in the amphidiploid and its wild progenitors and at 138 DAS in the genotypes of *A. hypogaea* L. (Figure 1).

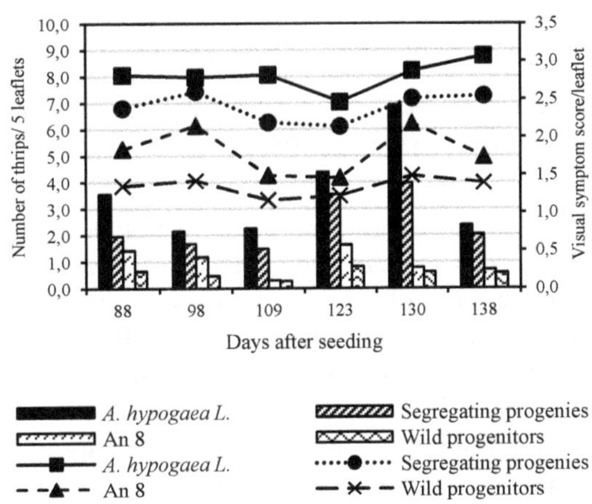

Figure 1. Population dynamics (bars) and visual symptom scores (lines) of *E. flavens* in genotypes of *A. hypogaea* L. (Runner IAC 888, IAC 503 and L. 335), segregating progenies in the F_3 generation, and amphidiploid (*A. magna* x *A. cardenasii*)^{4x} and wild progenitors (*A. magna* and *A. cardenasii*). Pindorama, State of São Paulo, 2013/2014.

It can be observed that the wild progenitors (*A. magna* and *A. cardenasii*) had the lowest infestation and the lowest scores for visual symptoms of *E. flavens*, followed by the amphidiploid (*A. magna* x *A.*

cardenasii)^{4x} and by the segregating progenies. These genotypes were therefore less infested and suffered less damage caused by the insect compared to the genotypes of *A. hypogaea* L., corroborating the research results of Janini et al. (2010) and Michelotto, Godoy, Fávero, Carrega, and Finoto (2013). The lower infestation and minor damage exhibited by the wild genotypes were observed in the first evaluation and remained throughout the plant cycle.

For the F_4 generation, the highest infestation of thrips was observed at 79 DAS and the highest scores of visual symptoms was at 107 DAS, in all genotypes (Figure 2).

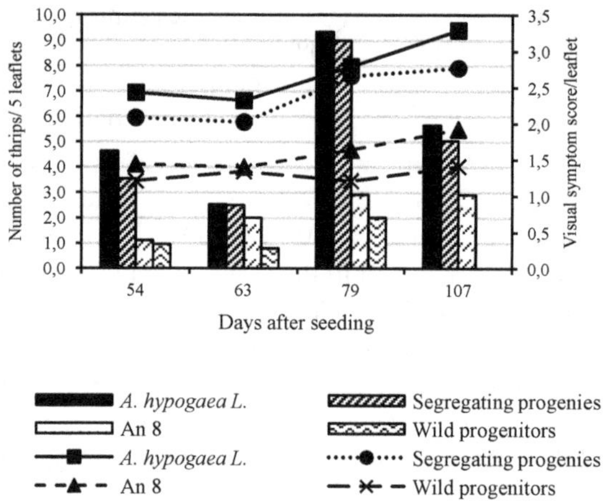

Figure 2. Population dynamics and visual symptom scores of *E. flavens* in genotypes of *A. hypogaea* L. (Runner IAC 888, IAC 503 and L. 335), segregating progenies in the F_4 generation, and amphidiploid (*A. magna* x *A. cardenasii*)^{4x} and wild progenitors (*A. magna* and *A. cardenasii*). Pindorama, State of São Paulo, 2014/2015.

The mean performance of the genotypes proved what was observed in the figures; in other words, the genotypes of *A. hypogaea* L. had higher infestation and damage rates, especially Runner IAC 886 with the highest mean values in both generations. Wild progenitors had the lowest mean values of infestation and damage, specifically the wild species *A. cardenasii* (GKP 10017) (Table 1).

For infestation by *E. flavens*, the segregating progenies were not significantly ($p > 0.05$) different from IAC 503 and line 335 in the F_3 generation, and from Runner IAC 886 and IAC 503 in the F_4 generation (Table 1). In relation to visual symptom scores, these progenies showed lower mean values of damage than the genotypes of *A. hypogaea* L. in the two generations, thus emphasizing that studies on the type of resistance presented by these genotypes should be conducted in the next stages of this study (Table 1).

Table 1. Number of thrips in 5 leaflets (NT) and visual symptom scores per leaflet (NS) in genotypes *A. hypogaea* L. (Runner IAC 888, IAC 503 and L. 335), segregating progenies, and amphidiploid (*A. magna* x *A. cardenasii*)^{4x} and wild progenitors (*A. magna* and *A. cardenasii*) in two generations, F_3 and F_4. Pindorama, State of São Paulo.

Genotypes	F_3		F_4	
	$NT^{1,2}$	NS^2	$NT^{1,3}$	NS^3
Progeny	2.77b	2.43c	5.05a	2.39b
Runner IAC 886	4.81a	2.90a	5.21a	2.73a
IAC 503	3.21b	2.88a	4.36a	2.71a
L. 335	2.54b	2.63b	3.73ab	2.58ab
An. 8	1.27c	1.81d	2.24bc	1.60c
V 13751	0.61c	1.39e	1.44c	1.37cd
GKP 10017	0.53c	1.29e	0.96c	1.21d
Mean	2.25	2.19	3.28	2.08
F-test	48.07**	181.79**	145.81**	107.73**
C.V. (%)	7.94	3.83	8.52	6.37

Means followed by different letters, in the same column, are significantly different by Tukey's test at 5% probability. [1] = data transformed into $(x+0.5)^{1/2}$, [2] = mean of 6 evaluations, [3] = mean of 4 evaluations. Runner IAC 886 and IAC 503 = commercial cultivars. L.335 = line of the IAC breeding program. An.8 = amphidiploid (*A. magna* x *A. cardenasii*)^{4x}, V = 13751 wild species *A. magna*. GKP 10017 = wild species *A. cardenasii*. ** = Significant at 1%, C.V. = coefficient of variation (%).

In the combined analysis of variance for the different traits (Table 2), there were significant differences ($p < 0.01$) in the two generations for most traits, except for visual symptom scores (NS), indicating the influence of generations. Additionally, significant effects were detected ($p < 0.01$ or $p < 0.05$) in the progenies for all traits, except for number of pods plant^{-1}, indicating genetic variability in the segregating populations for most traits.

Considering the interaction (Prog. x Gen.), the traits visual symptom score, pod production, grain production, and percentage of pods with two locules showed significance at 1%. For the number of thrips, significance was 5%, while for the traits number of pods and number of grains per plant, the interaction was not significant (Table 2).

The significance of this interaction indicates that the behavior of the progenies was non-coincident in the evaluated generations, namely, the best progenies in the F_3 generation were not necessarily the best in the F_4. Some authors describe the occurrence of this type of interaction (Lima, Ramalho, & Abreu, 2012).

In general, there is little agreement about the best progenies between generations, i.e., both the phenotypic and genotypic values were different between them (Table 3). This was observed for all traits, and these differences in behavior found in the progenies over the generations are possibly due to the interaction 'generation x progenies', and the existence of segregation difference between them.

Table 2. Summary of analysis of variance for the traits: number of thrips in 5 leaflets (NT), visual symptom score per leaflet (NS), pod production (PV) (grams plant^{-1}), grain production (PG) (grams plant^{-1}), number of pods per plant (NV), number of grains per plant (NG), and percentage of pods with two locules per plant (V2). Pindorama, State of São Paulo.

SV	DF	Mean Square						
		NT[1]	NS	PV (g)[2]	PG (g)[2]	NV[2]	NG[2]	V_2[3] (%)
Block	3	0.03[ns]	0.08[ns]	1.48[ns]	1.03[ns]	0.77[ns]	1.23[ns]	1.91[ns]
Test. vs Prog.	1	5.26**	3.79**	27.97**	7.53**	4.22*	3.96*	2838.17**
Generation	1	14.48**	0.09[ns]	6.34**	20.34**	5.01**	9.79**	1278.00**
Progeny	91	0.13**	0.09**	2.30**	1.69**	1.01[ns]	1.84*	178.38**
Prog. x Gen.	91	0.09*	0.06**	1.48**	1.20**	0.82[ns]	1.39[ns]	73.65**
Controls	5	1.89**	4.19**	23.79**	21.03**	6.04**	17.12**	7528.01**
Residual	39	0.05	0.03	0.73	0.51	0.72	1.08	14.43
		Means[4]						
Progeny		3.91	2.41	20.99	11.14	42.86	46.22	39.23
Runner IAC 886		5.01	2.82	37.01	24.82	69.38	76.71	77.37
IAC 503		3.79	2.80	25.87	17.05	43.01	62.67	70.38
L. 335		3.14	2.61	16.76	12.85	39.30	53.43	64.16
An. 8		1.76	1.71	5.66	2.52	22.95	33.39	0.71
V 13751		1.03	1.38	2.13	1.25	18.20	15.06	0.25
GKP 10017		0.75	1.25	1.17	0.45	12.98	8.48	0.08
C.V. (%)		11.13	7.34	25.06	27.56	22.96	23.72	10.47

SV = source of variation, DF = degrees of freedom, CV = coefficient of variation (%). [1] = data transformed into $(x+0.5)^{1/2}$, [2] = data transformed into Box-Cox, [3] = data transformed into arcsine $(\%/100)^{1/2}$. [4] = raw means *, ** = significant at 5 and 1%, respectively, by F-test F, [ns] = non-significant by F-test. Runner IAC 886 and IAC 503 = commercial cultivars. L.335 = line of the IAC breeding program. An.8 = amphidiploid (*A. magna* x *A. cardenasii*)^{4x}, V 13751 = wild species *A. magna*. GKP 10017 = wild species *A. cardenasii*.

Table 3. Phenotypic and genotypic values of the 20 best progenies for the traits number of thrips and visual symptom scores compared to the behavior of the controls, in the F_3 and F_4 generations. Pindorama, State of São Paulo.

Number of thrips/5 leaflets						Visual symptom score/leaflet					
	F_3			F_4			F_3			F_4	
Prog	MA	BLUP	Prog	MA	BLUP	Prog	MA	BLUP	Prog	MA	BLUP
21	0.75	-1.72	26	1.85	-1.62	80	1.77	-0.59	24	1.59	-0.41
23	0.75	-1.72	39	2.14	-1.45	88	1.77	-0.59	77	1.96	-0.25
57	0.75	-1.70	65	2.64	-1.28	24	1.81	-0.56	65	2.00	-0.17
54	0.92	-1.55	66	2.68	-1.25	67	1.92	-0.45	66	2.01	-0.17
71	1.00	-1.64	32	2.70	-1.11	77	1.93	-0.45	80	2.01	-0.22
64	1.08	-1.41	33	2.72	-1.10	41	1.98	-0.41	64	2.10	-0.12
68	1.16	-1.41	60	2.84	-1.16	46	2.01	-0.38	57	2.11	-0.11
26	1.16	-1.37	37	2.98	-0.95	63	2.02	-0.36	59	2.11	-0.11
28	1.16	-1.37	48	3.03	-1.05	50	2.06	-0.33	16	2.13	-0.13
31	1.16	-1.37	57	3.03	-1.05	70	2.07	-0.32	48	2.14	-0.10
77	1.33	-1.35	44	3.09	-0.88	19	2.10	-0.32	61	2.15	-0.09
5	1.42	-1.14	80	3.09	-1.16	75	2.13	-0.26	46	2.15	-0.12
17	1.42	-1.14	41	3.14	-0.85	35	2.14	-0.26	56	2.16	-0.09
69	1.42	-1.12	83	3.36	-1.00	14	2.17	-0.25	43	2.16	-0.12
81	1.50	-1.20	82	3.61	-0.85	71	2.17	-0.23	5	2.17	-0.11
24	1.50	-1.08	73	3.67	-0.81	74	2.17	-0.23	39	2.17	-0.11
55	1.58	-0.98	77	3.73	-0.78	76	2.17	-0.23	17	2.19	-0.10
56	1.58	-0.98	88	3.73	-0.78	39	2.18	-0.23	63	2.19	-0.07
2	1.59	-0.98	59	3.73	-0.63	61	2.19	-0.21	8	2.20	-0.09
14	1.59	-0.99	90	3.77	-0.75	10	2.20	-0.23	41	2.20	-0.10
MS	1.24	-1.31		3.08	-1.02		2.05	-0.34		2.09	-0.14
MP	2.77			5.05			2.43			2.39	
IAC 886[1]	4.81			5.21			2.90			2.73	
IAC 503[1]	3.21			4.36			2.88			2.71	
L. 335[1]	2.54			3.73			2.63			2.58	
An. 8[1]	1.27			2.24			1.81			1.60	
V 13751[1]	0.61			1.44			1.39			1.37	
GKP 10017[1]	0.53			0.96			1.29			1.21	

Prog = progenies selected to calculate the genetic gain; MA = adjusted mean; BLUP = best linear unbiased prediction; MS = mean of the selected progenies; MP = mean of the population; [1] = controls: cultivars Runner IAC 886 and IAC 503, line 335 of the IAC breeding program, amphidiploid An.8 (*A. magna* x *A. cardenasii*)^{4x}, wild progenitors V 13751 (*A. magna*) and GKP 10017 (*A. cardenasii*).

Regarding the number of insects, for the F_3 generation, the mean value of the selected progenies nearly equaled the mean value of the amphidiploid (*A. magna* x *A. cardenasii*)^{4x}, though still a 65% reduction in the occurrence of thrips compared to the genotypes of *A. hypogaea* L. (Runner IAC 886, IAC 503 and L. 335), especially progenies 21, 23, and 57. For the F_4 generation, this reduction was close to 30%, showing that such progenies may show resistance similar to that of wild genotypes.

(Table 3). As for visual symptom scores, progenies 80 and 88 stood out with the lowest mean (1.77) in the F_3 generation and the progeny 24 stood out with a mean of 1.59 in the F_4 generation. The selected progenies had a 27 and 22% reduction in insect attack compared to the cultivated genotypes of *A. hypogaea* L. (Table 3).

The mean production of pods and grains of the

selected progenies exceeded by 118% and 40% the mean production of the *A. hypogaea* L. genotypes. In the F_3 generation, progenies 49 produced 131.30 g $plant^{-1}$ of pods and 45 produced 61.79 g $plant^{-1}$ of grains. Mean production exceeded by 64% and 53% in the F_4 generation, highlighting the progeny 47 that produced 88.21 g $plant^{-1}$ of pods and 60.67 g $plant^{-1}$ of grains (Table 4).

Considering the number of pods and grains, the progeny 75 stood out with 259.00 pods $plant^{-1}$ and 239.00 grains $plant^{-1}$ in the F_3 generation, an increase of 117% and 55%, respectively, compared to the *A. hypogaea* L. genotypes. In the F_4 generation, the progeny 79 stood out with 167.03 pods $plant^{-1}$ and 186 grains $plant^{-1}$ with mean increases of 77% in the number of pods and 79% in the number of grains per plant (Table 5).

In the commercial species, *A. hypogaea* L. and, depending on the cultivar, the pods have two or more well-developed seeds. Many wild species have a large number of pods with only a very small seed, representing an important agronomic limitation (Santos et al., 2013). The amphidiploid (*A. magna* x *A. cardenasii*)^{4x} used in the cross of this study has these characteristics and a relatively low number of pods (Table 6).

Considering that wild species and amphidiploid have mostly unilocular pods (with one grain), this variable is a strong indicator of the level of the genotypes' agronomic suitability (Santos et al., 2013). The selected progenies obtained mean values close to the *A. hypogaea* L. genotypes; however, there were reductions of 11 and 14% in the percentage of pods with two locules for these genotypes. It is worth noting that the progeny 49 had 83.24% of its pods filled with two locules in the F_3 generation and the progeny 43 had 80.41% in the F_4 generation. Both had mean values higher than Runner IAC 886, with 79.38 and 75.36% in the respective generations, a cultivar considered to have excellent yield potential (Table 6).

Some segregating progenies that were selected as resistant, in both generations evaluated, that also showed good reproductive traits, but had a low level of agronomic suitability to the *A. hypogaea* L. genotypes, are as follows: 5, 17, 41, 56, 57, 63, 64, and 77 (Tables 3 to 6). These progenies are of great value because they indicate potential in crossings following the breeding.

Table 4. Phenotypic and genotypic values of the 20 best progenies for the traits production of pods and production of grains per plant (grams) compared to the behavior of the controls, in the F_3 and F_4 generations. Pindorama, State of São Paulo.

Production of pods (g)						Production of grains (g)					
F_3			F_4			F_3			F_4		
Prog	MA	BLUP	Prog	MA	BLUP	Prog	MA	BLUP	Prog	MA	BLUP
49	131.30	106.45	47	88.21	26.38	45	61.79	46.27	47	60.67	18.44
47	96.13	73.15	79	64.27	15.76	47	45.12	33.56	79	40.01	9.64
75	92.59	68.07	77	63.54	15.46	75	44.43	30.89	77	38.01	8.84
79	78.07	54.32	34	52.59	13.56	38	43.29	29.74	7	32.08	8.24
38	77.10	53.98	7	50.87	13.10	89	36.03	23.29	76	31.77	6.34
45	60.77	38.51	51	50.12	10.64	79	34.88	22.36	34	30.22	7.50
89	59.30	36.55	76	47.83	8.96	27	29.18	17.13	27	27.05	6.23
71	57.77	35.10	78	45.92	8.18	55	20.20	11.30	6	26.57	6.04
27	53.18	31.32	9	44.04	10.27	91	19.84	8.93	51	26.54	4.78
91	42.91	21.03	54	42.96	7.68	70	19.69	8.79	54	25.63	4.41
36	40.34	19.17	27	41.75	9.08	88	18.81	8.14	56	24.65	4.02
55	37.69	17.81	90	41.08	6.18	71	18.25	7.51	82	22.97	2.82
19	37.62	17.86	82	40.30	5.85	74	17.45	6.79	90	22.79	2.74
34	37.57	16.54	8	37.15	7.43	34	17.37	6.58	14	21.73	4.10
13	36.62	16.91	56	37.04	5.24	36	16.09	5.98	74	21.30	2.15
88	36.37	14.84	6	36.40	7.12	19	14.26	5.42	8	20.78	3.72
41	36.09	15.14	64	35.26	4.50	13	14.09	5.27	89	20.70	1.91
53	31.53	11.97	72	33.62	3.09	15	14.03	5.22	3	20.50	3.61
23	30.35	10.97	55	33.38	3.72	48	13.75	5.54	72	20.13	1.68
70	29.94	8.75	74	32.75	2.73	40	13.10	2.77	55	20.04	2.17
MS	55.16	33.42		45.95	9.25		25.61	14.57		27.70	5.47
MP	19.61			22.37			8.90			13.38	
IAC 886[1]	41.68			32.33			26.56			23.07	
IAC 503[1]	21.50			30.24			16.32			17.77	
L. 335[1]	12.64			20.88			12.10			13.60	
An. 8[1]	3.17			8.14			1.86			3.17	
V 13751[1]	1.34			2.91			0.80			1.70	
GKP 10017[1]	0.44			1.89			0.23			0.66	

Prog = progenies selected to calculate the genetic gain; MA = adjusted mean; BLUP = best linear unbiased prediction; MS = mean of the selected progenies; MP = mean of the population; [1] = controls: cultivars Runner IAC 886 and IAC 503, line 335 of the IAC breeding program, amphidiploid An.8 (*A. magna* x *A. cardenasii*)^{4x}, wild progenitors V 13751 (*A. magna*) and GKP 10017 (*A. cardenasii*).

Table 5. Phenotypic and genotypic values of the 20 best progenies for the traits number of pods and number of grains per plant (score) compared to the behavior of the controls, in the F_3 and F_4 generations. Pindorama, State of São Paulo.

	Number of pods plant^{-1} (score)						Number of grains plant^{-1} (score)				
	F_3			F_4			F_3			F_4	
Prog	MA	BLUP	Prog	MA	BLUP	Prog	MA	BLUP	Prog	MA	BLUP
75	259.00	195.89	79	167.03	96.64	75	289.00	158.62	79	186.68	107.35
79	155.00	101.58	77	143.89	76.70	79	183.00	112.18	47	154.75	83.89
49	144.46	99.87	81	116.14	52.78	38	165.59	99.91	27	137.34	74.61
38	125.81	77.33	47	113.61	56.20	45	146.59	84.15	77	135.37	64.58
45	120.81	72.79	82	113.14	50.20	47	135.97	87.02	6	120.42	61.75
47	115.46	73.67	76	112.31	49.48	27	120.59	62.59	81	116.91	49.19
27	109.81	62.82	78	109.81	47.33	89	120.00	59.94	34	111.31	52.92
71	106.00	57.15	34	105.41	52.99	36	104.59	49.32	82	110.74	44.05
89	104.00	55.33	9	92.92	45.05	13	100.44	50.25	9	101.64	46.10
41	85.81	41.06	64	88.86	34.87	71	99.00	42.53	7	93.52	39.33
34	83.81	39.24	74	87.92	28.46	88	86.00	31.75	58	90.82	30.70
36	82.81	28.34	27	81.80	32.65	91	83.00	29.26	74	90.68	27.32
13	82.73	41.18	75	81.52	22.95	34	79.59	28.59	54	89.25	29.39
88	76.00	29.94	54	78.30	25.77	74	77.00	24.28	76	88.57	25.57
23	71.73	31.21	51	77.53	25.11	19	74.44	28.69	51	87.08	27.58
19	69.73	29.40	58	72.86	21.08	23	74.44	28.69	78	86.91	24.18
74	66.00	20.88	90	71.97	14.72	41	70.59	21.13	64	83.25	24.39
91	63.00	18.16	6	71.53	26.62	15	65.44	21.23	56	83.15	24.31
78	54.00	9.99	56	70.96	19.44	78	65.00	14.33	55	80.81	22.36
83	53.00	9.09	72	69.14	12.88	64	54.97	19.85	90	73.57	13.06
MS	101.45	55.25		96.33	39.57		107.26	52.72		106.14	43.64
MP	38.76			46.96			40.92			51.52	
IAC 886[1]	61.85			76.90			90.09			63.32	
IAC 503[1]	41.18			44.83			63.42			61.91	
L. 335[1]	37.34			41.25			53.98			52.88	
An. 8[1]	16.09			29.81			31.59			35.19	
V 13751[1]	12.27			24.13			10.36			19.75	
GKP 10017[1]	3.90			22.06			3.13			13.83	

Prog = progenies selected to calculate the genetic gain; MA = adjusted mean; BLUP = best linear unbiased prediction; MS = mean of the selected progenies; MP = mean of the population; [1] = controls: cultivars Runner IAC 886 and IAC 503, line 335 of the IAC breeding program, amphidiploid An.8 (*A. magna* x *A. cardenasii*)^{4x}, wild progenitors V 13751 (*A. magna*) and GKP 10017 (*A. cardenasii*).

Table 6. Phenotypic and genotypic values of the 20 best progenies for the trait percentage of pods with two locules compared to the behavior of the controls, in the F_3 and F_4 generations. Pindorama, State of São Paulo.

	Percentage of pods with two locules				
	F_3			F_4	
Prog	MA	BLUP	Prog	MA	BLUP
49	83.24	46.36	43	80.41	34.64
17	79.79	43.17	49	77.39	31.01
57	73.96	37.26	47	74.83	28.65
55	66.23	29.68	57	65.34	19.92
2	65.64	29.30	3	63.34	19.01
5	64.08	27.77	17	61.52	17.33
47	63.58	27.08	37	61.01	16.79
38	63.19	27.25	72	60.82	17.19
59	62.12	25.65	56	60.44	15.41
18	62.01	25.74	59	59.85	14.87
56	59.57	23.15	55	59.38	14.44
4	59.15	22.93	84	57.38	14.02
32	52.02	23.16	58	57.17	12.40
91	57.14	21.30	5	56.88	13.06
53	56.10	19.74	45	56.73	12.85
54	56.10	19.74	7	56.06	12.31
51	55.21	18.87	1	55.63	11.91
15	54.21	18.38	6	55.47	11.76
63	52.81	16.52	4	55.39	11.69
14	52.01	15.93	18	55.29	11.60
MS	62.28	25.95		61.52	17.04
MP	35.64			42.82	
IAC 886[1]	79.38			75.36	
IAC 503[1]	68.43			72.32	
L. 335[1]	62.12			66.19	
An. 8[1]	0.00			1.42	
V 13751[1]	0.00			0.49	
GKP 10017[1]	0.00			0.16	

Prog = progenies selected to calculate the genetic gain; MA = adjusted mean; BLUP = best linear unbiased prediction; MS = mean of the selected progenies; MP = mean of the population; [1] = controls: cultivars Runner IAC 886 and IAC 503, line 335 of the IAC breeding program, amphidiploid An.8 (*A. magna* x *A. cardenasii*)^{4x}, wild progenitors V 13751 (*A. magna*) and GKP 10017 (*A. cardenasii*).

In addition, the best phenotypic values coincided with the highest predicted values (BLUP) for all traits in the two generations (Tables 3 to 6), corroborating the results obtained by Michelotto et al. (2015). BLUP is a function of phenotypic values that have the common objective of predicting true genetic values, and it has been used as a selection tool in various crops (Baldissera et al., 2012; Pinheiro et al., 2013; Regitano Neto, Ramos Júnior, Gallo, Freitas, & Azzini, 2013). In peanut crop, BLUP has been used to assist the selection of lines at the early stages of breeding programs (Luz, Santos, Silva Filho, & Mello Filho, 2010) and has helped to identify sources of resistance to aerial fungal diseases and in interspecific accessions and hybrids of the genus *Arachis* L. (Michelotto et al., 2015).

It can be inferred that, in general, the genetic variance was greater than the environmental variance for most of the traits evaluated in the two generations, except for production of pods and the production of grains in the F_4 generation, likely because these traits are highly influenced by the environment (Table 7). This ratio between the variances characterizes a situation favorable to the selection process.

Table 7. Estimates of genetic parameters of peanut progenies in the F_3 and F_4 generations. Pindorama, State of São Paulo.

Characteristics	Gen	Genetic parameters				
		σ_g^2	σ_e^2	h^2	G.S. (%)	I.C.
Number of thrips 5 leaflets^{-1}	F_3	1.46	0.23	0.86	-47.50	-47.63 a -47.37
	F_4	1.25	0.85	0.59	-23.01	-23.23 a -22.79
Visual symptom score leaflet^{-1}	F_3	0.08	0.01	0.91	-14.23	-14.29 a -14.24
	F_4	0.02	0.02	0.52	-6.53	-6.56 a -6.50
Production of pods (g plant^{-1})	F_3	514.25	28.82	0.95	172.23	170.75 a 173.71
	F_4	98.96	140.49	0.41	43.22	40.77 a 45.66
Production of grains (g plant^{-1})	F_3	109.97	13.13	0.89	167.10	166. 08 a 168.12
	F_4	38.17	57.18	0.40	42.81	41.27 a 44.34
Number of pods plant^{-1}	F_3	1488.50	152.99	0.91	147.18	143.79 a 150.57
	F_4	832.87	133.55	0.86	90.41	87.22 a 93.60
Number of grains plant^{-1}	F_3	1454.15	299.46	0.83	134.56	129.90 a 139. 22
	F_4	1019.75	203.57	0.80	88.00	83.75 a 92.25
% pods with two locules plant^{-1}	F_3	364.18	7.18	0.98	73.26	72.48 a 74.04
	F_4	183.63	15.92	0.92	40.18	39.06 a 41.30

Gen = generation, σ_g^2 = genetic variance, σ_e^2 = environmental variance, h^2 = broad-sense heritability, G.S. = gain with selection (%), I.C. = confidence interval.

One of the most useful parameters for breeders is the heritability coefficient (h^2). It allows one to predict for the possibility of a successful selection and reflects the proportion of phenotypic variation that can be inherited (Ramalho et al., 2012). In general, estimates of broad-sense heritability had medium and high magnitude for all the traits through the generations. In F_3, they ranged from 98% for the percentage of pods with two locules to 83% for the number of grains plant^{-1}. In F_4, they ranged from 92% for the percentage of pods with two locules to 40% for the production of grains per plant (Table 7).

With respect to the infestation of thrips on the F_4 peanut (*A. hypogaea* L.) progenies, Ekvised et al. (2006) obtained a h^2 ranging from 0 to 80% in the number of insects and from 0 to 88% in the symptoms of attack. For the number of pods, Luz et al. (2010) found a h^2 ranging from 25.8 to 31.69%. Sikinarum et al. (2007) detected higher h^2 values, from 55 to 99% for the same trait.

There was a reduction in the heritability coefficients from the F_3 to the F_4 generation (Table 7), indicating that it is not a constant measure and that several factors can change it, such as, for example, the decrease in genetic variability by increased inbreeding, a characteristic belonging to the genetic structure of populations of perfect autogamous plants such as peanuts and because of the high interaction 'progenies x generations' (Ramalho et al., 2012).

Other than the mean performance of the progenies, gains with the selection must be greatly considered in a breeding program. In general, gains were relatively greater in the F_3 generation (Table 7), mainly due to the higher heritability coefficients in this generation.

Importantly, for characteristics related to resistance, it is expected that the gain will decrease because the intention was to reduce the occurrence and the damage caused by the pest insect. In the F_3 generation, gains were -47.50 and -14.23% in the number and symptoms of thrips, and in the F_4 generation, gains were -23.01 and -6.53%. For reproductive traits, the greatest gains in the F_3 generation were for the number of pods (147.18%), number of grains (134.56%) and production of pods (172.23%). In the F_4 generation, the greatest gains for reproductive traits were achieved in the number of grains (88.00%) and number of pods (90.41%) (Table 7).

Along with the genetic gains, selection should also be based on the confidence interval, wherein the lower limits must be taken into consideration (Resende et al., 1995). Analyzing them, it can be observed that these were higher in the F_3 generation and showed smaller amplitudes for most of the evaluated characteristics, except for the number of pods and grains (Table 7). This indicates that selection in this generation enabled higher lower limits and less wide ranges, as well as the greatest genetic gain.

Conclusion

Along the evaluated generations, the segregating progenies selected demonstrated higher resistance to thrips, with the amphidiploid (*A. magna* x *A. cardenasii*)^{4x} as a source of resistance to the insect.

Some segregating progenies selected as resistant, in both generations evaluated, also showed good production traits, but with a low level of agronomic suitability to the *A. hypogaea* L. genotypes, were as follows: 5, 17, 41, 56, 57, 63, 64, and 77.

Acknowledgements

The authors are indebted to CNPq/Universal (n° 480921/2013-0) for financial support. To CAPES for the scholarship granted to the first author and CNPq for the productivity scholarship to the third author.

References

Baldissera, J. N. C., Bertoldo, J. G., Valentini, G., Coan, M. M. D., Rozeto, D. S., Guidolin, A. F., & Coimbra,

J. L. M. (2012). Uso do melhor preditor linear não-visado (BLUP) na predição de híbridos em feijão. *Bioscience Journal, 28*(3), 395-403.

Boiça Júnior, A. L., Chagas Filho, N. R., Godoy, I. J., Lourenção, A. L., & Souza, J. R. (2012). Avaliação de resistência de cultivares de amendoim de hábito rasteiro a *Enneothrips flavens* Moulton (Thysanoptera: Thripidae). *Arquivos do Instituto Biológico, 79*(1), 33-38.

Box, G. E. P., & Cox, D. R. (1964). An analysis of transformations. *Journal of the Royal Statistical Society. Series B (Methodological), 26*(2), 211-252.

Companhia Nacional de Abastecimento [CONAB]. (2015). *Levantamento safra de grãos 2015/2016*. Retrieved from http://www.conab.gov.br/

Chagas Filho, N. R., Boiça Júnior, A. L., Godoy, I. J., Lourenção, A. L., & Ribeiro, Z. A. (2008). Resistência de cultivares de amendoim de hábito de crescimento ereto a *Enneothrips flavens* Moulton (Thysanoptera: Thripidae). *Arquivos do Instituto Biológico, 75*(2), 149-156.

Ekvised, S., Jogloy, S., Akkasaeng, C., Keerati-kasikorn, M., Kesmala, T., Buddhasimma, I., & Patanothai, A. (2006). Heritability and correlation of thrips resistance and agronomic traits in peanut. *Asian Journal of Plant Sciences, 5*(6), 923-931. doi: 10.3923/ajps.2006.923.931

Fávero, A. P., Godoy, I. J., & Suassuna, T. M. F. (2011). Uso de espécies silvestres no pré-melhoramento do amendoim. In Lopes, M. A., Fávero, A. P., Ferreira, M. A. J. F., Faleiro, F. G., Folle, S. M., & Guimarães, E. P. (Eds.), *Pré-melhoramento de plantas: estado da arte e experiências de sucesso* (p. 266-291). Brasília, DF: Embrapa.

Fávero, A. P., Moraes, S. A., Garcia, A. A. F., Valls, J. F. M., & Vello, N. A. (2009). Characterization of rust, early and late leaf spot resistance in wild and cultivated peanut germplasm. *Scientia Agricola, 66*(1), 110-117.

Fávero, A. P., Simpson, C. E., Valls, J. F. M., & Vello, N. A. (2006). Study of the evolution of cultivated peanut through crossability studies among *Arachis ipäensis*, *A. duranensis*, and *A. hypogaea*. *Crop Science, 46*(4), 1546-1552.

Federer, W. T. (1956). Augmented (hoonuiaku) designs. *Hawaian Planters' Record, 55*(2), 191-208.

Gallo, D., Nakano, O., Silveira Neto, S., Carvalho, R. P. L., Batista, G. C., Berti Filho, E., ... Omoto, C. (2002). *Entomologia Agrícola*. Piracicaba, SP: Fundação de Estudos Agrários Luiz de Queiroz.

Godoy, I. J., Moraes, S. A., Santos, R. C., & Zanotto, M. (2005). Melhoramento do amendoim. In A. Borém (Ed.), *Melhoramento de espécies cultivadas* (2a ed., p. 54-95). Viçosa, MG: UFV.

Henderson, C. R., Kempthorne, O., Searle, S. R., & Krosigk, C. M. V. (1959). The estimation of environmental and genetic trends from records subject to culling. *Biometrics, 15*(2), 192-218.

Henderson, C. R. (1973). *Maximum likelihood estimation of variance components*. Unpublished manuscript. Ithaca, NY: Department of Animal Science, Cornell University, 14850.

Holbrook, C., & Stalker, T. H. (2003). Peanut breeding and genetic resources. In J. Janick (Ed.), *Plant breeding reviews* (v. 22, p. 297-356). Hoboken, NJ: John Wiley & Sons, Inc.

Janini, J. C., Boiça Júnior, A. L., Godoy, I. J., Michelotto, M. D., & Fávero, A. P. (2010). Avaliação de espécies silvestres e cultivares de amendoim para resistência a *Enneothrips flavens* Moulton. *Bragantia, 69*(4), 891-898.

Lima, L. K., Ramalho, M. A. P., & Abreu, A. F. B. (2012). Implications of the progeny x environment interaction in selection index involving characteristics of common bean. *Geneticand Molecular Reserch, 11*(4), 4093-4099. doi: 10.4238/2012.September.19.5

Luz, L. N., Santos, R. C., Silva Filho, J. L., & Melo Filho, P. A. (2010). Estimativas de parâmetros genéticos em linhagens de amendoim baseadas em descritores associados ao ginóforo. *Revista Ciência Agronômica, 41*(1), 132-138.

Michelotto, M. D., Godoy, I. J., Fávero, A. P., Carrega, W. C., & Finoto, E. L. (2013). Occurrence of *Enneothrips flavens*, Moulton, and *Stegasta bosquella*, Chambers, and its effects on agronomic traits of wild *Arachis* accessions. *Bioscience Journal, 29*(1), 115-124.

Michelotto, M. D., Barioni Jr., W., de Resende, M. D. V., Godoy, I. J., Leonardecz, E., & Fávero, A. P. (2015). Identification of fungus resistant wild accessions and interspecific hybrids of the genus *Arachis*. *PLoS ONE, 10*(6), 1-17.

Moraes, A. R. A.; Lourenção, A. L.; Godoy, I. J. & Teixeira, G. C. (2005). Infestation by *Enneothrips flavens* Moulton and yield of peanut cultivars. *Scientia Agricola, 62*(5), 469-472.

Patterson, H. D., & Thompson, R. (1971). Recovery of inter-block information when blocks sizes are unequal. *Biometrika, 58*(3), 545-554.

Perozini, A. C., Silva, P. S., & Goussain, R. C. S. (2014). Resistência ao tripes *Enneothrips flavens* Moulton (Thysanoptera: Thrpidae) e o potencial produtivo em genótipos de amendoim de hábito de crescimento rasteiro. *Vértices, 16*(2), 79-92.

Pinheiro, L. C. M., God, P. I. V. G., Faria, V. R., Oliveira, A. G., Hasui, A. A., Pinto, E. H. G., ... Moreira, M. A. (2013). Parentesco na seleção para a produtividade e teores de óleo e proteína em soja via modelos mistos. *Pesquisa Agropecuária Brasileira, 48*(9), 1246-1253.

R Development Core Team. (2013). *R: A language and environment for statistical computing*. Vienna, AU: R Foundation for Statistical Computing.

Ramalho, M. A. P., Abreu, A. F. B., Santos, J. B., & Nunes, J. A. R. (2012). *Aplicações da genética quantitativa no melhoramento de plantas autógamas*. Lavras, MG: UFLA.

Regitano Neto, A., Ramos Junior, E. U., Gallo, P. B., Freitas, J. G. & Azzini. (2013). Comportamento de genótipos de arroz de terras altas no estado de São Paulo. *Revista Ciência Agronômica, 44*(3), 512-519.

Resende, M. D. V., Araujo, A. J., Sampaio, P. T. B., & Wiecheteck, M. S. S. (1995). Acurácia seletiva, intervalos de confiança e variância de ganhos genéticos associados a 22 métodos de seleção em *Pinus caribaea* var. hondurensis. *Revista Floresta*, *25*(1/2), 3-16.

Santos, J. F., Godoy, I. J., Fávero, A. P., Moura, N. F., Michelotto, M. D., & Martins, A. L. M. (2011). Resistência à mancha preta em populações F_4 selecionadas de cruzamentos entre um amendoim cultivado e um anfidiploide de Arachis. *Bragantia*, *70*(3), 512-518.

Santos, J. F., Godoy, I. J., Michelotto, M. D., & Fávero, A. P. (2013). Resistência à mancha preta e qualidade agronômica de plantas RC_1F_2 de cruzamentos do híbrido anfidiploide (*Arachis ipaënsis* x *Arachis duranensis*) com o amendoim cultivado (*A. hypogaea*). *Bioscience Journal*, *29*(2), 280-287.

Sikinarum, J., Jaisil, P., Jogloy, S., Toomsam, B., Kesmala, T., & Patanothai, A. (2007). Heritability and correlation for nitrogen fixation and related traits in peanut (*Arachis hypogaea* L.). *Pakistan Journal of Biological Science*, *10*(12), 1956-1962.

Simpson, C. E. (1991). Pathways for introgression of pest resistance into *Arachis hypogaea* L. *Peanut Science*, *18*(1), 22-26.

Statistical Analyses System [SAS]. (2011). *The SAS System, release 9.3.* Cary, NC: SAS Institute Inc.

Wynne, J. C., & Halward, T. M. (1989). Germplasm enhancement in peanut. In H. T. Stalker, & C. Chapman (Eds.), *Science management of germplasm: Characterization, evaluation and enhancement* (p. 155-174). Rome, IT: International Board for Plant Genetic Resources.

Permissions

All chapters in this book were first published in ASA, by Editora da Universidade Estadual de Maringá; hereby published with permission under the Creative Commons Attribution License or equivalent. Every chapter published in this book has been scrutinized by our experts. Their significance has been extensively debated. The topics covered herein carry significant findings which will fuel the growth of the discipline. They may even be implemented as practical applications or may be referred to as a beginning point for another development.

The contributors of this book come from diverse backgrounds, making this book a truly international effort. This book will bring forth new frontiers with its revolutionizing research information and detailed analysis of the nascent developments around the world.

We would like to thank all the contributing authors for lending their expertise to make the book truly unique. They have played a crucial role in the development of this book. Without their invaluable contributions this book wouldn't have been possible. They have made vital efforts to compile up to date information on the varied aspects of this subject to make this book a valuable addition to the collection of many professionals and students.

This book was conceptualized with the vision of imparting up-to-date information and advanced data in this field. To ensure the same, a matchless editorial board was set up. Every individual on the board went through rigorous rounds of assessment to prove their worth. After which they invested a large part of their time researching and compiling the most relevant data for our readers.

The editorial board has been involved in producing this book since its inception. They have spent rigorous hours researching and exploring the diverse topics which have resulted in the successful publishing of this book. They have passed on their knowledge of decades through this book. To expedite this challenging task, the publisher supported the team at every step. A small team of assistant editors was also appointed to further simplify the editing procedure and attain best results for the readers.

Apart from the editorial board, the designing team has also invested a significant amount of their time in understanding the subject and creating the most relevant covers. They scrutinized every image to scout for the most suitable representation of the subject and create an appropriate cover for the book.

The publishing team has been an ardent support to the editorial, designing and production team. Their endless efforts to recruit the best for this project, has resulted in the accomplishment of this book. They are a veteran in the field of academics and their pool of knowledge is as vast as their experience in printing. Their expertise and guidance has proved useful at every step. Their uncompromising quality standards have made this book an exceptional effort. Their encouragement from time to time has been an inspiration for everyone.

The publisher and the editorial board hope that this book will prove to be a valuable piece of knowledge for researchers, students, practitioners and scholars across the globe.

List of Contributors

Marcio Zilio and Adelina Ferreira Michels
Programa de Pós-graduação em Ciências Agrárias (Produção Vegetal), Centro de Ciências Agroveterinárias, Universidade do Estado de Santa Catarina, Lages, Santa Catarina, Brazil

Clovis Arruda Souza, Cileide Maria Medeiros Coelho and David José Miquelluti
Departamento de Agronomia, Centro de Ciências Agroveterinárias, Universidade do Estado de Santa Catarina, Av. Luís Camões, 2090, 88520-000, Lages, Santa Catarina, Brazil

Rodrigo Moreira Ribeiro, Antônio Teixeira do Amaral Júnior, Guilherme Ferreira Pena, Marcelo Vivas and Railan Nascimento Kurosawa
Centro de Ciências e Tecnologia Agropecuária, Universidade Estadual do Norte Fluminense Darcy Ribeiro, Av. Alberto Lamego, 2000, 28013- 602, Parque Califórnia, Campos dos Goytacazes, Rio de Janeiro, Brazil

Leandro Simões Azeredo Gonçalves
Departamento de agronomia, Universidade Estadual de Londrina, Londrina, Paraná, Brazil

Sara Regina Silvestrin Rovaris, Cássio Egídio Cavenaghi Prete, Valquíria Sheila Zago and Lucas José Ferreira da Silva
Departamento de Agronomia, Centro de Ciências Agrárias, Universidade Estadual de Londrina, Rod. Celso Garcia Cid, PR-445, Km 380, 86051-990, Londrina, Paraná, Brazil

Pedro Mário de Araújo and Deoclécio Domingos Garbuglio
Área de Melhoramento Genético Vegetal, Diretoria Técnico Cientifica, Instituto Agronômico do Paraná, Londrina, Paraná, Brazil

Verônica de Jesus Boaventura, Rudiney Ringenberg and Carlos Alberto da Silva Ledo
Embrapa Mandioca e Fruticultura, Parque Estação Biológica, s/n.°, 70770-901, Brasília, Distrito Federal, Brasil

Magali Ferrari Grando, Madelaine Lurdes Varnier, Marilia Rodrigues de Silva and Marilei Suzin
Laboratório de Biotecnologia Vegetal, Faculdade de Agronomia e Medicina Veterinária, Universidade de Passo Fundo, BR-285, Km 171, 99052-900, Campus I, São José, Passo Fundo, Rio Grande do Sul, Brazil

Beatriz Marti Emydio and Luiz Ricardo Pereira
Empresa Brasileira de Pesquisa Agropecuária, Embrapa Trigo, Passo Fundo, Rio Grande do Sul, Brazil

Thiago Gledson Rios Terra
Universidade Federal do Tocantins, Gurupi, Tocantins, Brazil

Haroldo Silva Rodrigues
Departamento de Fitotecnia, Universidade Federal de Viçosa, Viçosa, Minas Gerais, Brazil

Paulo Hideo Nakano Rangel
Embrapa Arroz e Feijão, Santo Antônio de Goiás, Goiás, Brazil

Rafael Simões Tomaz
Universidade Estadual Paulista, Dracena, São Paulo, Brazil

Cosme Damião Cruz
Departamento de Biologia Geral, Universidade Federal de Viçosa, Viçosa, Minas Gerais, Brazil

Aluízio Borém
Departamento de Fitotecnia, Universidade Federal de Viçosa, Viçosa, Minas Gerais, Brazil

Edmar Soares de Vasconcelos, Lívia Maria Lemos Hoepers, Rafaela Goularte Amaral, Vanessa Aline Egewarth and Andressa Strenske
Departamento de Agronomia, Centro de Ciências Agrárias, Universidade Estadual do Oeste do Paraná, Rua Pernambuco, 1777, 85960-000, Marechal Cândido Rondon, Paraná, Brazil

Sara Regina Silvestrin Rovaris, André Luís Bombonato Oliveira, Eduardo Sawazaki and Maria Elisa Ayres Guidetti Zagatto Paterniani
Instituto Agronômico em Campinas, Centro de Grãos e Fibras, Fazenda Santa Elisa, Av. Theodureto de Almeida Camargo, 1500, 13012- 970, Vila Nova, Campinas, São Paulo, Brazil

Paulo Boller Gallo
Agência Paulista de Tecnologia dos Agronegócios, Pólo Regional de Desenvolvimento do Nordeste Paulista, Mococa, São Paulo, Brazil

Willame Rodrigues do Nascimento Sousa and Ana Paula Peron
Programa de Pós-graduação em Genética e Melhoramento, Centro de Ciências Agrárias, Universidade Federal do Piauí, 64049-550, Bairro Iningá, Teresina, Piauí, Brazil

Angela Celis de Almeida Lopes, Reginaldo de Carvalho and Regina Lúcia Ferreira Gomes
Departamento de Biologia, Universidade Federal Rural de Pernambuco, Recife, Pernambuco, Brazil

Júlio César DoVale, Roberto Fritsche-Neto and Glauco Vieira Miranda
Departamento de Fitotecnia, Universidade Federal de Viçosa, Av. Peter Henry Rolfs, s/n, 36570-000, Viçosa, Minas Gerais, Brazil

Ciro Maia
Departamento de Biologia Geral, Universidade Federal de Viçosa, Viçosa, Minas Gerais, Brazil

Paulo Cezar Cavatte
Departamento de Biologia Vegetal, Universidade Federal de Viçosa, Viçosa, Minas Gerais, Brazil

Haroldo Silva Rodrigues, Cosme Damião Cruz and Aluízio Borém
Departamento de Fitotecnia, Universidade Federal de Viçosa, Av. Peter Henry Rolfs, s/n, 36570-900, Campus Universitário, Viçosa, Minas Gerais, Brazil

Jeferson Luis Vasconcelos de Macêdo and Ricardo Lopes
Embrapa Amazônia Ocidental, Manaus, Amazonas, Brazil

Marcos Deon Vilela de Resende
Embrapa Florestas, Colombo, Paraná, Brazil

Andréa Ferreira Costa, José Aires Ventura and Hélcio Costa
Instituto Capixaba de Pesquisa, Assistência Técnica e Extensão Rural, Estação Experimental Mendes da Fonseca, BR-262, km 94, Domingos Martins, Espírito Santo, Brazil

Nilton Rocha Leal and Antonio Teixeira do Amaral Júnior
Centro de Ciências e Tecnologias Agropecuárias, Laboratório de Melhoramento Genético Vegetal, Universidade Estadual do Norte Fluminense "Darcy Ribeiro", Campos dos Goytacazes, Rio de Janeiro, Brazil

Leandro Simões Azeredo Gonçalves
Centro de Ciências Agrárias, Departamento de Agronomia, Universidade Estadual de Londrina, Londrina, Paraná, Brazil

Éder Matsuo, Tuneo Sediyama and Luiz Renato Cadore
Departamento de Fitotecnia, Universidade Federal de Viçosa, Campus Universitário, 36570-000, Viçosa, Minas Gerais, Brazil

Cosme Damião Cruz
Departamento de Biologia Geral, Universidade Federal de Viçosa, Viçosa, Minas Gerais, Brazil

Rita de Cássia Teixeira Oliveira
Bacuri Pesquisa e Melhoramento, Viçosa, Minas Gerais, Brazil

Allan Emanoel Mezzomo Zemolin, Nerinéia Dalfollo Ribeiro, Cleiton Renato Casagrande, Maykon Júnior da Silva and Fernanda Daltrozo Arns
Departamento de Fitotecnia, Centro de Ciências Rurais, Universidade Federal de Santa Maria, Avenida Roraima, 1000, 97105-900, Santa Maria, Rio Grande do Sul, Brazil

Amanda Nogueira de Albuquerque, Marco Antônio Aparecido Barelli, Leonarda Grillo Neves, Vinícius Ribeiro Arantes and Kleberson Luiz Miranda da Silva
Departamento de Agronomia, Universidade do Estado do Mato Grosso, Av. Tancredo Neves, 1095, Cavalhada II, 78200-000, Cáceres, Mato Grosso, Brazil

Marisol Galicia-Juárez, Leopoldo Ernesto Mendoza-Onofre, Víctor Arturo González-Hernández, Ignacio Benítez-Riquelme and Leobigildo Córdova-Téllez
Colegio de Postgraduados, Campus Montecillo, Estado de México, México

María Eugenia Cisneros-López
Campo Experimental Río Bravo, Instituto Nacional de Investigaciones Forestales, Agrícolas y Pecuarias, km 61, Carretera Matamoros-Reynosa, Río Bravo, Tams, México

Alessandro de Lucca e Braccini
Programa de Pós-graduação em Genética e Melhoramento, Departamento de Agronomia, Centro de Ciências Agrárias, Universidade Estadual de Maringá, Av. Colombo, 5790, 87020-900, Maringá, Paraná, Brazil

Leandra Regina Texeira
Programa de Pós-graduação em Genética e Melhoramento, Departamento de Agronomia, Centro de Ciências Agrárias, Universidade Estadual de Maringá, Av. Colombo, 5790, 87020-900, Maringá, Paraná, Brazil
Cooperativa Central de Pesquisa Agrícola, BR 467, km 98, Cascavel, Paraná, Brazil

Elisa Serra Negra Vieira, Polyana Kelly Martins and Ivan Schuster
Cooperativa Central de Pesquisa Agrícola, BR 467, km 98, Cascavel, Paraná, Brazil

Bruno Grimaldo Martinho Churata
Departamento de Estatística, Universidade Federal do Paraná, Curitiba, Paraná, Brazil

Elisângela Knoblauch Viega de Andrade, Valter Carvalho de Andrade Júnior, José Sebastião Cunha Fernandes and Altino Junior Mendes Oliveira
Departamento de Agronomia, Universidade Federal dos Vales do Jequitinhonha e Mucuri, Campus JK, Rodovia MGT-367, km 583, 5000, Alto da Jacuba, Diamantina, Minas Gerais, Brazil

Marcelo Luiz de Laia
Departamento de Engenharia Florestal, Universidade Federal dos Vales do Jequitinhonha e Mucuri, Diamantina, Minas Gerais, Brazil

Alcinei Mistico Azevedo
Universidade Federal de Minas Gerais, Instituto de Ciências agrárias, Montes Claros, Minas Gerais, Brasil

Rodrigo Oliboni, Marcos Ventura Faria, Mikael Neumann, Guilherme Mendes Battistelli, Rafael Gallo Tegoni and Juliano Tadeu Vilela de Resende
Departamento de Agronomia, Universidade Estadual do Centro Oeste, R. Simeão Camargo Varella de Sá, 3, Guarapuava, Paraná, Brazil

Fernando Higino de Lima e Silva, Alexandre Pio Viana, Eileen Azevedo Santos, Jôsie Cloviane de Oliveira Freitas, Daniele Lima Rodrigues and Antonio Teixeira do Amaral Júnior
Laboratório de Melhoramento Genético Vegetal, Universidade Estadual do Norte Fluminense Darcy Ribeiro, Av. Alberto Lamego, 2000, 28013-602, Parque Califórnia, Campos dos Goytacazes, Rio de Janeiro, Brazil

Nerison Luís Poersch, Nerinéia Dalfollo Ribeiro, Daniele Piano Rosa and Sandra Maria Maziero
Centro de Ciências Rurais, Departamento de Fitotecnia, Universidade Federal de Santa Maria, Av. Roraima, 1000, 97105-900, Santa Maria, Rio Grande do Sul, Brazil

Evandro Jost
Instituto Federal de Educação, Ciência e Tecnologia Farroupilha, São Vicente do Sul, Rio Grande do Sul, Brazil

Dielli Aparecida Didoné, Marilia Rodrigues de Silva, Cássia Canzi Ceccon, Tiago Teixeira, Marilei Suzin and Magali Ferrari Grando
Laboratório de Biotecnologia de Plantas, Universidade de Passo Fundo, BR-285, 99052-900, Passo Fundo, Rio Grande do Sul, Brazil

Juliana Araújo Santos Martins
Instituto Federal de Educação, Ciência e Tecnologia do Triângulo Mineiro, Fazenda Sobradinho, s/n, 38400-974, Uberlândia, Minas Gerais, Brazil

Fernando Cézar Juliatti
Instituto de Ciências Agrárias, Universidade Federal de Uberlândia, Uberlândia, Minas Gerais, Brazil

Melina Zacarelli Pirotta, Fabiana Mota da Silva and Sandra Helena Unêda-Trevisoli
Departamento de Produção Vegetal, Universidade Estadual Paulista, Via de Acesso Prof. Paulo Donato Castellane, s/n., 14884-900, Jaboticabal, São Paulo, Brazil

Marcos Doniseti Michelotto
Agência Paulista de Tecnologia dos Agronegócios, Polo Regional Centro Norte, Pindorama, São Paulo, Brazil

Alessandra Pereira Fávero
Empresa Brasileira de Pesquisa Agropecuária, Embrapa Pecuária Sudeste, São Carlos, São Paulo, Brazil

Ignácio José de Godoy
Instituto Agronômico de Campinas, Campinas, São Paulo, Brazil

Index

www.ingramcontent.com/pod-product-compliance
Lightning Source LLC
LaVergne TN
LVHW080649210926
848485LV00003B/352